Advanced Real-Time On-Site Sensing Technologies in Food and Environment Analysis

Advanced Real-Time On-Site Sensing Technologies in Food and Environment Analysis

Guest Editors

Zhenbo Wei
Shanshan Qiu

Basel • Beijing • Wuhan • Barcelona • Belgrade • Novi Sad • Cluj • Manchester

Guest Editors

Zhenbo Wei
Zhejiang University
Hangzhou
China

Shanshan Qiu
Hangzhou Dianzi University
Hangzhou
China

Editorial Office
MDPI AG
Grosspeteranlage 5
4052 Basel, Switzerland

This is a reprint of the Special Issue, published open access by the journal *Chemosensors* (ISSN 2227-9040), freely accessible at: https://www.mdpi.com/journal/chemosensors/special_issues/65NAJ5L10Z.

For citation purposes, cite each article independently as indicated on the article page online and as indicated below:

Lastname, A.A.; Lastname, B.B. Article Title. *Journal Name* **Year**, *Volume Number*, Page Range.

ISBN 978-3-7258-2975-0 (Hbk)
ISBN 978-3-7258-2976-7 (PDF)
https://doi.org/10.3390/books978-3-7258-2976-7

© 2025 by the authors. Articles in this book are Open Access and distributed under the Creative Commons Attribution (CC BY) license. The book as a whole is distributed by MDPI under the terms and conditions of the Creative Commons Attribution-NonCommercial-NoDerivs (CC BY-NC-ND) license (https://creativecommons.org/licenses/by-nc-nd/4.0/).

Contents

Rocío Cánovas, Nick Sleegers, Alexander L.N. van Nuijs and Karolien De Wael
Tetracycline Antibiotics: Elucidating the Electrochemical Fingerprint and Oxidation Pathway
Reprinted from: *Chemosensors* **2021**, *9*, 187, https://doi.org/10.3390/chemosensors9070187 . . . 1

Nan Jiang, Tao Liang, Chunlian Qin, Qunchen Yuan, Mengxue Liu, Liujing Zhuang and Ping Wang
A Microphysiometric System Based on LAPS for Real-Time Monitoring of Microbial Metabolism
Reprinted from: *Chemosensors* **2022**, *10*, 177, https://doi.org/10.3390/chemosensors10050177 . . 18

Jin Liu, Shanshan Qiu and Zhenbo Wei
Real-Time Measurement of Moisture Content of Paddy Rice Based on Microstrip Microwave Sensor Assisted by Machine Learning Strategies
Reprinted from: *Chemosensors* **2022**, *10*, 376, https://doi.org/10.3390/chemosensors10100376 . . 28

Jin Tong, Chengxin Song, Tianjian Tong, Xuanjie Zong, Zhaoyang Liu, Songyang Wang, et al.
Design and Optimization of Electronic Nose Sensor Array for Real-Time and Rapid Detection of Vehicle Exhaust Pollutants
Reprinted from: *Chemosensors* **2022**, *10*, 496, https://doi.org/10.3390/chemosensors10120496 . . 44

Zhongyuan Zhang, Shanshan Qiu, Jie Zhou and Jingang Huang
Monitoring of MSW Incinerator Leachate Using Electronic Nose Combined with Manifold Learning and Ensemble Method
Reprinted from: *Chemosensors* **2022**, *10*, 506, https://doi.org/10.3390/chemosensors10120506 . . 56

Zsanett Bodor, Mariem Majadi, Csilla Benedek, John-Lewis Zinia Zaukuu, Márta Veresné Bálint, Éva Csajbókné Csobod and Zoltan Kovacs
Detection of Low-Level Adulteration of Hungarian Honey Using near Infrared Spectroscopy
Reprinted from: *Chemosensors* **2023**, *11*, 89, https://doi.org/10.3390/chemosensors11020089 . . . 72

Chaiyong Wu, Dengfeng Li, Qianli Jiang and Ning Gan
A Paper-Chip-Based Phage Biosensor Combined with a Smartphone Platform for the Quick and On-Site Analysis of *E. coli* O157:H7 in Foods
Reprinted from: *Chemosensors* **2023**, *11*, 151, https://doi.org/10.3390/chemosensors11020151 . . 87

Marta Mesías, Juan Diego Barea-Ramos, Jesús Lozano, Francisco J. Morales and Daniel Martín-Vertedor
Application of an Electronic Nose Technology for the Prediction of Chemical Process Contaminants in Roasted Almonds
Reprinted from: *Chemosensors* **2023**, *11*, 287, https://doi.org/10.3390/chemosensors11050287 . . 99

Dimitra Kourti, Michailia Angelopoulou, Panagiota Petrou and Sotirios Kakabakos
Optical Immunosensors for Bacteria Detection in Food Matrices
Reprinted from: *Chemosensors* **2023**, *11*, 430, https://doi.org/10.3390/chemosensors11080430 . . 110

Nils Schlatter, Bernd G. Lottermoser, Simon Illgner and Stefanie Schmidt
Utilising Portable Laser-Induced Breakdown Spectroscopy for Quantitative Inorganic Water Testing
Reprinted from: *Chemosensors* **2023**, *11*, 479, https://doi.org/10.3390/chemosensors11090479 . . 142

Article

Tetracycline Antibiotics: Elucidating the Electrochemical Fingerprint and Oxidation Pathway

Rocío Cánovas [1,2], Nick Sleegers [1,2], Alexander L.N. van Nuijs [3] and Karolien De Wael [1,2,*]

[1] AXES Research Group, Bioscience Engineering Department, University of Antwerp, Groenenborgerlaan 171, 2020 Antwerp, Belgium; rocio.canovasmartinez@uantwerpen.be (R.C.); nick.sleegers@uantwerpen.be (N.S.)
[2] NANOlab Center of Excellence, University of Antwerp, Groenenborgerlaan 171, 2020 Antwerp, Belgium
[3] Toxicological Center, University of Antwerp, Universiteitsplein 1, 2610 Antwerp, Belgium; alexander.vannuijs@uantwerpen.be
* Correspondence: karolien.dewael@uantwerpen.be

Abstract: Herein, a complete study of the electrochemical behavior of the most commonly used tetracycline antibiotics (TCs) on unmodified carbon screen-printed electrodes (SPEs) is presented. In addition, the oxidation pathway of TCs on SPE is elucidated, for the first time, with liquid chromatography-quadrupole time-of-flight mass spectrometry (LC-QTOF-MS). Square wave voltammetry (SWV) was used to study the electrochemical fingerprint (EF) of the antibiotics shaping the different oxidation processes of the TCs in a pH range from 2 to 12. Their characteristic structure and subsequent EF offer the possibility of distinguishing this class of antibiotics from other types. Under the optimized parameters, calibration curves of tetracycline (TET), doxycycline (DOXY), oxytetracycline (OXY), and chlortetracycline (CHL) in a Britton Robinson buffer solution (pH 9) exhibited a linear range between 5 and 100 μM with excellent reproducibilities (RSD_{TET} = 3.01%, RSD_{DOXY} = 3.29%, RSD_{OXY} = 9.78% and RSD_{CHL} = 6.88% at 10 μM, N = 3) and limits of detection (LOD) of LOD_{TET} = 4.15 μM, LOD_{DOXY} = 2.14 μM, LOD_{OXY} = 3.07 μM and LOD_{CHL} = 4.15 μM. Furthermore, binary, tertiary, and complex mixtures of all TCs were analyzed with SWV to investigate the corresponding EF. A dual pH screening (pH 4 and pH 9), together with the use of a custom-made Matlab script for data treatment, allowed for the successful confirmation of a single presence of TCs in the unknown samples. Overall, this work presents a straightforward study of the electrochemical behavior of TCs in SPE, allowing for the future on-site identification of residues of tetracycline antibiotics in real samples.

Keywords: tetracycline antibiotics; electrochemical fingerprint; liquid chromatography mass spectrometry; oxidation pathway

1. Introduction

Tetracycline antibiotics (TCs) produced by *Streptomyces* are broad-spectrum agents characterized by containing four condensed aromatic rings known as naphthacene cores (see Figure S1). TCs are especially effective against a broad variety of Gram-positive and Gram-negative bacteria, such as *Staphylococcus*, *Streptococcus*, *Pneumococcus*, *Gonococcus*, *Cholera*, *Dysentery bacillis*, *Pertussis*, *Rickettsia*, *Chlamydia*, and *Mycoplasma* [1,2]. Since their first introduction into medicine in the 1940s, antibiotics have been widely used in livestock farming—as veterinary medicines, feed additives, growth promoters [3], or to prevent/treat mastitis and metritis in cows [4]—and in healthcare for the treatment of many different infections, such as respiratory tract infections, urethritis and severe acne, or even malaria [5,6]. TCs can be divided into four main types: tetracycline (TET), doxycycline (DOXY), oxytetracycline (OXY), and chlortetracycline (CHL), which are most often applied to livestock animals (including honeybees) because of their affordability [5]. However, during the last decades, the generation of new TCs has increased considerably, with more than nine different types of TC [1]. The action mechanism of TC lies in its active transport into the cells of the susceptible bacteria, for the inhibition of protein biosynthesis after binding to the 30S ribosomal subparticle [1,7].

In spite of its unquestionably antibacterial clinical application, the excessive use of TCs exhibits important disadvantages and concerns in our society. First, some examples are the possible allergic sensibilization of exposed individuals (raising intracranial pressure and skin infections, such as rosacea or perioral dermatitis) [8] and adverse effects, including gastrointestinal disturbances (due to the selective pressure that antimicrobial drug residues may exert over human gut microflora) and renal dysfunction [6,9]. Moreover, TCs have been described as hepatotoxic, more dangerous during pregnancy [8,10], and carcinogenic agents, therefore, exposure to them should be reduced to the minimum possible level [11]. Finally, the overuse of these antibiotics promotes resistance genes in bacteria [12]. These resistant bacteria may spread from animals to humans via the food chain [13]. Thus, pathogens could possibly become resistant to these drugs leading to a failure risk of the antibiotic treatment, decreasing its efficiency and triggering potential negative effects for both human and animal health [5,13,14].

According to the World Health Organization (WHO), more than half of the global production of antibiotics is used in farm animals [15]. As TCs have been widely used for animal feed, meat, milk, and fish production, the analytical methods developed are mainly focused on their determination in food samples. Although a maximum residue limit (MRL) has been set for these compounds in food-safety control programs in many countries (see Table S1, MRLs established by European Union), a fast and sensitive analytical method is still needed for achieving the lowest possible detection limit [5].

On the other hand, as TCs are difficult to completely decompose in the body of animals and people, significant quantities of the toxic TC are discharged into the environment [16]. In this way, residues of TCs can also be heavily adsorbed into environmental materials, as well as in waste effluents from hospitals and pharmaceutical industry, where they keep their activity, leading to bacterial resistance [17]. For example, it has been demonstrated that OXY remains undegraded for more than ten months when it is adsorbed in marine sediments and soils [17]. A few studies have reported that the concentration of TCs in groundwater and surface water is as high as 0.2–10 nM, which may be sufficient to cause serious pollution of the aquatic environment [16]. However, the possible contamination of ground water is largely unknown [18], and the exhaustive control of these residues in the environment should be addressed in the near future [19,20]. Hence, accurate monitoring, based on highly sensitive and selective user-friendly sensors for in situ application, is urgently required.

Electrochemical-based detection is of particular interest due to its remarkable advantages, such as low cost, excellent sensitivity, portability, and fast response in comparison to laboratory-based equipment such as high-performance liquid chromatographic (HPLC), spectrophotometry, or capillary electrophoresis, which require relatively expensive instrumentation, have long analysis times, and entail trained personnel [6,21]. Moreover, electrochemical sensors manufactured using screen-printing technology offer huge advantages, utilizing the progressive drive towards miniaturized, sensitive, and portable device and making them user-friendly and disposable [22].

During the last decades, the electrochemical detection of several types of TCs has been pursued using different approaches. Biorecognition is one of the most common strategies and uses aptamer-based sensors [23–32], which are the most recent and commonly found in the literature, followed by enzyme-linked immunoassay (ELISA) functionalization [14,15,33,34], and sensors based on molecular imprinted polymers (MIPs) [35,36] or antibodies [8,37,38] (see Table S2). Furthermore, the oxidation of TCs has been reported using diverse types of electrodes and materials, ranging from metallic nanoparticles [39–45], graphene oxide [46–50], composites [43,51–55], gold [3,11,56], boron-doped diamond electrode (BDDE) [57–60], ruthenium oxide-hexacyanor-uthenate (RuO-RuCN) [61], or multi-wall carbon nanotubes (MWNTs) [17,21] (Table S3). Interestingly, several cases have shown a combination of electrochemical detection (ED) with other analytical methods, such as

HPLC [59,62,63], as well as flow-injection (FI) [6,57,58,64–66] (see Table S3). However, to date, few studies have explored and reported an understanding of the basis of the electrochemistry of TCs per se. One example is the work reported by Hou et al., where the authors studied the electrochemical behavior of four different TCs by cyclic voltammetry and differential pulse voltammetry at electrochemically pre-treated glassy carbon electrodes [67]. Furthermore, there is a lack of information and publications specifically showing the oxidation pathways and their corresponding oxidation products.

Herein, a meaningful study of the electrochemical behavior of the most commonly used TCs (TET, DOXY, OXY, and CHL) on carbon screen-printed electrodes (SPE), as well as the elucidation of the oxidation pathway on the SPE via liquid chromatography-quadrupole time-of-flight mass spectrometry (LC-QTOF-MS), is for the first time presented. First, the voltammetric response of the four TCs at different pHs (ranging from 2 to 12) was measured, to explore their characteristic electrochemical fingerprints (EF). Afterwards, the selective identification of single TCs and the influence of binary and complex mixtures on the EF of TCs was studied by square wave voltammetry (SWV), using a dual pH strategy (pH 4 and pH 9). Moreover, a custom-made data treatment based on Matlab software was used to enhance the peak separation, thus facilitating the identification of the TCs [68,69]. In parallel, the oxidation mechanism of TCs was studied by analyzing partially electrolyzed samples with LC-QTOF-MS, aiming to understand the redox processes at the carbon SPE. Overall, the characteristic EF of TCs shown in this work enables a clear discrimination between them and other antibiotics, due to their specific electrooxidation processes. This rapid and low-cost profiling approach will ultimately allow the identification of TET, DOXY, OXY, and CHL in a decentralized manner.

2. Materials and Methods

2.1. Reagents

Tetracycline antibiotic with >91% purity was obtained from Alfa Aesar Thermo Fisher (Kandel) Germany, doxycycline hyclate purity >98% and chlortetracycline hydrochloride purity >89.5% were purchased from Acros Organics (Geel, Belgium), and oxytetracycline hydrochloride purity 95% was acquired from TCI Tokyo Chemical Industry. Phenol with purity 99% was obtained from J&K scientific GmbH, Germany. Benz[b]anthracene, N,N-dimethylcyclohexylamine and 6,11-dihydroxy-5,12-naphthacenedione, as well as all analytical grade salts of potassium chloride, sodium phosphate, sodium acetate, sodium borate, and potassium hydroxide were purchased from Sigma-Aldrich (Overijse, Belgium). All TCs stocks were prepared in 18.2 MΩ cm^{-1} doubly deionized water, except TET stock, which was prepared in ethanol, all in a concentration of 10 mM. Electrochemical measurements were performed in Britton Robinson buffer at 20 mM ionic strength with a supporting electrolyte 100 mM KCl by applying 50 µL of the buffer onto the SPE.

2.2. Instrumentation and Apparatus

All solutions were prepared in 18.2 MΩ cm^{-1} doubly deionized water (Sartorius, Arium® Ultrapure Water Systems). The pH was measured using a 913 pH meter consisting of a glass pH electrode (6.0262.100) with a reference electrolyte 3 M of KCl from Metrohm (The Netherlands).

All SWV measurements were performed using a MultiPalmSens4 or EmStat Blue potentiostats (PalmSens, The Netherlands) with PSTrace/MultiTrace or PStouch software, respectively. Disposable ItalSens IS-C graphite screen-printed electrodes (SPE) (provided by PalmSens, Utrecht, The Netherlands), containing a graphite working electrode (Ø = 3 mm), a carbon counter electrode, and a (pseudo) silver reference electrode were used for all measurements. The optimized SWV parameters: potential range of 0–1.5 V, frequency 10 Hz, 35 mV amplitude, and 5 mV step potential were used (see Supplementary Material). All the voltammograms were background corrected using the "moving average iterative background correction" (peak width = 1) tool in PSTrace software.

A custom-made script (Matlab R2018b, MathWorks, Natick, MA, USA) [68,69] was used after the analysis by SWVs to enhance peak identification. In brief, the script removes the background signal and applies a top-hat filter that provides an enhanced separation of overlapped peaks, which permits a successful identification of the TCs at pH 9 [70].

The chromatography-mass spectrometry experiments were performed on an liquid chromatograph coupled to a quadrupole time-of-flight mass spectrometer (LC-QTOF-MS) using electrospray ionization (ESI) in positive mode. The apparatus consisted of a 1290 Infinity LC (Agilent Technologies, Wilmington, DE, United States) connected to a 6530 Accurate-Mass QTOF-MS (Agilent Technologies) with a heated-ESI source (Jet-Stream ESI). Further information on the LC-QTOF-MS conditions can be found in the Supplementary Material.

3. Results and Discussions

3.1. Electrochemical Behavior of Tetracyclines via SWV

The electrochemical behavior of TCs was studied on carbon SPE in the whole pH range (2–12) using SWV. SWV was chosen because it is considered a faster and more sensitive technique, which allows for a better resolution and separation between peaks of the oxidation processes in comparison with cyclic voltammetry (CV) [71]. It is important to highlight that TCs are derived from a system of four six membered rings arranged linearly with characteristic double bonds and they have several functional groups, resulting in strong complexing properties [72]. The high similarity between structures (i.e., all TCs contain the same phenolic and di(methyl)amino substituents on position 10 and 4, respectively [73]) (Figure S1a–e) allows for an easy electrochemical discrimination from other types of antibiotics, such as amoxicillin or penicillin (see Figure 1). Therefore, it is expected that the electrochemical oxidation of the TCs will occur through the aforementioned moieties. Accordingly, two main oxidation peaks were observed in all TCs during the pH screening, with the exception of the appearance of a third oxidation process at basic pH or the merge of all the peaks at neutral pH (Figure 2). Figure 2 shows the EF of TCs (black lines), exhibiting similar electrochemical oxidation processes. Few noteworthy similarities were found between the EF in the couples TET–OXY and DOXY–CHL at pHs below 7, whereas the similarity among the couples becomes the opposite at higher pHs (TET–CHL and DOXY–OXY). As reported in the literature, TCs have variable charges on different sites, depending on the pH of the solution, because they contain three ionizable groups (tricarbonyl, dimethylammonium, and phenolic-diacetone) [10,74]. Hence, when the pH is under 4, TC exists as a cation (TCH^{3+}), due to the protonation of the dimethylammonium group (pK$_a$ = 9.7). At a pH between 3.5 and 7.5, TC exists as a zwitterion (TCH$_2^0$), due to the loss of a proton from the phenolic diketone moiety (pK$_a$ = 7.7). At a pH higher than 7, TC exists as an anion (TCH$^-$ or TC^{2-}), due to the loss of protons from the tri-carbonyl system (pK$_a$ = 3.3) and phenolic diketone moiety [75].

Taking into account the three ionizable groups, whole pH screening (2–12) was subsequently carried out using another four chemical compounds with similar structures (see Figure S1f–i) in order to confirm the origin of the oxidation peaks. Figure 2 summarizes the EF of the four TCs (black lines) and the other chemical compounds using the entire range of pH, and Figure S2 shows those afterward selected as optimal, pH 4 and pH 9. Benz[b]anthracene (Figure S1f) was identified as the naphthacene core shared by all TCs, thus, it was used as negative control. As expected, no oxidation peaks were observed in the EF of Figure 2 and Figure S2 (grey dashed line). Phenol (Figure S1g) directly corresponds with the phenolic substituent in position 10 (see Figure S1a), showing similar electrochemical behaviors among the EFs, as can be seen in Figure 2 and Figure S2 (blue dashed line). It is important to point out that the overlap between the EFs of phenol and the TCs was not complete in some examples, such as DOXY and CHL (see Figure 2b,d). This could be due to the presence of a chlorine atom in position 7 in the case of CHL and the absence of the hydroxyl group (–OH) at position 6 in the case of DOXY, which influences the oxidation, causing a shift towards higher potentials. Except for this minor shift, a clear visual correlation can be observed between both EFs (phenol and TCs), which mostly coincides with the first oxidation peak of TCs in the pH screening. Subsequently, N,N-dimethylcyclohexylamine (Figure S1h) was used in the experiment, because of its tertiary amine ionizable group, to verify the origin of the second peak. As can be seen in Figure 2 and Figure S2 (red dashed line), both EFs from the corresponding TC and the N,N-dimethylcyclohexylamine overlap at pHs above 8, since the tertiary amine is protonated bellow pH 7, making its oxidation difficult [68,76]. Last but not least, the origin of a third oxidation process at higher potentials and basic pHs could be associated with the oxidation of some hydroxyl groups present in the chemical structures of TCs. To corroborate this, 6,11-dihydroxy-5,12-naphthacenedione (Figure S1i) was used, as it contains hydroxyl groups in a naphthacene core. Interestingly, the suggested oxidation of a hydroxyl group was shown as a peak at basic pHs (i.e., pH 9 to 12) (Figure 2 and Figure S2 green dashed line). Notably, the concentration of 6,11-dihydroxy-5,12-naphthacenedione was increased until 1 mM to lighten the oxidation peaks in the voltammogram (green dashed line, Figure 2 and Figure S2), while the rest of pH screenings were performed with a concentration of 10 µM. Overall, the EF of TC is enriched at basic pH, exhibiting multiple oxidation peaks attributed to the aforementioned oxidizable groups. A clear example is shown for the TC at pH 9, were the obtained EF shows three overlapped oxidation peaks. At this point, from a electrochemical point of view, the origin of the oxidation peaks exhibited in the EF of the four TCs can be confirmed.

According to the EF, the best-resolved anodic signals for oxidation of TCs on SPE were obtained at basic pHs due to the deprotonation of the oxidizable groups. However, the possibility of a dual pH approach for a selective detection of TC was parallelly considered. In particular, pH 9 was chosen as one optimal option, due to the similarities between the couples TET–CHL and DOXY–OXY, which yielded EFs with three distinguishable peaks ($P1_{TET} = 0.65$ V, $P2_{TET} = 0.74$ V, $P3_{TET} = 0.84$ V and $P1_{CHL} = 0.64$ V, $P2_{CHL} = 0.71$ V, $P3_{CHL} = 0.81$ V) and two peaks ($P1_{DOXY} = 0.65$ V, $P2_{DOXY} = 0.77$ V and $P1_{OXY} = 0.65$ V, $P2_{OXY} = 0.78$ V), respectively. These results differ from the previous studies reported in the literature, where the more favorable peaks always appeared at acidic pHs [54]. However, it is important to clarify that the oxidative processes and the oxidation potential may vary depending on the type of electrode, which hinders a direct comparison.

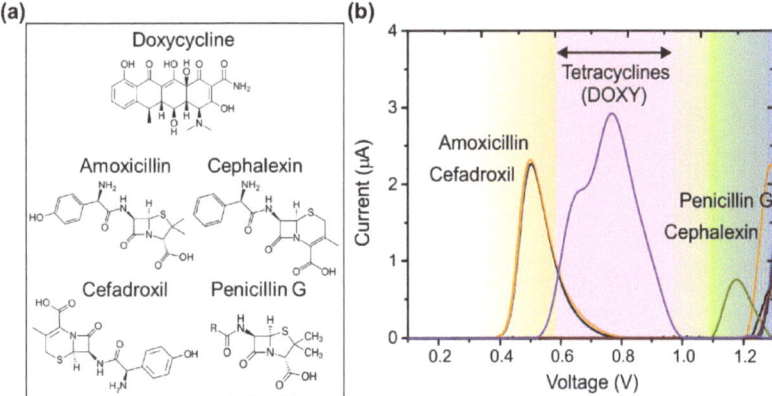

Figure 1. Comparison between (**a**) chemical structures and (**b**) peak potential range of the characteristic electrochemical fingerprints of the tetracyclines (Doxycycline, pink region) with other type of antibiotics, such as Amoxicillin and Cefadroxil (orange and grey region) or Cephalexin and Penicillin G (green and blue region). Moving average corrected square wave voltammogram using a concentration of 100 µM for each antibiotic, except for Penicillin G (1 mM).

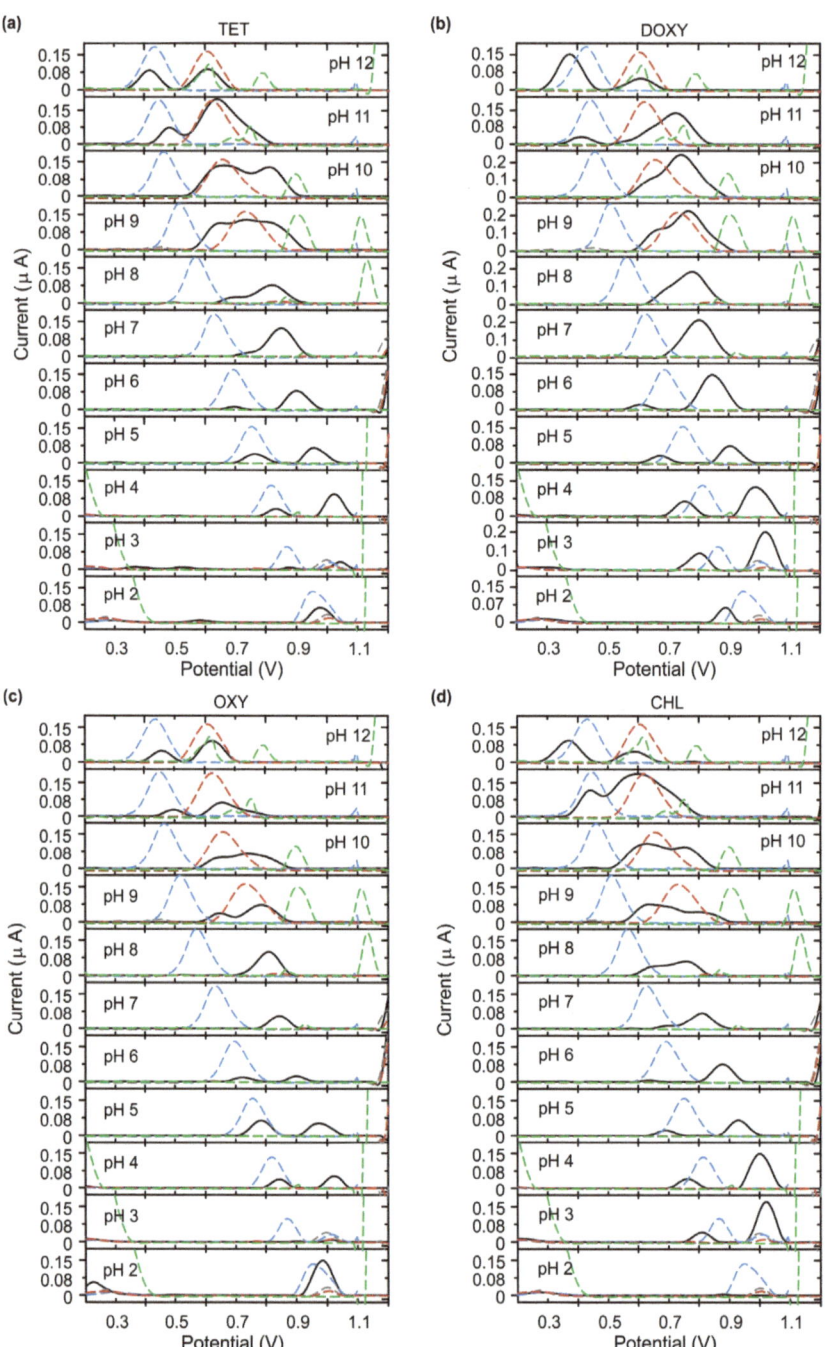

Figure 2. Square wave voltammograms (baseline-corrected) obtained in the whole pH screening range (2–12) using Britton Robinson buffer of all TCs (black line) (**a**) TET, (**b**) DOXY, (**c**) OXY, and (**d**) CHL at 10 µM concentration each. Phenol (blue dashed line), benz[b]anthracene (grey dashed line), and N,N-dimethylcyclohexylamine (red dashed line) in a concentration of 10 µM, and 6,11-dihydroxy-5,12-naphthacenedione (green dashed line) in a concentration of 1 mM.

Figure S3 shows the shift in the oxidation peak potential of TC oxidation as a function of pH (from pH 2 to 12) for the first peak corresponding to the phenol group (P1, blue squares) and the second peak corresponding to the tertiary amine group (P2, red dots). Indeed, the E_p for P1 and P2 shifts negatively with the increase of pH, which corresponds to the oxidation process of phenolic compounds and tertiary amines [73]. When a regular shift in the peak potential with pH is observed, this indicates the involvement of protons during TC oxidation reaction [40]. Some authors stated that the first peak is due to the phenol oxidation, whereas the second peak can be attributed to the oxidation of the intermediates generated during the first reaction (redox process from hydroquinone to benzoquinone) and that this can be justified because of the linear regions along the pH increment [40,73]. In this case, the linear relationship (from pH 2 to pH 12) follows a pseudo-Nernstian response in the case of the first peak (Ep (V)$_{TET}$ = −0.047 pH + 1.04, Ep (V)$_{DOXY}$ = −0.039 pH + 0.95, Ep (V)$_{OXY}$ = −0.049 pH + 1.09, Ep (V)$_{CHL}$ = −0.044 pH + 0.97), probably caused by the different overlapping of both peaks between pH 7 and 10 (see Figure S3), and a closer to Nernstian response in the case of the second peak (Ep (V)$_{TET}$ = −0.053 pH + 1.22, Ep (V)$_{DOXY}$ = −0.039 pH + 1.10, Ep (V)$_{OXY}$ = −0.049 pH + 1.2, Ep (V)$_{CHL}$ = −0.059 pH + 1.22). Interestingly, the secondary peak of CHL reaches a plateau from pH 10 to pH 12, likely due to the deprotonated tertiary amine (pK_a = 8.61). The linear relationship suggests the equal transfer of proton and electrons (2e$^-$/2H$^+$), matching the elucidated oxidation pathway (see Scheme 1). Interestingly, the breaks of linearity evidence that the pK_a of the different moieties was reached during the pH screening [54]. It is important to highlight that the electrodes and electrochemical approaches of the previously reported studies were different from our conditions, making it difficult to compare between the observed oxidation processes. Therefore, the elucidation of the oxidation pathway on the SPE by LC-QTOF-MS is crucial for understanding and confirming the oxidation processes (see Section 3.2).

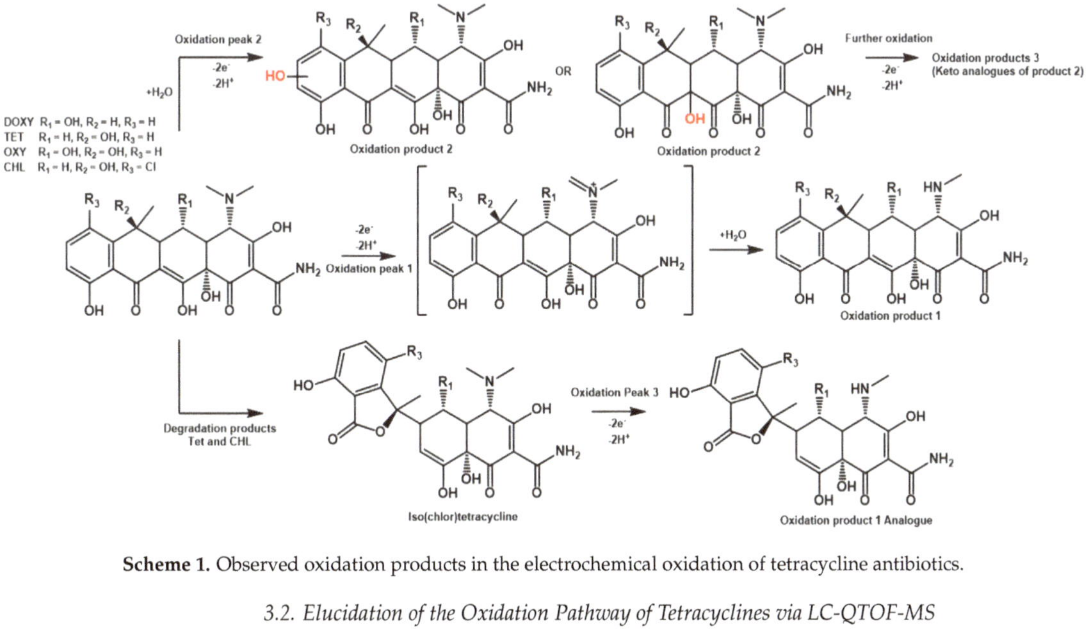

Scheme 1. Observed oxidation products in the electrochemical oxidation of tetracycline antibiotics.

3.2. Elucidation of the Oxidation Pathway of Tetracyclines via LC-QTOF-MS

Understanding the oxidation processes taking place during the voltammetric scans can play an important role in the development of efficient detection strategies [69]. Several authors have tried to explain the oxidation of different TCs through voltammetric studies; some of these are very well summarized in the work reported by Calixto et al. [54].

However, to the best of our knowledge, a profound analysis focussing on the identification of oxidation products has not yet been reported. Therefore, in order to gain an insight into the oxidation processes at the SPE and to identify possible oxidation products, TC solutions were partially electrolyzed on SPE and subsequently analysed using LC-QTOF-MS. A solution of 200 µM TCs was electrolyzed in Britton Robinson buffer pH 9 at 0.66 V and 0.85 V and in pH 4 at 0.8 V and 1.0 V. After 60 min the electrolyzed samples were diluted 1:5 with ultrapure water and directly injected into the equipment. The obtained chromatograms are compared against the standards of TCs in the same concentration, diluted 1:5. Figure 3 shows the results for DOXY, OXY, TET, and CHL (top, in black) and the three main oxidations product, all information is summarized in Table S4.

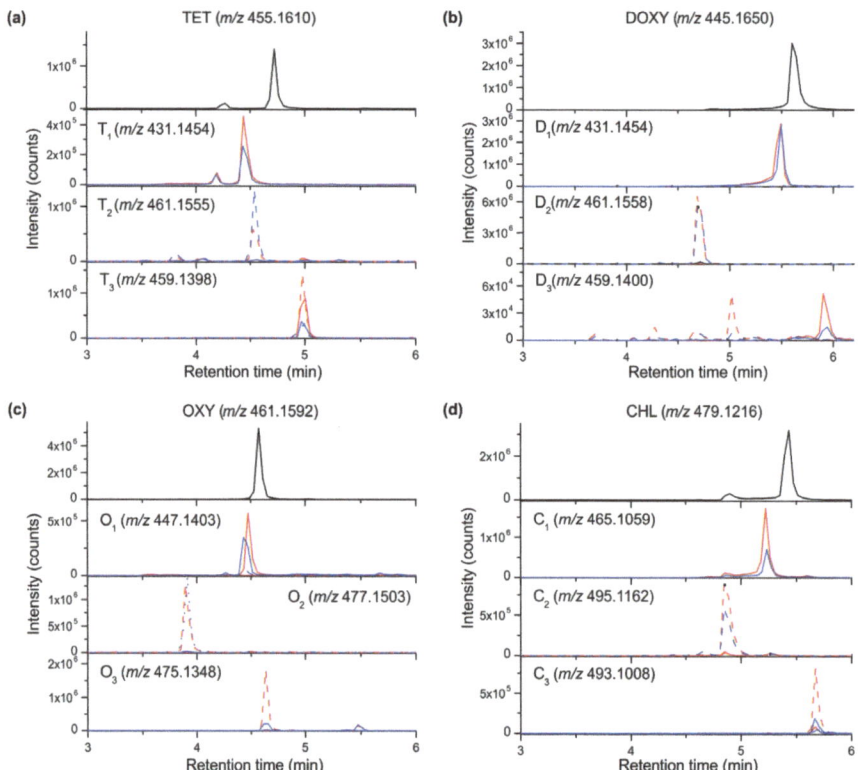

Figure 3. Extracted ion chromatogram of (**a**) TET, (**b**) DOXY, (**c**) OXY, and (**d**) CHL (diluted 1:5) and the oxidation products found after electrolysis in both Britton Robinson buffer pH 9: 0.66 V (red line) and 0.85 V (red dashed line), and pH 4: 0.8 V (blue line) and 1.0 V (blue dashed line) for 60 min.

One product was principally formed for all four TCs during the electrolysis at lower potentials, respectively 0.66 V for pH 9 and 0.80 V for pH 4. The structural information needed to suggest the structures of the oxidation products lies in these MS/MS fragmentation patterns. The key fragment in the case of the tetracyclines is "F2" (m/z 154.0498), which includes the electrochemically active tertiary amine in the tetracycline structure. In product 1 (D1, O1, T1 and C1), this F2 was not found as m/z 154.0498, but rather as 140.0340, which elutes just before the remaining non-oxidized TC with a clear loss of −14 Da, indicating the loss of a methyl-group. Surprisingly, the first electrochemical oxidation signal is not related to the phenolic-moiety of the TCs, as was previously shown by the SWV results, but rather the oxidation of the tertiary-amine group. During the oxidation process, the tertiary amine is oxidatively converted to a secondary amine. This was confirmed by comparing the MSMS spectra of DOXY (Figure S4a) and the oxidation products D_1 and D_2 (Figure S4b,c). Not only the mother-ion $[M + H]^+$ of product D1 exhibits a clear loss of the 14 Da compared to DOXY (*m/z* 431.1454 *versus m/z* 445.1603, see Table S4), but it can also be seen in the typical fragment of *m/z* 154.0498 of DOXY observed in product D1 as *m/z* 140.0430 (Figure S4a,b), corresponding with the loss of a methyl-group. Moreover, the fragment that indicates the presence of the tertiary amine group, *m/z* 58.0657 ($[C_3H_8N]^+$), could not be observed for D_1, while other fragments such as *m/z* 321.075 ($[C_{19}H_{13}O_5]^+$) and *m/z* 267.065 ($[C_{16}H_{11}O_4]^+$) are identical to DOXY as they do not contain any nitrogen atoms.

For the electrolysis at higher potentials, namely 0.85 V for pH 9 and 1.00 V pH 4, a second main oxidation product was formed (D_2, O_2, T_2 and C_2, Figure 3 and Table S4), while the products of the first oxidation peak had almost completely disappeared. When the molecular ion masses of the products are compared to the corresponding non-oxidized TC, a clear addition of 16 Da is observed for all TCs, indicating that an additional oxygen is incorporated into the structure. Based on the MSMS fragmentation pattern of D_2 this addition does not take place close to the tertiary amine group, as the fragment *m/z* 154.0498 can still be observed. Therefore, this product is formed during the oxidation of the phenolic-moiety in the DOXY, and the same conclusion is attributed to the rest of the TCs. On the other hand, it is extremely difficult to assign a specific position to the additional hydroxyl-group in this product based on the LC-QTOF-MS analysis alone. In the literature, for oxidative degradation of tetracyclines this is sometimes suggested to be at either the para, meta, or ortho-position of the phenolic-group (OH groups indicated in red in Scheme 1), whereas other articles suggest a similar product as for the enzymatic oxidation of tetracyclines [16,77–79]. Herein, it is known that the oxidative addition takes place at the 11a position. At these higher potentials the newly formed incorporated hydroxyl-group can be even oxidized, resulting in a keto analogue of the second oxidation product, which corresponds to the remaining oxidation products D_3, O_3, T_3, and C_3 (Figure 3, Table S4). However, this does not explain the appearance of the third oxidation peak previously shown only for TET and CHL (Figure 2). However, the answer lies in the occurrence of a second peak in the chromatogram at 4.72 min (*m/z* 445.1605) and 5.44 min (*m/z* 479.1233) of the standards of TET and CHL, respectively. These products possess the same *m/z*-value as the TC itself, meaning that they are rearranged degradation products specific to these TC. It has been reported for TET and CHL that these two TCs are particularly sensitive to the formation of iso(chlor)tetracycline [80]. Therefore, the formation of isotetracycline is the most likely explanation for this part of the study and was also shown to be electroactive by the HPLC coupled to an electrochemical detector (Figure S5). These results correlate with the third oxidation peak observed throughout the SWV analysis. Hence, the redox pathway suggested for the oxidation of tetracycline antibiotics is exhibited in Scheme 1.

3.3. Calibration Curves

A preliminary optimization was carried out before performing the calibration curve at pH 9 to test the analytical performance of the sensor towards different TCs (Figure 4). First, the influence of supporting electrolytes (potassium chloride, KCl) on the electrochemical response was initially studied at pH 2 (Figure S6), showing an oxidation peak at ca. 0.95 V corresponding to the oxidation of the phenolic moiety. Hence, 100 mM KCl was selected as an optimal concentration for further experiments. Moreover, the optimization of the SWV parameters—step potential (E step), amplitude, and frequency—was carried out (see Figure S7). As a result, an E step of 5 mV, amplitude of 35 mV, and frequency of 10 Hz were chosen as optimal conditions to obtain the best performance regarding the peak intensity, better signal-to-noisy ratio, and/or the best adaptation to the specific software for data treatment (Matlab script). Figure 4a–d displays the SW voltammograms of the TCs upon increasing concentrations. Accordingly, Figure 4e–h exhibits the corresponding calibration curves, with the calculated analytical parameters shown in Table 1. Subsequently, the reproducibility ($N = 3$) of the SWV of the TCs at pH 9 in a concentration of 10 µM, as well as the stability of the stock solutions (from 1 to 3 weeks) at pH 9 to test its possible degradation over time in the alkaline solution, were evaluated (Figures S8 and S9, respectively). The results show excellent reproducibility, with RSD < 10% (Table 1). Moreover, negligible degradation of DOXY, OXY, and CHL solution stocks in pH 9 was observed over time (more than 3 weeks period), showing RSD values ($N = 3$) of 5.04%, 5.72%, and 5.15%, respectively. On the other hand, TET exhibits higher degradation after the first week with an RSD of 20.23% ($N = 3$), consistent with the change in color (from yellow to red) and the appearance of precipitates.

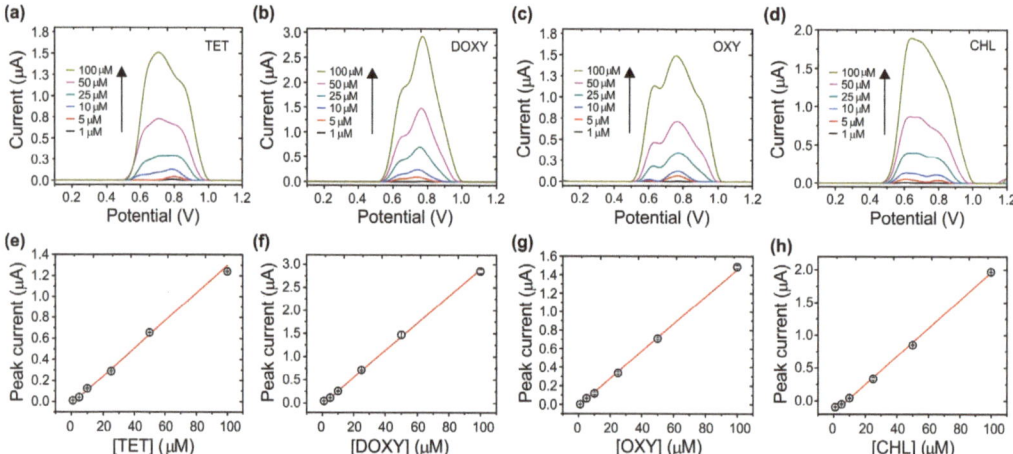

Figure 4. Baseline-corrected SWV for (**a**) TET, (**b**) DOXY, (**c**) OXY, and (**d**) CHL at pH 9 in Britton Robinson buffer in a range from 1 to 100 µM using carbon SPE. Corresponding calibration curves of the four TCs (**e**) TET, (**f**) DOXY, (**g**) OXY, and (**h**) CHL showing the peak current versus concentration of the different TCs ($N = 3$).

Table 1. Analytical parameters obtained from calibrations curves of TET, DOXY, OXY, and CHL in a range from 1 to 100 µM concentration and RDS obtained from the reproducibility study (Figure S8).

	TET	DOXY	OXY	CHL
Sensitivity (µA µM^{-1})	0.013	0.030	0.015	0.020
R-squared	0.998	0.999	0.999	0.999
Linear range (µM)	5–100	5–100	5–100	5–100
Limit of detection (µM)	4.17	2.14	3.07	2.49
RSD (%) at 10 µM, $N = 3$	3.01	3.29	9.78	6.88

3.4. Data Treatment towards an Enhanced Peak Analysis

A Matlab script was designed to improve the identification of the oxidation peaks in the EF [68,69]. In particular, for the cases in which the overlapping of signals creates shoulders and tails in the peak, the aforementioned tool improves the peak separation, and thus the peak identification, allowing for a trustworthy compound identification based on the characteristic oxidation processes of the EF [70]. Hence, the script was implemented for all the EF obtained during the pH screening. After studying the general behavior of the TCs at different pHs, two pHs exhibited a characteristic enrichment of the EFs: (*i*) pH 9 showed high peak intensity and interesting EF similarities (same E_p) between the couples TET–CHL (with three peaks, $P1_{TET} = 0.65$ V, $P2_{TET} = 0.74$ V, $P3_{TET} = 0.84$ V and $P1_{CHL} = 0.64$ V, $P2_{CHL} = 0.71$ V, $P3_{CHL} = 0.81$ V) and DOXY–OXY (in contrast showing two peaks, $P1_{DOXY} = 0.65$ V, $P2_{DOXY} = 0.77$ V and $P1_{OXY} = 0.65$ V, $P2_{OXY} = 0.78$ V, respectively), and (*ii*) pH 4 due to the easy identification between the peak potentials of two different couples of TCs, TET–OXY ($P1_{TET} = 0.84$ V, $P2_{TET} = 1.02$ V, $P1_{OXY} = 0.85$ V, $P2_{OXY} = 1.02$ V) and DOXY–CHL ($P1_{DOXY} = 0.75$ V, $P2_{DOXY} = 0.99$ V, $P1_{CHL} = 0.76$ V, $P2_{CHL} = 1$ V). Therefore, a dual pH strategy might permit the selective determination of specific TCs based on the EF in an unknown sample.

Figure 5 summarizes the comparison of SWVs obtained from the potentiostat software (i.e., baseline corrected by moving average correction), and after employing data analysis (i.e., tailor-made script) of the four TCs at pH 4 and pH 9. The script did not improve the peak separation for pH 4 (Figure 5a–b). On the contrary, the script was crucial to elucidate the number of peaks at pH 9 that potentially correspond to different oxidation processes. Figure 5c shows the SWVs of TCs that clearly exhibit shoulders on oxidation peaks corresponding to partially overlaying signals of different electroactive groups. After data analysis (Figure 5d), an improved peak separation was accomplished, allowing for an easy identification of the specific EF for each TC, showing three peaks in the case of TET and CHL, and only two peaks in the case of DOXY and OXY. Therefore, at this point, a two-step protocol was established for the determination of TET, DOXY, OXY, and CHL: (*i*) a preliminary screening at pH 4 to obtain a first identification between two different couples (TET–OXY or CHL–DOXY), followed by (*ii*) a confirmatory screening at pH 9 to elucidate the specific TC present in the sample, taking advantage of the similarities between the different couples (three peaks in the couple TET–CHL and two peaks in DOXY–OXY).

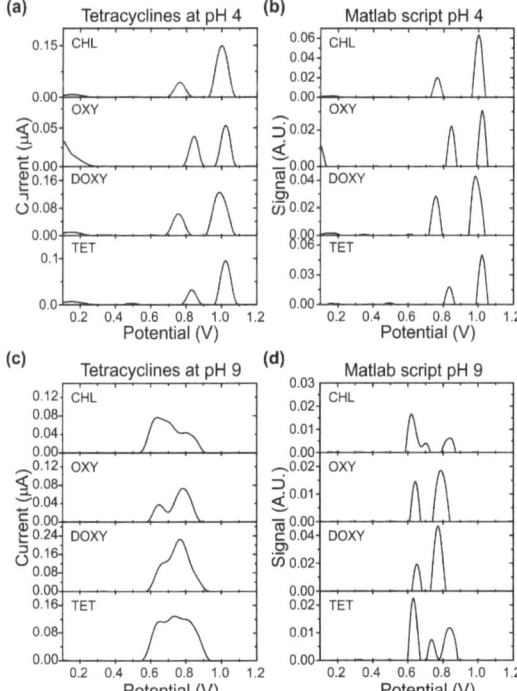

Figure 5. Data treatment with Matlab script to improve peak separation/identification. (**a**) Baseline-corrected SWVs of tetracyclines at pH 4, (**b**) output signal after the application of the script at pH 4, (**c**) baseline-corrected SWVs of TCs at pH 9, where overlapping of peaks exists, and (**d**) output signal after the application of the script at pH 9.

3.5. Single, Binary, Tertiary, and Complex Mixtures of TCs

The aim of the proposed approach was not just being able to distinguish TCs among other antibiotics (which was successfully shown in Figure 1), but also between themselves, and thus to identify possible overlapping and suppression effects, which could lead to false positive and false negative results, respectively. Therefore, apart from being able to detect single TCs in an unknown sample, the next step was to analyze if different TCs could be identified in binary, tertiary, or quaternary mixtures. Figure S10 shows another example where all TCs in different mixtures at pH 9 can be compared between the SWVs obtained from the potentiostat software (i.e., moving average correction), and after employing data analysis (i.e., tailor-made script), which facilitates the separation between peaks. It is worth mentioning that the treated signal after the script does not correspond to the current intensity of the SWVs, thus producing signals for a qualitative analysis.

Figure 6 displays the whole summary of single, binary, tertiary, and quaternary combinations of the four studied TCs at pH 4 (Figure 6a) and pH 9 (Figure 6b). A dashed line color code has been added to facilitate the identification of the TC present in the corresponding mixture, i.e., orange for TET, green for DOXY, blue for OXY, and purple for CHL. As can be seen from the figure, if a single TC is present in the sample, the identification can be successfully achieved (green tick) by the dual pH strategy and the Matlab script (only required at pH 9, see Figure S10). As a practical example, if an EF showing two peaks at 0.82 V and 1.1 V is obtained at pH 4, one could easily identify the presence of TET or OXY. Afterwards, a second SWV at pH 9 is carried out to confirm which TC is present in the sample. Therefore, the appearance of two peaks (after Matlab script) at 0.63 V and 0.79 V means that the TC present in the sample is OXY.

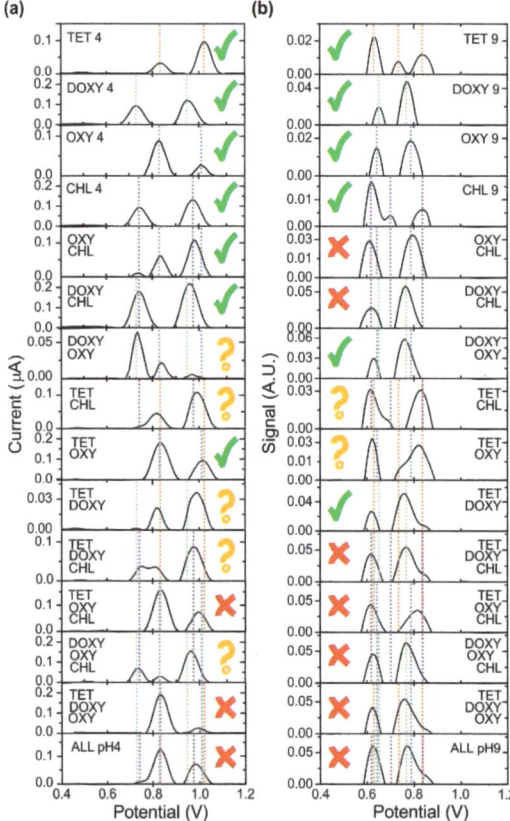

Figure 6. Electrochemical fingerprint of TCs in all type of combinations and mixtures at (**a**) pH 4, SWVs raw data (moving average correction) and (**b**) pH 9, output signal after the application of the script. The successful recognition of the TC is represented in the figure with a green tick (✓), followed by an orange question mark (?) or a red cross (✗) for inconclusive and unsuccessful identifications, respectively.

After the binary mixtures are introduced, the complexity of the exercise increases, making possible a proper identification in some cases (i.e., DOXY + OXY and TET + DOXY, Figure 6b). Other mixtures lead to inconclusive results (orange question mark), such as TET + CHL and TET + OXY, where the middle peaks corresponding to CHL and TET appear with a slight shoulder after applying the Matlab script, instead of being clearly separated. Unfortunately, tertiary and quaternary mixtures of TCs were very difficult to determine (red cross). Future steps will consider a deeper exploration and optimization in both the electrochemical setup and the Matlab script. Nevertheless, it is important to again point out that although the identification becomes challenging upon increasing the complexity of the mixtures, it is still possible to distinguish the entire class of TCs from other types of antibiotics (Figure 1).

4. Conclusions

This work reveals, for the first time, a complete study of the electrochemistry of four different tetracyclines (TC) antibiotics: tetracycline (TET), doxycycline (DOXY), oxytetracycline (OXY), and chlortetracycline (CHL). The aim of the present manuscript was to collect detailed information about the oxidation processes of these TCs from both electrochemical and chromatographic points of view, in order to clarify the reactions and pave the way towards the development of more accurate and reliable detection platforms. Furthermore, a broad overview of the previously reported electrochemical sensors and biosensors based on surface modification, aptamers, antibodies, enzymes, or MIP for TC determination was included. From the electrochemical section, the electrochemical fingerprint of the tetracyclines in the whole pH range (from 2 to 12) was studied via SWV on unmodified carbon SPEs. Moreover, the main oxidation processes were attributed to the corresponding electroactive groups (i.e., phenol, tertiary amine, and hydroxyl groups) by comparing the electrochemical behavior of TCs with different chemical compounds sharing similar moieties. In addition, a successful protocol for the identification of a single TC in unknown samples was developed based on a dual pH strategy and an innovative custom-made script designed to enhance the separation between oxidation peaks. Subsequently, from a chromatographic perspective, the oxidation mechanism of the TCs was elucidated for the first time by LC-QTOF-MS and different oxidation products were clearly detected and identified after the electrolysis at SPE. Overall, this work encompasses relevant information about the electrochemical behavior of TCs and points out a simple, rapid, and reliable electrochemical approach for the identification of tetracycline antibiotics in combination with a custom Matlab script. The present work aims to pave the way and bring new insights towards the development of future analytical platforms for the reliable determination of antibiotics in decentralized settings.

Supplementary Materials: The following are available online at https://www.mdpi.com/article/10.3390/chemosensors9070187/s1, LC-QTOF-MS Conditions, Figure S1. Chemical structure of the tetracycline antibiotics, Figure S2. pH screening (2–12) for all tetracyclines, Figure S3. Peak potential *versus* pH, Figure S4. MS/MS spectra of DOXY, Figure S5. HPLC–ECD chromatograms of TCs at 10 µM, Figure S6. Electrochemical influence of the supporting electrolyte, Figure S7. Optimization of the main parameters used in SWV, Figure S8. Reproducibility study, Figure S9. Stability study, Figure S10. Output signal after the utilization of the script, Table S1. Maximum residue limit (MRL) for tetracyclines in animal tissues, Table S2. Summary of electrochemical biosensors based on aptamers, antibodies, molecular imprinted polymers and enzyme-linked immunoassays for tetracyclines (TCs) detection, Table S3. Summary of electrochemical approaches based on modified electrodes for tetracycline determination, Table S4. LC-QTOF data for all four TCs and the three main oxidation products.

Author Contributions: R.C.: Conceptualization, Data curation, Formal analysis, Investigation, Methodology, Validation, Visualization, Writing—original draft, drafting the article and revising it critically for important intellectual content. N.S.: Data curation, Formal analysis, Investigation. A.L.N.v.N.: Resources, Writing—final approval of the version to be submitted. K.D.W.: Funding acquisition, Methodology, Resources, Supervision, Project administration, Writing—final approval of the version to be submitted. All authors have read and agreed to the published version of the manuscript.

Funding: FWO (Research Foundation—Flanders) grant number G054819N, 2018.

Institutional Review Board Statement: Not applicable.

Informed Consent Statement: Not applicable.

Data Availability Statement: Not applicable.

Acknowledgments: The authors would like to acknowledge the financial support from FWO (Research Foundation—Flanders, grant number G054819N, 2018) and the financial support from the University of Antwerp.

Conflicts of Interest: The authors declare no conflict of interest.

References

1. Oka, H.; Ito, Y.; Matsumoto, H. Chromatographic analysis of tetracycline antibiotics in foods. *J. Chromatogr. A* **2000**, *882*, 109–133. [CrossRef]
2. Masawat, P.; Slater, J.M. The determination of tetracycline residues in food using a disposable screen-printed gold electrode (SPGE). *Sens. Actuators B Chem.* **2007**, *124*, 127–132. [CrossRef]
3. Le, T.H.; Pham, V.P.; La, T.H.; Phan, T.B.; Le, Q.H. Electrochemical aptasensor for detecting tetracycline in milk. *Adv. Nat. Sci. Nanosci. Nanotechnol.* **2016**, *7*. [CrossRef]
4. Zhao, F.; Zhang, X.; Gan, Y. Determination of tetracyclines in ovine milk by high-performance liquid chromatography with a coulometric electrode array system. *J. Chromatogr. A* **2004**, *1055*, 109–114. [CrossRef]
5. Liu, X.; Huang, D.; Lai, C.; Zeng, G.; Qin, L.; Zhang, C.; Yi, H.; Li, B.; Deng, R.; Liu, S.; et al. Recent advances in sensors for tetracycline antibiotics and their applications. *TrAC Trends Anal. Chem.* **2018**, *109*, 260–274. [CrossRef]
6. Palaharn, S.; Charoenraks, T.; Wangfuengkanagul, N.; Grudpan, K.; Chailapakul, O. Flow injection analysis of tetracycline in pharmaceutical formulation with pulsed amperometric detection. *Anal. Chim. Acta* **2003**, *499*, 191–197. [CrossRef]
7. Brodersen, D.E.; Clemons, W.M.; Carter, A.P.; Morgan-Warren, R.J.; Wimberly, B.T.; Ramakrishnan, V. The structural basis for the action of the antibiotics tetracycline, pactamycin, and hygromycin B, on the 30S ribosomal subunit. *Cell* **2000**, *103*, 1143–1154. [CrossRef]
8. Liu, X.; Zheng, S.; Hu, Y.; Li, Z.; Luo, F.; He, Z. Electrochemical Immunosensor Based on the Chitosan-Magnetic Nanoparticles for Detection of Tetracycline. *Food Anal. Methods* **2016**, *9*, 2972–2978. [CrossRef]
9. Pellegrini, G.E.; Carpico, G.; Coni, E. Electrochemical sensor for the detection and presumptive identification of quinolone and tetracycline residues in milk. *Anal. Chim. Acta* **2004**, *520*, 13–18. [CrossRef]
10. Kim, Y.J.; Kim, Y.S.; Niazi, J.H.; Gu, M.B. Electrochemical aptasensor for tetracycline detection. *Bioprocess Biosyst. Eng.* **2010**, *33*, 31–37. [CrossRef]
11. Casella, I.G.; Fabio, P. Determination of tetracycline residues by liquid chromatography coupled with electrochemical detection and solid phase extraction. *J. Agric. Food Chem.* **2009**, *57*, 8735–8741. [CrossRef]
12. Rudnicki, K.; Sipa, K.; Brycht, M.; Borgul, P.; Skrzypek, S.; Poltorak, L. Electrochemical sensing of fluoroquinolone antibiotics. *TrAC Trends Anal. Chem.* **2020**, *128*, 115907. [CrossRef]
13. Shen, G.; Guo, Y.; Sun, X.; Wang, X. Electrochemical Aptasensor Based on Prussian Blue-Chitosan-Glutaraldehyde for the Sensitive Determination of Tetracycline. *Nano-Micro Lett.* **2014**, *6*, 143–152. [CrossRef]
14. Pastor-Navarro, N.; Morais, S.; Maquieira, Á.; Puchades, R. Synthesis of haptens and development of a sensitive immunoassay for tetracycline residues. Application to honey samples. *Anal. Chim. Acta* **2007**, *594*, 211–218. [CrossRef]
15. Faridah, S.; Azura, N.; Hazana, R.; Gayah, A.R.; Norzaili, Z.; Azima, A.; Zamri, I. Electrochemical sensors for detection of tetracycline antibiotics Unbound free tetracycline and tetracycline conjugates were removed during the washing step Direct competitive ELISA method Carbon working electrode was connected to the electrochemical ayse. *Malays. Soc. Anim. Prod.* **2012**, *15*, 67–80.
16. Zhi, D.; Qin, J.; Zhou, H.; Wang, J.; Yang, S. Removal of tetracycline by electrochemical oxidation using a Ti/SnO$_2$–Sb anode: Characterization, kinetics, and degradation pathway. *J. Appl. Electrochem.* **2017**, *47*, 1313–1322. [CrossRef]
17. Vega, D.; Agüí, L.; González-Cortés, A.; Yáñez-Sedeño, P.; Pingarrón, J.M. Voltammetry and amperometric detection of tetracyclines at multi-wall carbon nanotube modified electrodes. *Anal. Bioanal. Chem.* **2007**, *389*, 951–958. [CrossRef]
18. Zhu, J.; Snow, D.D.; Cassada, D.A.; Monson, S.J.; Spalding, R.F. Analysis of oxytetracycline, tetracycline, and chlortetracycline in water using solid-phase extraction and liquid chromatography-tandem mass spectrometry. *J. Chromatogr. A* **2001**, *928*, 177–186. [CrossRef]
19. Hayat, A.; Marty, J.L. Aptamer based electrochemical sensors for emerging environmental pollutants. *Front. Chem.* **2014**, *2*, 1–9. [CrossRef] [PubMed]
20. Dos Santos, A.J.; Kronka, M.S.; Fortunato, G.V.; Lanza, M.R.V. Recent advances in electrochemical water technologies for the treatment of antibiotics: A short review. *Curr. Opin. Electrochem.* **2021**, *26*, 100674. [CrossRef]

21. Wong, A.; Scontri, M.; Materon, E.M.; Lanza, M.R.V.; Sotomayor, M.D.P.T. Development and application of an electrochemical sensor modified with multi-walled carbon nanotubes and graphene oxide for the sensitive and selective detection of tetracycline. *J. Electroanal. Chem.* **2015**, *757*, 250–257. [CrossRef]
22. Hayat, A.; Marty, J.L. Disposable screen printed electrochemical sensors: Tools for environmental monitoring. *Sensors* **2014**, *14*, 10432–10453. [CrossRef]
23. Feng, Y.; Yan, T.; Wu, T.; Zhang, N.; Yang, Q.; Sun, M.; Yan, L.; Du, B.; Wei, Q. A label-free photoelectrochemical aptasensing platform base on plasmon Au coupling with MOF-derived In_2O_3@g-C_3N_4 nanoarchitectures for tetracycline detection. *Sens. Actuators B Chem.* **2019**, *298*, 126817. [CrossRef]
24. Guo, Y.; Wang, X.; Sun, X. A label-free electrochemical aptasensor based on electrodeposited gold nanoparticles and methylene blue for tetracycline detection. *Int. J. Electrochem. Sci.* **2015**, *10*, 3668–3679.
25. Alawad, A.; Istamboulié, G.; Calas-Blanchard, C.; Noguer, T. A reagentless aptasensor based on intrinsic aptamer redox activity for the detection of tetracycline in water. *Sens. Actuators B Chem.* **2019**, *288*, 141–146. [CrossRef]
26. Taghdisi, S.M.; Danesh, N.M.; Ramezani, M.; Abnous, K. A novel M-shape electrochemical aptasensor for ultrasensitive detection of tetracyclines. *Biosens. Bioelectron.* **2016**, *85*, 509–514. [CrossRef]
27. Piaopiao, C.; Yichen, X.; Xiaoxiao, C.; Shan, Z.; Yang, L.; Chaobiao, H. A "Signal On" Photoelectrochemical Aptasensor For Tetracycline Detection Based On Semiconductor Polymer Quantum Dots. *J. Electrchem. Soc.* **2020**, *167*, 067516. [CrossRef]
28. Zarei, M. Sensitive visible light-driven photoelectrochemical aptasensor for detection of tetracycline using ZrO_2/g-C_3N_4 nanocomposite. *Sens. Int.* **2020**, *1*, 100029. [CrossRef]
29. Yang, Y.; Yan, W.; Guo, Y.; Wang, X.; Zhang, F.; Yu, L.; Guo, C.; Fang, G. Sensitive and selective electrochemical aptasensor via diazonium-coupling reaction for label-free determination of oxytetracycline in milk samples. *Sens. Actuators Rep.* **2020**, *2*, 100009. [CrossRef]
30. Hu, X.; Xu, Y.; Cui, X.; Li, W.; Huang, X.; Li, Z.; Shi, J.; Zou, X. Fluorometric and electrochemical dual-mode nanoprobe for tetracycline by using a nanocomposite prepared from carbon nitride quantum dots and silver nanoparticles. *Microchim. Acta* **2020**, *187*, 1–10. [CrossRef]
31. Zhang, J.; Wu, Y.; Zhang, B.; Li, M.; Jia, S.; Jiang, S.; Zhou, H.; Zhang, Y.; Zhang, C.; Turner, A.P.F. Label-Free Electrochemical Detection of Tetracycline by an Aptamer Nano-Biosensor. *Anal. Lett.* **2012**, *45*, 986–992. [CrossRef]
32. Zhou, L.; Li, D.J.; Gai, L.; Wang, J.P.; Li, Y. Bin Electrochemical aptasensor for the detection of tetracycline with multi-walled carbon nanotubes amplification. *Sens. Actuators B Chem.* **2012**, *162*, 201–208. [CrossRef]
33. Jeon, M.; Kim, J.; Paeng, K.J.; Park, S.W.; Paeng, I.R. Biotin-avidin mediated competitive enzyme-linked immunosorbent assay to detect residues of tetracyclines in milk. *Microchem. J.* **2008**, *88*, 26–31. [CrossRef]
34. Zhang, Y.; Lu, S.; Liu, W.; Zhao, C.; Xi, R. Preparation of anti-tetracycline antibodies and development of an indirect heterologous competitive enzyme-linked immunosorbent assay to detect residues of tetracycline in milk. *J. Agric. Food Chem.* **2007**, *55*, 211–218. [CrossRef] [PubMed]
35. Devkota, L.; Nguyen, L.T.; Vu, T.T.; Piro, B. Electrochemical determination of tetracycline using AuNP-coated molecularly imprinted overoxidized polypyrrole sensing interface. *Electrochim. Acta* **2018**, *270*, 535–542. [CrossRef]
36. Wang, H.; Zhao, H.; Quan, X.; Chen, S. Electrochemical Determination of Tetracycline Using Molecularly Imprinted Polymer Modified Carbon Nanotube-Gold Nanoparticles Electrode. *Electroanalysis* **2011**, *23*, 1863–1869. [CrossRef]
37. Jampasa, S.; Pummoree, J.; Siangproh, W.; Khongchareonporn, N.; Bgamrojanavanich, N.; Chailapakul, O.; Chaiyo, S. Chemical "Signal-On" electrochemical biosensor based on a competitive immunoassay format for the sensitive determination of oxytetracycline. *Sens. Actuators B. Chem.* **2020**, *320*, 128389. [CrossRef]
38. Starzec, K.; Cristea, C.; Tertis, M.; Feier, B.; Wieczorek, M.; Koscielniak, P.; Kochana, J. Employment of electrostriction phenomenon for label-free electrochemical immunosensing of tetracycline. *Bioelectrochemistry* **2020**, *132*, 107405. [CrossRef]
39. El Alami El Hassani, N.; Baraket, A.; Boudjaoui, S.; Taveira Tenório Neto, E.; Bausells, J.; El Bari, N.; Bouchikhi, B.; Elaissari, A.; Errachid, A.; Zine, N. Development and application of a novel electrochemical immunosensor for tetracycline screening in honey using a fully integrated electrochemical Bio-MEMS. *Biosens. Bioelectron.* **2019**, *130*, 330–337. [CrossRef]
40. Kushikawa, R.T.; Silva, M.R.; Angelo, A.C.D.; Teixeira, M.F.S. Construction of an electrochemical sensing platform based on platinum nanoparticles supported on carbon for tetracycline determination. *Sens. Actuators B Chem.* **2016**, *228*, 207–213. [CrossRef]
41. Luo, Y.; Xu, J.; Li, Y.; Gao, H.; Guo, J.; Shen, F.; Sun, C. A novel colorimetric aptasensor using cysteamine-stabilized gold nanoparticles as probe for rapid and specific detection of tetracycline in raw milk. *Food Control* **2015**, *54*, 7–15. [CrossRef]
42. Chen, B.; Zhang, Z.; Sun, X.; Kuang, Y.; Mao, X.; Wang, X.; Yan, Z.; Li, B.; Xu, Y.; Yu, M.; et al. Biallelic Mutations in PATL2 Cause Female Infertility Characterized by Oocyte Maturation Arrest. *Am. J. Hum. Genet.* **2017**, *101*, 609–615. [CrossRef] [PubMed]
43. Xu, Q.; Liu, Z.; Fu, J.; Zhao, W.; Guo, Y.; Sun, X.; Zhang, H. Ratiometric electrochemical aptasensor based on ferrocene and carbon nanofibers for highly specific detection of tetracycline residues. *Sci. Rep.* **2017**, *7*, 14729. [CrossRef] [PubMed]
44. Zhan, X.; Hu, G.; Wagberg, T.; Zhan, S.; Xu, H.; Zhou, P. Electrochemical aptasensor for tetracycline using a screen-printed carbon electrode modified with an alginate film containing reduced graphene oxide and magnetite (Fe_3O_4) nanoparticles. *Microchim. Acta* **2016**, *183*, 723–729. [CrossRef]
45. Shi, Z.; Hou, W.; Jiao, Y.; Guo, Y.; Sun, X.; Zhao, J.; Wang, X. Ultra-sensitive aptasensor based on IL and Fe_3O_4 nanoparticles for tetracycline detection. *Int. J. Electrochem. Sci.* **2017**, *12*, 7426–7434. [CrossRef]

46. Jahanbani, S.; Benvidi, A. Comparison of two fabricated aptasensors based on modified carbon paste/oleic acid and magnetic bar carbon paste/Fe$_3$O$_4$@oleic acid nanoparticle electrodes for tetracycline detection. *Biosens. Bioelectron.* **2016**, *85*, 553–562. [CrossRef]
47. Munteanu, F.D.; Titoiu, A.M.; Marty, J.L.; Vasilescu, A. Detection of antibiotics and evaluation of antibacterial activity with screen-printed electrodes. *Sensors* **2018**, *18*, 901. [CrossRef]
48. Han, Q.; Wang, R.; Xing, B.; Chi, H.; Wu, D.; Wei, Q. Label-free photoelectrochemical aptasensor for tetracycline detection based on cerium doped CdS sensitized BiYWO6. *Biosens. Bioelectron.* **2018**, *106*, 7–13. [CrossRef]
49. Lorenzetti, A.S.; Sierra, T.; Domini, C.E.; Lista, A.G.; Crevillen, A.G.; Escarpa, A. Electrochemically Reduced Graphene Oxide-Based Screen-Printed Electrodes for Total Tetracycline Determination by Adsorptive Transfer Stripping. *Sensors* **2020**, *20*, 76. [CrossRef]
50. Mohammad-Razdari, A.; Ghasemi-Varnamkhasti, M.; Rostami, S.; Izadi, Z.; Ensafi, A.A.; Siadat, M. Development of an electrochemical biosensor for impedimetric detection of tetracycline in milk. *J. Food Sci. Technol.* **2020**, *57*, 4697–4706. [CrossRef] [PubMed]
51. Tang, Y.; Liu, P.; Xu, J.; Li, L.; Yang, L.; Liu, X.; Liu, S.; Zhou, Y. Electrochemical aptasensor based on a novel flower-like TiO$_2$ nanocomposite for the detection of tetracycline. *Sens. Actuators B Chem.* **2018**, *258*, 906–912. [CrossRef]
52. Calixto, C.M.F.; Cavalheiro, É.T.G. Determination of Tetracycline in Bovine and Breast Milk Using a Graphite–Polyurethane Composite Electrode. *Anal. Lett.* **2017**, *50*, 2323–2334. [CrossRef]
53. Abraham, T.; Gigimol, M.G.; Priyanka, R.N.; Susan, M.; Korah, B.K.; Mathew, B. In-situ fabrication of Ag$_3$PO$_4$ based binary composite for the efficient electrochemical sensing of tetracycline. *Mater. Lett.* **2020**, *279*, 128502. [CrossRef]
54. Calixto, C.M.F.; Cervini, P.; Cavalheiro, É.T.G. Determination of tetracycline in environmental water samples at a graphite-polyurethane composite electrode. *J. Braz. Chem. Soc.* **2012**, *23*, 938–943. [CrossRef]
55. Calixto, C.M.F.; Cavalheiro, É.T.G. Determination of Tetracyclines in Bovine and Human Urine using a Graphite-Polyurethane Composite Electrode. *Anal. Lett.* **2015**, *48*, 1454–1464. [CrossRef]
56. Que, X.; Chen, X.; Fu, L.; Lai, W.; Zhuang, J.; Chen, G.; Tang, D. Platinum-catalyzed hydrogen evolution reaction for sensitive electrochemical immunoassay of tetracycline residues. *J. Electroanal. Chem.* **2013**, *704*, 111–117. [CrossRef]
57. Wangfuengkanagul, N.; Siangproh, W.; Chailapakul, O. A flow injection method for the analysis of tetracycline antibiotics in pharmaceutical formulations using electrochemical detection at anodized boron-doped diamond thin film electrode. *Talanta* **2004**, *64*, 1183–1188. [CrossRef]
58. Charoenraks, T.; Palaharn, S.; Grudpan, K.; Siangproh, W.; Chailapakul, O. Flow injection analysis of doxycycline or chlortetracycline in pharmaceutical formulations with pulsed amperometric detection. *Talanta* **2004**, *64*, 1247–1252. [CrossRef] [PubMed]
59. Treetepvijit, S.; Preechaworapun, A.; Praphairaksit, N.; Chuanuwatanakul, S.; Einaga, Y.; Chailapakul, O. Use of nickel implanted boron-doped diamond thin film electrode coupled to HPLC system for the determination of tetracyclines. *Talanta* **2006**, *68*, 1329–1335. [CrossRef] [PubMed]
60. Allahverdiyeva, S.; Yardım, Y.; Senturk, Z. Electrooxidation of tetracycline antibiotic demeclocycline at unmodified boron-doped diamond electrode and its enhancement determination in surfactant-containing media. *Talanta* **2021**, *223*, 121695. [CrossRef] [PubMed]
61. Loetanantawong, B.; Suracheep, C.; Somasundrum, M.; Surareungchai, W. Electrocatalytic Tetracycline Oxidation at a Mixed-Valent Ruthenium Oxide-Ruthenium Cyanide-Modified Glassy Carbon Electrode and Determination of Tetracyclines by Liquid Chromatography with Electrochemical Detection. *Anal. Chem.* **2004**, *76*, 2266–2272. [CrossRef] [PubMed]
62. Charoenraks, T.; Chuanuwatanakul, S.; Honda, K.; Yamaguchi, Y.; Chailapakul, O. Analysis of tetracycline antibiotics using HPLC with pulsed amperometric detection. *Anal. Sci.* **2005**, *21*, 241–245. [CrossRef] [PubMed]
63. Pizan-Aquino, C.; Wong, A.; Aviles-Felix, L.; Sotomayor, M.D.P.T. Evaluation of the performance of selective M-MIP to tetracycline using electrochemical and HPLC-UV method. *Mater. Chem. Phys.* **2020**, *245*, 122777. [CrossRef]
64. Ji, H.; Wang, E. Flow injection amperometric detection based on ion transfer across a water—Solidified nitrobenzene interface for the determination of tetracycline and terramycin. *Analyst* **1988**, *113*, 1541–1543. [CrossRef]
65. Agüí, L.; Guzman, A.; Pedrero, M.; Yáñez-Sedeño, P.; Pingarrón, J.M. Voltametric and flow injection determination of oxytetracycline residues in food samples using carbon fiber microelectrodes. *Electroanalysis* **2003**, *15*, 601–607. [CrossRef]
66. Oungpipat, W.; Southwell-Keely, P.; Alexander, P.W. Flow injection detection of tetracyclines by electrocatalytic oxidation at a nickel-modified glassy carbon electrode. *Analyst* **1995**, *120*, 1559–1565. [CrossRef]
67. Hou, W.; Wang, E. Liquid chromatographic determination of tetracycline antibiotics at an electrochemically pre-treated glassy carbon electrode. *Analyst* **1989**, *114*, 699–702. [CrossRef]
68. Felipe Montiel, N.; Parrilla, M.; Beltran, V.; Nuyts, G.; Van Durme, F.; De Wael, K. The opportunity of 6-monoacetylmorphine to selectively detect heroin at preanodized screen printed electrodes. *Talanta* **2021**, *226*, 122005. [CrossRef] [PubMed]
69. Schram, J.; Parrilla, M.; Sleegers, N.; Samyn, N.; Bijvoets, S.M.; Heerschop, M.W.J.; van Nuijs, A.L.N.; De Wael, K. Identifying electrochemical fingerprints of ketamine with voltammetry and LC-MS for its detection in seized samples Identifying electrochemical fingerprints of ketamine with voltammetry and LC-MS for its detection in seized samples. *Anal. Chem.* **2020**, *92*, 13485–13492. [CrossRef]
70. Van Echelpoel, R.; de Jong, M.; Daems, D.; Van Espen, P.; De Wael, K. Unlocking the full potential of voltammetric data analysis: A novel peak recognition approach for (bio)analytical applications. *Talanta* **2021**, *233*, 122605. [CrossRef]

71. Mirceski, V.; Skrzypek, S.; Stojanov, L. Square-wave voltammetry. *ChemTexts* **2018**, *4*, 1–14. [CrossRef]
72. Wang, H.; Zhao, H.; Quan, X. Gold modified microelectrode for direct tetracycline detection. *Front. Environ. Sci. Eng. China* **2012**, *6*, 313–319. [CrossRef]
73. Dang, X.; Hu, C.; Wei, Y.; Chen, W.; Hu, S. Sensitivity improvement of the oxidation of tetracycline at acetylene black electrode in the presence of sodium dodecyl sulfate. *Electroanalysis* **2004**, *16*, 1949–1955. [CrossRef]
74. Turku, I.; Sainio, T.; Paatero, E. Thermodynamics of tetracycline adsorption on silica. *Environ. Chem. Lett.* **2007**, *5*, 225–228. [CrossRef]
75. Ghadim, E.E.; Manouchehri, F.; Soleimani, G.; Hosseini, H.; Kimiagar, S.; Nafisi, S. Adsorption properties of tetracycline onto graphene oxide: Equilibrium, kinetic and thermodynamic studies. *PLoS ONE* **2013**, *8*, e79254. [CrossRef]
76. Schram, J.; Parrilla, M.; Sleegers, N.; Van Durme, F.; Van Den Berg, J.; van Nuijs, A.L.N.; De Wael, K. Electrochemical profiling and LC-MS characterization of synthetic cathinones: From methodology to detection in forensic samples. *Drug Test. Anal.* **2021**. [CrossRef]
77. Wu, J.; Zhang, H.; Oturan, N.; Wang, Y.; Chen, L.; Oturan, M.A. Application of response surface methodology to the removal of the antibiotic tetracycline by electrochemical process using carbon-felt cathode and DSA (Ti/RuO_2-IrO_2) anode. *Chemosphere* **2012**, *87*, 614–620. [CrossRef]
78. Wang, J.; Zhi, D.; Zhou, H.; He, X.; Zhang, D. Evaluating tetracycline degradation pathway and intermediate toxicity during the electrochemical oxidation over a Ti/Ti_4O_7 anode. *Water Res.* **2018**, *137*, 324–334. [CrossRef] [PubMed]
79. Forsberg, K.J.; Patel, S.; Wencewicz, T.A.; Dantas, G. The Tetracycline Destructases: A Novel Family of Tetracycline-Inactivating Enzymes. *Chem. Biol.* **2015**, *22*, 888–897. [CrossRef]
80. Volkers, G.; Petruschka, L.; Hinrichs, W. Recognition of drug degradation products by target proteins: Isotetracycline binding to tet repressor. *J. Med. Chem.* **2011**, *54*, 5108–5115. [CrossRef]

Article

A Microphysiometric System Based on LAPS for Real-Time Monitoring of Microbial Metabolism

Nan Jiang [1,2,3], Tao Liang [4], Chunlian Qin [1], Qunchen Yuan [1], Mengxue Liu [1], Liujing Zhuang [1,2,3] and Ping Wang [1,2,3,*]

1. Biosensor National Special Laboratory, Key Laboratory for Biomedical Engineering of Education Ministry, Department of Biomedical Engineering, Zhejiang University, Hangzhou 310027, China; 21915031@zju.edu.cn (N.J.); lotusqin@zju.edu.cn (C.Q.); sumyuan@zju.edu.cn (Q.Y.); liumx15015@zju.edu.cn (M.L.); liujing123@zju.edu.cn (L.Z.)
2. The MOE Frontier Science Center for Brain Science & Brain-Machine Integration, Zhejiang University, Hangzhou 310058, China
3. State Key Laboratory of Transducer Technology, Chinese Academy of Sciences, Shanghai 200050, China
4. Research Center for Quantum Sensing, Zhejiang Lab, Hangzhou 310000, China; cooltao@zju.edu.cn
* Correspondence: cnpwang@zju.edu.cn

Abstract: Macronutrients (carbohydrates, fat and protein) are the cornerstones of daily diet, among which carbohydrates provide energy for the muscles and central nervous system during movement and exercise. The breakdown of carbohydrates starts in the oral cavity, where they are primarily hydrolyzed to glucose and then metabolized to organic acids. The end products may have an impact on the oral microenvironment, so it is necessary to monitor the process of microbial metabolism and to measure the pH change. Although a pH meter has been widely used, it is limited by its sensitivity. We then introduce a light addressable potentiometric sensor (LAPS), which has been used in extracellular acidification detection of living cells with the advantages of being objective, quantitative and highly sensitive. However, it is difficult to use in monitoring bacterial metabolism because bacteria cannot be immobilization on the LAPS chip as easily as living cells. Therefore, a microphysiometric system integrated with Transwell insert and microfluidic LAPS chip was designed and constructed to solve this problem. The decrease in pH caused by glucose fermentation in *Lactobacillus rhamnosus* was successfully measured by this device. This proves the feasibility of the system for metabolism detection of non-adhere targets such as microorganisms and even 3D cells and organoids.

Keywords: microphysiometer; LAPS; pH measurement; biosensor; microbial metabolism

1. Introduction

Macronutrients (carbohydrates, fat and protein) are the main components of most foods and beverages. As essential energy sources for humans, the breakdown of macronutrients starts in the oral cavity. The primary hydrolysis of the ingested biomolecules is catalyzed by digestive enzymes supplied by host and microbial sources [1]. Then, after complex metabolic activities, the ecology and biology of the oral microenvironment changes due to a series of end products. Carbohydrates, for example, are primarily hydrolyzed into monosaccharides such as glucose, which can be further catabolized by microorganisms in the oral cavity. Organic acids and other substances produced by metabolism can impact the balance of the local oral environment and can even lead to diseases such as tooth decay. Hence, it is necessary to monitor the process of microbial metabolism.

The microbial ecology of the oral cavity is very complex, with over 700 species of bacteria colonizing on mucosal and tooth surfaces [2]. Numerous studies have shown that lactic acid bacteria (LAB) exist in human oral cavity, especially in saliva, as normal flora [3,4], by which monosaccharides can be fermented to lactic acid and other organic acids [5]. By contrast, sugar substitutes that are increasingly used in food to replace sucrose cannot be broken down by LAB. The purpose of the present study is to detect the pH change in glucose and sugar substitutes solutions caused by metabolism of LAB and to provide the basis for studying the change in oral microenvironment.

It has been reported that a pH meter was used to detect the pH-lowering potential caused by acid production of *Lactobacillus rhamnosus* [3], but this method has low sensitivity and is time-consuming. In contrast, light addressable potentiometric sensor (LAPS) is more suitable for monitoring metabolic activity of microorganisms due to its outstanding superiority with its high sensitivity toward pH [6,7]. However, LAPS was commonly used to detect the metabolism of living cells, which can be immobilized [8,9]. Only a few studies have investigated the acidification of non-adherent targets such as microorganism directly in suspension [10–12]. In this work, a microphysiometer based on the LAPS principle and a microfluidic system combined with an innovative structure fabricated by a Transwell insert was constructed for real-time and long-term monitoring of fermentation process conducted by *Lact. rhamnosus*. Compared with the traditional microphysiometer [13], the combination of Transwell and LAPS is a simpler and lower-cost structure, which can be promoted as a model system for detecting the metabolism of microorganisms and other non-adherent targets.

2. Methods and Experiments
2.1. Principle
2.1.1. Biological Basis of Metabolism in LAB

According to the pathways of glucose fermentation with different end-products, LAB can be categorized into three main types: homofermentative, heterofermentative and facultative heterofermentative. *Lact. rhamnosus* belongs to the third type [3], which can either produce several end products such as lactic acid and acetic acid similar to heterofermentative bacteria or only produce lactic acid such as homofermentative bacteria. Figure 1 illustrates the principal steps involved in the fermentation process of *Lact. rhamnosus*, in which protons are generated and released using glucose as a carbon source. In the homolactic fermentation, one molecule of glucose is transformed to fructose-1,6-diphophate and then converted to two pyruvate molecules through the Embden-Meyerhof-Parnas (EMP) pathway. Lactic acid is formed by the reduction of pyruvate. In contrast, during the process of heterolactic fermentation, glucose is converted to xylulose-5-phosphate which is then catalyzed by phosphoketolase to glyceraldehyde-3-phosphate and acetyl phosphate. The former is transformed into lactic acid, while the latter is transformed into acetic acid [5,14,15]. In conclusion, regardless of the kind of fermentation, a large amount of acidic substance is produced, so the physiological state and activity of LAB can be reflected by monitoring the pH change during the metabolism process.

2.1.2. The Principle of LAPS Based Bacterial Metabolism Detection

The principle of detecting acidic metabolites of LAB by LAPS was similar to that of cellular acidification detection described previously, Si_3N_4 was still used as the H^+ sensitive material [8,9]. The H^+ concentration in the solution can affect the surface potential of the LAPS chip, which can affect the width of the depletion layer formed between the silicon layer and the insulating layer when a DC bias voltage is applied onto the reference electrode (RE) of the sensor. Due to the frequency-modulated illumination at the bottom of the sensor, hole–electron pairs and, subsequently, an AC photocurrent with the same frequency are generated [16]. The back-side illumination was chosen instead of front-side because the solution with bacteria is a suspension and bacteria are difficult to be immobilized on the surface of sensor chip. Since the light needs to pass through the whole silicon layer to reach the depletion layer, the LAPS chip was partially thinned to improve the signal-to-noise ratio of the sensor. A schematic illustration of the thinned LAPS chip for detecting acidic metabolites is depicted in Figure 2a.

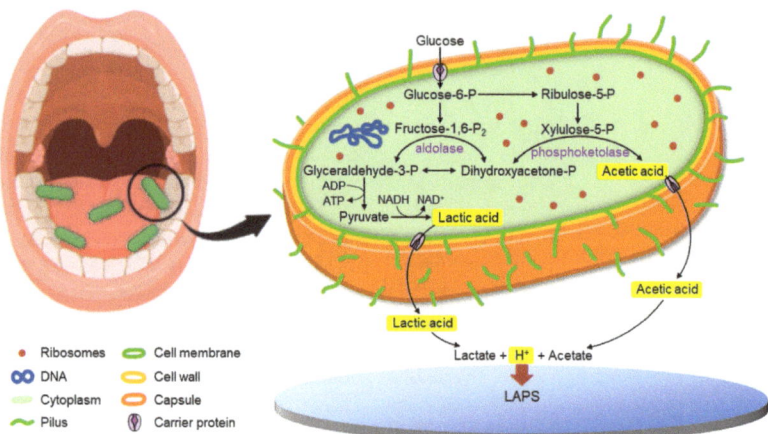

Figure 1. Schematic of the main steps of glucose metabolism in lactic acid bacteria in human oral cavity. The major intermediate and end products are shown in yellow background, of which lactic acid and acetic acid are the key substance causing pH change in the solution.

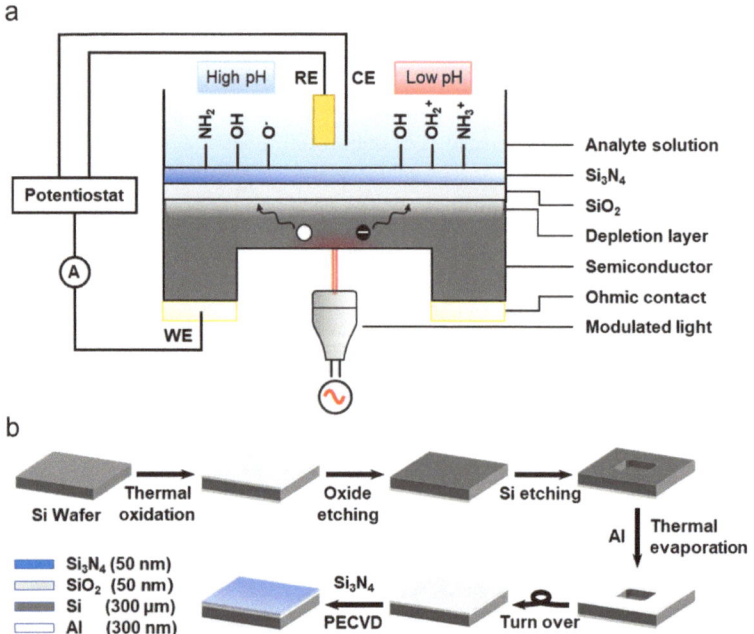

Figure 2. The detection principle and fabrication of the thinned LAPS chip. (**a**) The principle of acidic metabolites detection by the thinned LAPS chip. (**b**) The processing flow of the thinned LAPS chip.

2.2. Sensor Preparation

2.2.1. LAPS Chip Fabrication

The LAPS chip was partially thinned to achieve a backside illumination. Silicon was used as the sensor substrate, SiO_2 was used as the insulating layer, and Si_3N_4 was used as the sensing material for pH detection. The procedures of the LAPS chip fabrication are demonstrated in Figure 2b, and the 300 μm single-polished silicon wafer (4″, <100>) was thermally oxidized to grow 50 nm SiO_2 layers on both sides. Then, the SiO_2 on the rough side was removed by 5% HF etching, and the central part of the Si substrate was thinned to

100 µm. After deionized water washing and nitrogen drying, a 300 nm aluminum layer was thermally evaporated outside the thinned part to form an ohmic contact. Finally, a 50 nm Si_3N_4 layer was deposited by plasma-enhanced chemical vapor deposition (PECVD) to achieve a pH sensor. The processed LAPS chip was cut into 1.7 cm × 1.7 cm with a thinned central part of 7 mm × 7 mm and then in turn washed by acetone, ethanol and deionized water for storage and use. The thicknesses of the Si_3N_4 and SiO_2 layer and the thinned central part were confirmed using a field emission scanning electron microscope (Zeiss, Sigma300, Jena, Germany) and a surface profiler (KLA–Tencor, Alpha Step D–100, San Jose, CA, USA), respectively (Figure S1).

2.2.2. Manufacture of Sensor Unit with Transwell Insert

The Transwell insert with a 0.4 µm pore polycarbonate membrane (6.5 mm, Corning, CLS3413, New York, NY, USA), which can block the LAB while allowing liquids to pass through, was used innovatively to solve the problem that bacteria are difficult to adhere to the surface of LAPS chip. The polydimethylsiloxane (PDMS) chamber was made by mold casting, and the diameter of the through hole in the center was 9.5 mm to fix the Transwell insert. Then, the PDMS chamber was bonded with the LAPS chip by O_2 plasma treatment, and an O-ring was placed at the bottom of the chamber to maintain a constant space between the Transwell insert and the sensor chip. The LAB was placed within the scope of the O-ring. The PDMS plug in the Transwell insert was also achieved by the casting method, with two holes used as the inlet and outlet made by a puncher (0.5 mm, WPI). Then, two stainless-steel catheters (20 ga × 15 mm, Instech laboratories, Plymouth Meeting, PA, USA) were fixed into the inlet and outlet connecting the microbore tubings (0.031″ ID × 0.094″ OD, Cole-Parmer, Vernon Hills, IL, USA). The plug (with the stainless-steel catheters) was put into the upper compartment of the Transwell insert, and a small amount of uncured PDMS was used to encapsulate the edges to prevent liquid leakage. It is worth noting that the PDMS stopper was in the shape of a truncated cone, with the bottom edge closely attached to the inner wall of the Transwell insert to prevent contamination of the polycarbonate membrane from the uncured PDMS. The backside of the chip was pasted on a PCB pad with a hole using conductive silver glue to expose the thinned area and connected to the working electrode (WE). The counter electrode (CE) was connected to the stainless-steel catheter of the outlet. The reference electrode was held in a reservoir made of a PE tube, and the hole on the side wall was made to maintain the liquid volume. Figure 3a,b describe the composition of the sensor unit with Transwell insert, and the physical map is shown in Figure 3c.

2.3. LAPS System Set-Up

A 3D-printed resin scaffold was used to fix the optical path and the sensor unit (Figure 3d). The modulated laser with a wavelength at 685 nm was connected to the adjustable focusing collimator via pigtail. The parallel light after collimation (spot of about 1 mm) passed through the 45° optical mirror and illuminated vertically to the thinned area on the back of the LAPS chip. The optical mirror was fixed in a groove at the bottom of the resin scaffold so that the front and back positions could be fine-tuned when necessary. The sensor unit was fixed by the pin header and female header on the PCB pad, which was convenient for installation and disassembly. The scaffold and the sensor unit were placed together in a 37 °C thermostatic incubator for experiment. The medium was delivered to the chamber by the syringe pump (TYD02-10, Lead Fluid Technology Co., Ltd., Baoding, China). The other parts of the detection system were similar to those in a previous study [9].

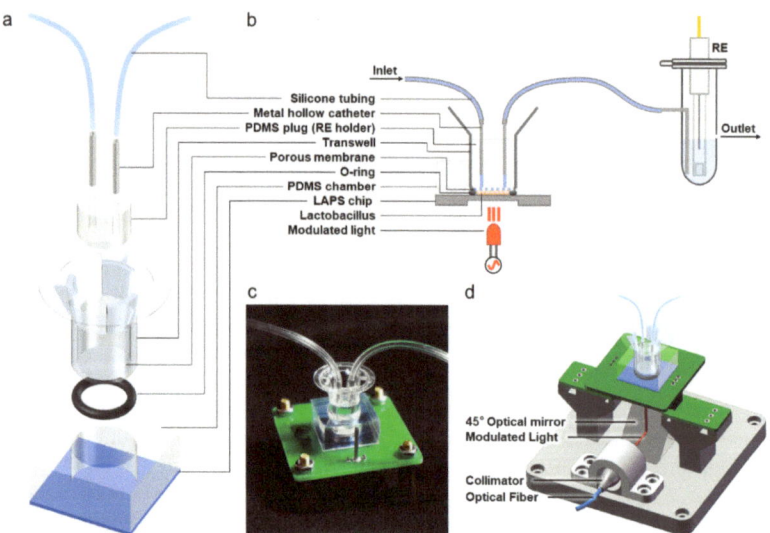

Figure 3. Schematic of microphysiometer sensor unit with the Transwell insert. (**a**) Schematic of the detection chamber fabricated by the Transwell insert with a microporous membrane. (**b**) Schematic of the microphysiometer sensor unit with a microfluidic structure. (**c**) Photography of the device. (**d**) Schematic of the 3D-printed resin scaffold used to fix the optical path and the sensor unit.

2.4. Lactobacillus rhamnosus Cultivation

The *Lactobacillus rhamnosus* (ATCC 7469) strain was purchased from the China general microbiological culture collection center (CGMCC). Before the experiment, the *Lact. rhamnosus* was inoculated into MRS (Solarbio, M8540, Beijing, China) broth medium, which was sterilized at 121 °C for 15 min. The bacteria were cultivated in an incubator (37 °C) for two generation to restore its activity and were identified by colony morphology and Gram staining. An optical density (OD) measurement (λ = 600 nm, OD_{600}) was carried out to evaluate the growth density and to obtain the growth curve of *Lact. rhamnosus* (SpectraMax Paradigm, Molecule Device, Sunnyvale, CA, USA).

2.5. Preparation of Test Solutions

Glucose (Macklin, G6172), which is commonly used in food, as well as a commercial sugar substitute (Shandong Youlezi Biotechnology Co., Ltd., Zibo, China) wer selected. Glucose is about 0.75 times sweeter than sucrose, while the substitute sweetener, which is a mixture of sucralose and erythritol, is about twice as sweet as sucrose. Most commercially available sugary beverages have 8–12% (w/v) added sugar, so the concentration of glucose solution was chosen as 12% (w/v) and the concentration of substitute sweetener of corresponding sweetness was 4.5% (w/v). All sugar solutions were prepared with artificial saliva (Dongguan Xinheng Technology Co., Ltd., Dongguan, China) to mimic the human oral environment while providing supporting electrolytes. The artificial saliva containing the same inorganic components as actual saliva was prepared according to ISO 10271 and was sterilized. The artificial saliva without sugar was used as a blank control group. Moreover, the 2.4% (w/v) glucose solution, which has a similar concentration to the glucose in the MRS broth medium, was selected for the experiment as well to verify that the acidification rate of glucose metabolism was different in glucose solutions of different concentrations. In addition, 1% (w/v) tryptone (Solarbio, T8490) was added to each solution to ensure the survival of *Lact. Rhamnosus*.

2.6. Bacterial Metabolism Monitoring

At the beginning of each measurement, the bacterial suspension was centrifuged at 3000 r/min for 10min and washed three times with artificial saliva, and the pellet was then resuspended in about 40 μL of artificial saliva. Subsequently, *Lact. rhamnosus* in suspension was pipetted into the sensor chamber formed by the Transwell insert, sensor chip and O-ring, where the bacteria were entrapped by the microporous membrane, while the nutrients and metabolites can pass through freely. The microbore tubing at the inlet was connected to the syringe pump and that at the outlet was connected to the PE tube. Glucose and substitute sweetener solutions were injected at 6 min intervals by the syringe pump to provide nutrients for the bacteria. The medium temperature was maintained at a constant 37 °C using a thermostatic water bath for the duration of the experiment. It is worth noting that all the materials such as O-ring, silicone tubing and metal hollow catheter were autoclaved prior to use and kept sterile until needed in order to avoid biofouling, which may affect the accuracy of the results.

3. Results and Discussion

3.1. Sensor Unit Characteristic Test

Phosphate buffer saline (PBS) was used in the experiment for sensor calibration, and the pH in a series of PBS solutions was adjusted between 3.2 and 7.2 using 0.1 mol/L HCl and NaOH solutions. The measurement parameters were as follows: the bias voltage between RE and WE ranged from −5500 mV to −1000 mV, with a step voltage of 10 mV. The modulated light frequency was 10 kHz with a wavelength at 685 nm. Figure 4a shows the detected I–V (photocurrent vs. bias voltage) curves in solutions with different pH gradients. It can be observed that the I–V curve shifts to the left as the pH decreases. The photocurrent value in the greatest slope was selected as the working point [17]. In this work, the goal was to detect the bacterial acidification; the pH change was small and did not exceed the linear part of I–V curve. Therefore, the constant voltage mode was chosen and the sensitivity at the working point was 93.73 nA/pH, as shown in Figure 4b.

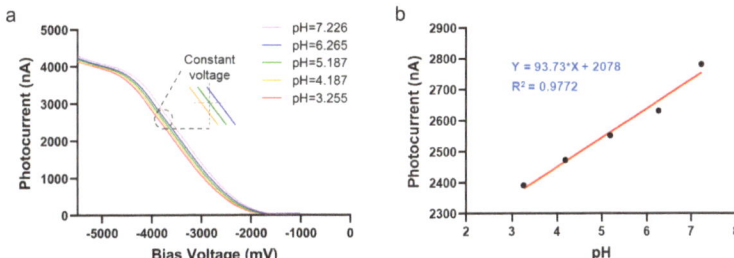

Figure 4. Performance evaluation of sensor unit. (**a**) I–V curves in different gradient pH solutions. (**b**) Sensor calibration curve in constant voltage mode (U = −3760 mV).

In addition, the stability and repeatability of the LAPS sensor was verified by PBS solution (pH = 7.4). In constant voltage mode, the photocurrent was continuously recorded for more than 30 min at the working point voltage. In three replicate experiments, photocurrent showed good stability, and the similar results also demonstrated good repeatability (Figure S2a). The fluid exchange capability of the designed microfluidic system with porous membrane was also verified. The photocurrent curve was obtained by alternately injecting PBS and 10-fold diluted PBS into the sensor chamber because the photocurrent amplitude was affected by the impedance of the solution (Figure S2b). The photocurrent value varied with the solution concentration and showed good consistency when the solution was repeatedly injected. Thus, the microfluidic system with porous membrane presents good fluid exchange capability, which can be used for bacterial metabolism monitoring.

3.2. Lactobacillus rhamnosus Cultivation and Identification

Figure 5a,b showed that the colony of *Lact. rhamnosus* was light yellow, smooth, moist, round with neat edges, slightly raised center, and about 1 mm in diameter. Microscopically, the bacteria had a shape of a slender, flexuous rod and yielded a positive result in Gram stain test (Figure 5c), which was consistent with the characteristics of *Lact. rhamnosus*.

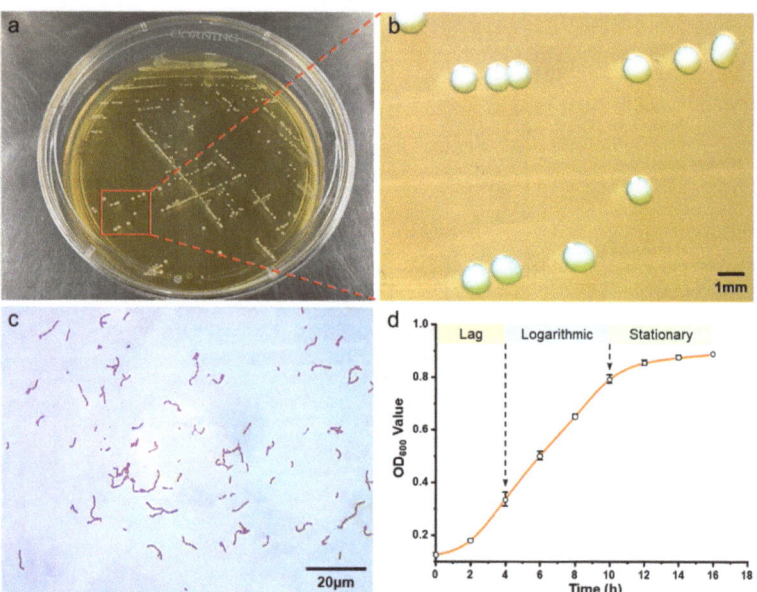

Figure 5. Morphological characteristics and growth curve of Lactobacillus rhamnosus. (**a**) The growth state of *Lact. rhamnosus* on MRS agar plate. (**b**) Observation of single colony under an optical microscope. (**c**) Bacterial morphology with Gram staining (×100 magnification). (**d**) Growth curve of *Lact. rhamnosus*.

In order to create a growth curve of *Lact. rhamnosus*, the OD_{600} value of the bacterial suspension was detected every two hours until the growth of the bacteria became visibly slow. As shown in Figure 5d, the curve was S-shaped, indicating that *Lact. rhamnosus* underwent various stages from rapid proliferation to steady growth. The bacteria entered the logarithmic phase at 2–4 h and the number began to increase exponentially. Moreover, the curve tended to be flat after 10 h, demonstrating that the bacteria entered a stationary phase. The bacteria at post-log phase, that is, about 8 h after inoculation into MRS broth medium, were chosen because of the strong reproduction ability and vigorous metabolism, which was conductive to the experiment.

3.3. Lactobacillus rhamnosus Metabolism Monitoring by the Microphysiometer

The sensor unit was always placed in the 37 °C thermostatic incubator during the experiment to maintain the strain vitality, and all of the medium was placed in the 37 °C thermostatic water bath. Initially, the 2.4% glucose solution was injected until there was liquid flowing out of the hole on the side wall of the PE tube. The bias voltage of the working point was applied, and the photocurrent was recorded continuously under constant voltage mode. The solution was injected at 6 min intervals by the syringe pump, and the photocurrent–time curve was obtained. After five flow–stop cycles, 12% glucose solution was injected, replacing the solution in the whole device completely, and the acidity change was also recorded, followed by the 4.5% sugar substitute solution and artificial saliva. The blank group had the same condition as the experimental group except the absence of *Lact. rhamnosus*.

According to the formula in Figure 4b, the photocurrent value was converted to a pH value to obtain the pH–time curves, which demonstrated the pH changes in the process of *Lact. rhamnosus* metabolism (Figure 6a). In the 2.4% glucose solution, the pH decreased because *Lact. rhamnosus* used glucose to produce acid substances such as lactic acid. After the glucose concentration was increased to 12%, a steeper decrease in pH occurred due to the enhanced metabolism. However, sucralose and erythritol could not be utilized as a carbon source by *Lact. rhamnosus*, so the pH the solution hardly changed, which was similar to the results of artificial saliva.

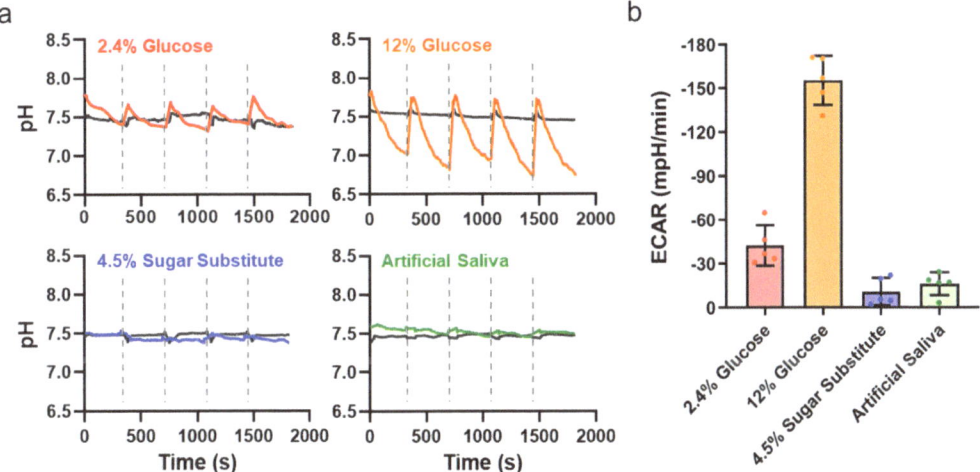

Figure 6. The pH change and ECAR values of sweetener metabolism of *Lactobacillus rhamnosus*. (**a**) The pH changed with time, and a fresh medium was injected every 360s; the 2.4% (*w*/*v*) glucose solution, the 12% (*w*/*v*) glucose solution, the 4.5% (*w*/*v*) sugar substitute solution and artificial saliva were injected successively. (**b**) The average ECAR values of different sweetener solutions and artificial saliva.

Due to the addition of fresh solution, the peak of the photocurrent appeared at the beginning of each cycle. It could be seen from the blank group that the photocurrent decreased slightly after 80 s in each cycle; thus, it could be inferred that the change in pH after 80 s was caused by bacterial metabolism. Therefore, the difference in pH change between the experimental group and the blank group 80 s after the beginning of each cycle was defined as ΔpH. Then, the extracellular acidification rate (ECAR) was obtained by the ratio of the ΔpH value to the time interval (280 s), and the average ECAR values of each group were calculated and shown in Figure 6b. The average of ECAR in the 2.4% glucose solution was -42.38 mpH/min but changed to -155.41 mpH/min in the 12% glucose solution. Additionally, the averages of ECAR in the 4.5% sugar substitute solution and artificial saliva were -10.79 mpH/min and -16.12 mpH/min, respectively. It can be concluded that the glucose fermentation in *Lact. rhamnosus* caused the pH of the solution to decrease, and with the increase in glucose concentration, the acidification rate also significantly increased. However, the pH of the sugar substitute solution and artificial saliva almost did not change, which confirmed that sugar substitute combined with sucralose and erythritol could not be utilized by *Lact. rhamnosus*.

In a previous study, the 24 h pH drop caused by glucose fermentation of LAB include *Lact. rhamnosus* was detected using a pH meter [3]. However, it has obvious limitations, such as the accuracy and sensitivity not being high and achieving real-time detection being difficult, which were overcome in this study with the newly designed microphysiometric system. In addition, there also exist several studies that use LAPS-based differential sensors to detect the extracellular acidification of bacteria [10,11]. Since the solution cannot be renewed, there are still problems in realizing continuous long-term monitoring. The results of this study confirmed that the microphysiometric system integrated with the Transwell insert and LAPS could monitor the metabolism of microorganisms such as *Lact. rhamnosus* in real time and with high sensitivity. The microfluidic system ensured continuous renewal of the medium to achieve long-term monitoring, and the stable illumination avoided disturbances in the large photocurrent fluctuation. In this way, long-term monitoring for metabolism of living microorganisms can be achieved.

3.4. Verification of Activity of Lactobacillus rhamnosus after Experiment

After the experiment, the bacterial suspension was collected and streaked on a MRS agar plate. After 48 h of incubation at 37 °C, single colonies of *Lact. rhamnosus* but no other types of colonies were observed (Figure S3). This indicated that the bacteria remained alive after a long time of detection, and there was no other bacterial contamination during the experiment. However, it could be found that the colony density was not high, which indicated that the number of viable bacteria decreased, or the ability of growth and reproduction reduced. It may be because the bacteria had left the most suitable medium environment for a long time, and sugar substitute and artificial saliva could not be used as carbon sources. According to the results, it can be concluded that the pH change in the glucose solution was caused by the metabolism of *Lact. rhamnosus*.

4. Conclusions

Although LAPS is widely used in extracellular acidification detection of living cells, it is rarely used to investigate non-adherent targets directly in suspension. In this study, a microphysiometric system with a novel structure fabricated by a Transwell insert and a microfluidic LAPS chip was used to detect the pH change in glucose and sugar substitutes solutions in the presence of *Lact. rhamnosus*, respectively. All of the solutions were prepared with artificial saliva to mimic the human oral environment. A polydimethylsiloxane (PDMS) chamber was manufactured for bacterial culture, and microfluidic flow paths were designed for medium delivery. A Transwell insert with a 0.4 µm pore polycarbonate membrane was used innovatively to solve the problem that adhering bacteria to the surface of a LAPS chip is difficult, which not only prevents bacteria from being washed away but also allows nutrients and metabolites to pass through. Additionally, the existence of O-ring restricts the bacteria to the effective detection area and avoids contact between the Transwell and the LAPS chip. Similar to previous studies, the experiment was carried out in several flow–stop cycles and compared with the control groups without the presence of bacteria. *Lact. rhamnosus* caused a decrease in the pH of the glucose solution, and the change became more significant when the glucose concentration increased, while the pH of other solutions hardly changed. The results presented that this microphysiometric system performed well in real-time monitoring of microbial metabolism. The functions of a traditional microphysiometer were achieved with simpler structure devices. Although changes in acidity can be evaluated by sensory evaluation, the pH change can be measured more objectively and quantitatively by this microphysiometric system. Further work can be carried out to improve the sensitivity of the sensor, such as using Al_2O_3 [18] or Ta_2O_5 [19] layers deposited on an insulator. In summary, our work presents a model system for non-adhere targets such as microorganisms and even 3D cells and organoids.

Supplementary Materials: The following supporting information can be downloaded at https://www.mdpi.com/article/10.3390/chemosensors10050177/s1, Figure S1: Verification of the thickness of the thinned LAPS chip; Figure S2: Verification of stability, repeatability and fluid exchange capability of the system; Figure S3: Verification of the activity of *Lactobacillus rhamnosus* after experiment.

Author Contributions: Conceptualization, N.J., T.L. and L.Z.; methodology, N.J., T.L. and Q.Y.; validation, N.J., T.L. and C.Q.; formal analysis and investigation, N.J. and T.L.; resources, N.J., C.Q. and M.L.; data curation, N.J. and T.L.; writing—original draft preparation, N.J. and T.L.; writing—review and editing, L.Z. and P.W. All authors have read and agreed to the published version of the manuscript.

Funding: This research was funded by the National Natural Science Foundation of China (No. 62120106004), National Key Research and Development Program of China (2021YFF1200803, 2021YFB3200801 and SQ2021AAA010032), Natural Science Foundation of Zhejiang Province (No. LBY21H180001 and LY21C100001) and China Postdoctoral Science Foundation Funded Project (No. 2020M671728 and 2021T140605).

Institutional Review Board Statement: Not applicable.

Informed Consent Statement: Not applicable.

Data Availability Statement: Not applicable.

Conflicts of Interest: The authors declare no conflict of interest.

References

1. Barbour, A.; Elebyary, O.; Fine, N.; Oveisi, M.; Glogauer, M. Metabolites of the oral microbiome: Important mediators of multikingdom interactions. *FEMS Microbiol. Rev.* **2022**, *46*, fuab039. [CrossRef] [PubMed]
2. Kilian, M.; Chapple, I.L.; Hanning, M.; Marsh, P.D.; Meuric, V.; Pedersen, A.M.L.; Tonetti, M.S.; Wade, W.S.; Zaura, E. The oral microbiome—An update for oral healthcare professionals. *Br. Dent. J.* **2016**, *221*, 657–666. [CrossRef] [PubMed]
3. Badet, M.C.; Richard, B.; Dorignac, C. An in vitro study of the pH-lowering potential of salivary lactobacilli associated with dental caries. *J. Appl. Microbiol.* **2001**, *90*, 1015–1018. [CrossRef] [PubMed]
4. Ahrné, S.; Nobaek, S.; Jeppsson, B.; Adlerberth, I.; Wold, A.; Molin, G. The normal Lactobacillus flora of healthy human rectal and oral mucosa. *J. Appl. Microbiol.* **1998**, *85*, 88–94. [CrossRef] [PubMed]
5. Kandler, O. Carbohydrate metabolism in lactic acid bacteria. *Antonie Van Leeuwenhoek* **1983**, *49*, 209–224. [CrossRef] [PubMed]
6. Werner, C.F.; Krumbe, C.; Schumacher, K.; Groebel, S.; Spelthahn, H.; Stellberg, H.; Wagner, T.; Yoshinobu, T.; Selmer, T.; Keusgen, M.; et al. Determination of the extracellular acidification of *Escherichia coli* by a light-addressable potentiometric sensor. *Phys. Status Solidi* **2011**, *208*, 1340–1344. [CrossRef]
7. Dantism, S.; Takenaga, S.; Wagner, T.; Wagner, P.; Schöning, M.J. Differential imaging of the metabolism of bacteria and eukaryotic cells based on light-addressable potentiometric sensors. *Electrochim. Acta* **2017**, *246*, 234–241. [CrossRef]
8. Hu, N.; Wu, C.; Ha, D.; Wang, T.; Liu, Q.; Wang, P. A novel microphysiometer based on high sensitivity LAPS and microfluidic system for cellular metabolism study and rapid drug screening. *Biosens. Bioelectron.* **2013**, *40*, 167–173. [CrossRef] [PubMed]
9. Liang, T.; Gu, C.; Gan, Y.; Wu, Q.; He, C.; Tu, J.; Pan, Y.; Qiu, Y.; Kong, L.; Wan, H.; et al. Microfluidic chip system integrated with light addressable potentiometric sensor (LAPS) for real-time extracellular acidification detection. *Sens. Actuators B* **2019**, *301*, 127004. [CrossRef]
10. Dantism, S.; Takenaga, S.; Wagner, P.; Wagner, T.; Schöning, M.J. Determination of the extracellular acidification of *Escherichia coli* K12 with a multi-chamber-based LAPS system. *Phys. Status Solidi* **2016**, *213*, 1479–1485. [CrossRef]
11. Dantism, S.; Röhlen, D.; Wagner, T.; Wagner, P.; Schöning, M.J. A LAPS-Based Differential Sensor for Parallelized Metabolism Monitoring of Various Bacteria. *Sensors* **2019**, *19*, 4692. [CrossRef] [PubMed]
12. Dantism, S.; Röhlen, D.; Dahmen, M.; Wagner, T.; Wagner, P.; Schöning, M.J. LAPS-based monitoring of metabolic responses of bacterial cultures in a paper fermentation broth. *Sens. Actuators B* **2020**, *320*, 128232. [CrossRef]
13. Hafner, F. Cytosensor®Microphysiometer: Technology and recent applications. *Biosens. Bioelectron.* **2000**, *15*, 149–158. [CrossRef]
14. Eiteman, M.A.; Ramalingam, S. Microbial production of lactic acid. *Biotechnol. Lett.* **2015**, *37*, 955–972. [CrossRef] [PubMed]
15. Wang, Y.; Wu, J.; Lv, M.; Shao, Z.; Hungwe, M.; Wang, J.; Bai, X.; Xie, J.; Wang, Y.; Geng, W. Metabolism Characteristics of Lactic Acid Bacteria and the Expanding Applications in Food Industry. *Front. Bioeng. Biotechnol.* **2021**, *9*, 612285. [CrossRef] [PubMed]
16. Owicki, J.C.; Bousse, L.J.; Hafeman, D.G.; Kirk, G.L.; Olson, J.D.; Wada, H.G.; Parce, J.W. The Light-Addressable Potentiometric Sensor: Principles and Biological Applications. *Annu. Rev. Biophys. Biomol. Struct.* **1994**, *23*, 87–114. [CrossRef] [PubMed]
17. Stein, B.; George, M.; Gaub, H.; Behrends, J.; Parak, W. Spatially resolved monitoring of cellular metabolic activity with a semiconductor-based biosensor. *Biosens. Bioelectron.* **2003**, *18*, 31–41. [CrossRef]
18. Schöning, M.J.; Tsarouchas, D.; Beckers, L.; Schubert, J.; Zander, W.; Kordoš, P.; Lüth, H. A highly long-term stable silicon-based pH sensor fabricated by pulsed laser deposition technique. *Sens. Actuators B* **1996**, *35*, 228–233. [CrossRef]
19. Yoshinobu, T.; Ecken, H.; Poghossian, A.; Lüth, H.; Iwasaki, H.; Schöning, M. Alternative sensor materials for light-addressable potentiometric sensors. *Sens. Actuators B* **2001**, *76*, 388–392. [CrossRef]

Article

Real-Time Measurement of Moisture Content of Paddy Rice Based on Microstrip Microwave Sensor Assisted by Machine Learning Strategies

Jin Liu [1], Shanshan Qiu [1,*] and Zhenbo Wei [2,*]

[1] College of Materials and Environmental Engineering, Hangzhou Dianzi University, Hangzhou 310018, China
[2] Department of Biosystems Engineering, Zhejiang University, 866 Yuhangtang Road, Hangzhou 310058, China
* Correspondence: qiuss@hdu.edu.cn (S.Q.); weizhb@zju.edu.cn (Z.W.)

Abstract: Moisture content is extremely imoprtant to the processes of storage, packaging, and transportation of grains. In this study, a portable moisture measuring device was developed based on microwave microstrip sensors. The device is composed of three parts: a microwave circuit module, a real-time measurement module, and software to display the results. This work proposes an improvement measure by optimizing the thickness of paddy rice samples (8–13 cm) and adding the ambient temperatures and the moisture contents (13.66–27.02% w.b.) at a 3.00 GHz frequency. A random forest, decision tree, k-nearest neighbor, and support vector machine were applied to predict the moisture content in the paddy rice. Microwave characteristics, phase shift, and temperature compensation were selected as the input variables to the prediction models, which have achieved high accuracy. Among those prediction models, the random forest model yielded the best performance with highest accuracy and stability (R^2 = 0.99, RMSE = 0.28, MAE = 0.26). The device showed a relatively stable performance (the maximum average absolute error was 0.55%, the minimum absolute error was 0.17%, the mean standard deviation was 0.18%, the maximum standard deviation was 0.41%, and the minimum standard deviation was 0.08%) within the moisture content range of 13–30%. The instrument has the advantages of real-time, simple structure, convenient operation, low cost, and portability. This work is expected to provide an important reference for the real-time in situ measurement of agricultural products, and to be of great significance for the development of intelligent agricultural equipment.

Keywords: microstrip microwave sensor; microwave attenuation; phase shift; moisture content; random forest

Citation: Liu, J.; Qiu, S.; Wei, Z. Real-Time Measurement of Moisture Content of Paddy Rice Based on Microstrip Microwave Sensor Assisted by Machine Learning Strategies. *Chemosensors* **2022**, *10*, 376. https://doi.org/10.3390/chemosensors10100376

Academic Editors: Manel del Valle and Chunsheng Wu

Received: 29 June 2022
Accepted: 13 September 2022
Published: 20 September 2022

Publisher's Note: MDPI stays neutral with regard to jurisdictional claims in published maps and institutional affiliations.

Copyright: © 2022 by the authors. Licensee MDPI, Basel, Switzerland. This article is an open access article distributed under the terms and conditions of the Creative Commons Attribution (CC BY) license (https://creativecommons.org/licenses/by/4.0/).

1. Introduction

Moisture content is an important index that reflects the freshness, preservation state, and internal quality of agricultural products [1], and it is a key measurement and control parameter in the processes of purchasing, processing, storing, and transporting. High moisture content leads to grain mildew, shortened storage time, and effects on the food taste [2]. Grain moisture content is a crucial factor for the management of harvesting and postharvest operations and is a basis for pricing in the grain trade [3].

The grain moisture content generally is measured by direct-contact methods or indirect-noncontact methods [4]. The direct-contact methods usually employ desiccation or a chemical method to measure the moisture content [5], such as the oven-drying method and the Karl Fisher titration method, which have the advantages of simple steps and high accuracy of the detection result. However, the processes of sample preparation and the long operation of inspection equipment would be occurred, which is not suitable for the real-time or field measurements there [6]. By detecting the variable parameters related to the water content of materials, indirect-noncontact methods, such as the neutron gauge [7], resistance detection [8], the capacitance method [9], near-infrared spectroscopy [10], and the microwave-drying method [11], serve to determine the moisture content in grains. Indirect-contact methods generally have a high measurement speed but usually have their own disadvantages, respectively. For near-infrared spectroscopy, the absorption spectrum varies with the moisture content of the material, and the infrared penetration depth is on the micron level, which means the detection area will be quite small on the surface of samples; however, for agricultural products, the measurement area should be as large as possible, thus results in limited application of the near-infrared method. The capacitance method regards the detected objects as a dielectric material and measures the dielectric constant of the measured objects [12]; however, the results of this method are greatly affected by the environment's temperature, and the device needs to be recalibrated after a long period of use. The capacitance method demonstrates poor performance for materials with high moisture content and is not a perfect solution for on-line measurement because the capacitive sensor takes a long time to introduce samples between the electrodes [13].

The principle of the microwave method is similar to that of the capacitance method, and it has the advantages of non-destructive detection, high speed, and real-time monitoring [14]. For moisture measurement based on microwaves, the dominant detecting method can be mainly divided into microwave reflection and microwave transmission [15]. One of the advantages of microwave reflection detection is that the integration of signal transmission and reception can simplify the installation; however, the advantage can also turn into a disadvantage, as it causes a low resolution and smaller reflected signals from individual antennae, which need special auxiliary hardware to amplify the signals and improve precision [16]. Applications of microwave transmission on the moisture measurements of peanuts, corn, and soybeans [17–19] have been achieved, but the detection equipment based on horn antenna were large and the experiments were inconvenience when carrying on-site. With the development of microstrip antenna sensors in recent years, it is possible to use microstrip antenna sensors instead of a horn antenna to realize microwave signal receiving and transmitting [20] and to design a set of portable real-time detection devices for grain moisture content.

The water molecule in the detected material is a strong dipole, which plays a dominant role in the impact on the dielectric properties compared to the dry matter. Therefore, the moisture content in the detected material can be measured by analyzing the dielectric properties, but the detecting equipment is expensive and complicated. This study converts to directly measure the parameters related to the dielectric properties, such as microwave attenuation and phase shift, to detect the moisture content of the material and to make the detecting equipment portable.

In general, very limited optimization work has been carried out regarding the realization of the on-line measurement of the moisture content of agricultural and food products. In this context, the aims of the present study are: (1) to develop a low-cost portable device for the real-time measurement of paddy rice with a moisture content range of generally 13–30%; (2) to determine the optimal measurement parameters according to experimental objects (microwave characteristics, phase shift, and temperature compensation); and (3) to develop an effective moisture content prediction model by comparing different regression algorithms. This work is expected to provide an important reference for the real-time in situ measurement of other agricultural products and to be of great significance for the development of intelligent agricultural equipment.

2. The Design of the Portable Moisture Content Detecting Device

2.1. The Principle of Microwave Transmission Detection

As a non-contact detecting method, microwave transmission detection has the advantages of a fast detection speed, high applicability, and strong practicability. The detection principle of the microwave transmission method is as follows:

The relationships between attenuation and phase shift and the dielectric properties are shown in Figure 1. The attenuation and phase shift of the microwave signal can be obtained as follows:

$$\Delta A = P_{inc} - P_{tra} - P_{ref} \quad (1)$$

$$\Delta \varphi = \varphi_{tra} - \varphi_{inc} - 2\pi n \quad (2)$$

where P_{inc} is the incident signal power, dBm; P_{tra} is the transmitted signal power, dBm; P_{ref} is the reflected signal power, dBm; φ_{inc} is the phase of the incident signal, rad; φ_{tra} is the phase of the transmitted signal, rad; and N is an integer determined by the thickness of the test sample.

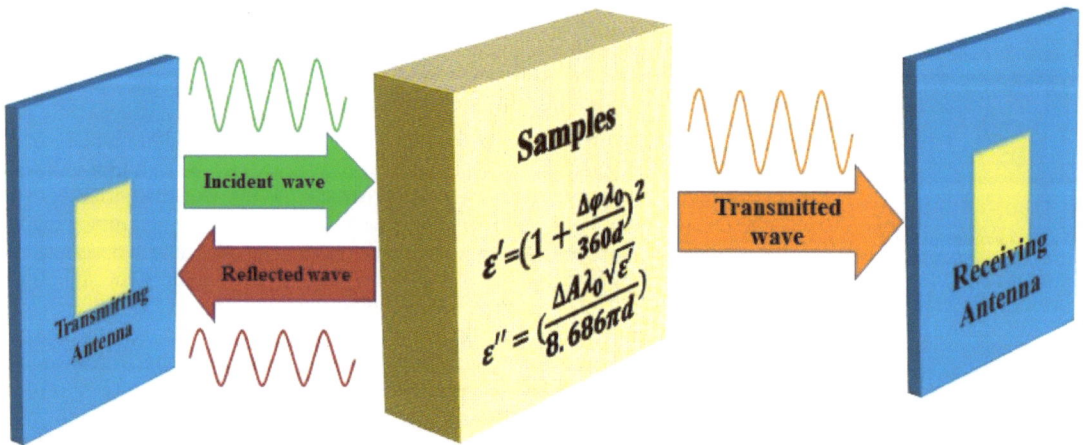

Figure 1. Detection principle of microwave transmission method: ε' is the dielectric constant; ε'' is the loss factor; $\Delta \varphi$ is the phase shift, rad; λ_0 is the wavelength, m; D is the sample thickness, m; ΔA is the attenuation, dB.

2.2. The Portable Detection Device

2.2.1. The Structure and the Principle

A schematic diagram of the detection device is shown in Figure 2. The device is divided into three sub-modules: (a) the microwave signal transmission-generation module, MSTGM; (b) the real-time measurement module (RTMM) with two microstrip microwave sensors (transmitting and receiving); (c) and the signal receiving–processing module, SRPM. The MSTGM sends a microwave signal at the specified frequency, which goes into the power distributor by the splitter. Two of the same microwave signals are generated after the power splitter. One microwave signal, processed by the low-noise amplifier 2 is translated to the microstrip transmitting antenna in RTMM as the incident signal. The incident signal perpendicularly irradiates the grain samples, a part of which is reflected by the sample surface to form a reflected signal. Due to the reverse cut-off function caused by the isolator, the reflected signal is extremely reduced, and the signal generator is protected. The other part of the microwave incident signal turns into the transmission signal after penetrating the grain samples. The transmission signal, which is received by the microstrip receiving antenna, is transmitted to the radio-frequency port of the in-phase and quadrature mixer (IQ mixer) as the input radio-frequency (RF) input signals. The other signal from the power splitter is amplified by the low-noise amplifier 1 and then enters the local oscillator (LO) of the IQ mixer as the input reference signal.

The microstrip transmitting antenna and microstrip receiving antenna were in close contact with the tested sample during measurement, and the distance between the two antennas changed as the sample cell distance changed. The microwave attenuation data, phase-shift data, and environment temperature data were transmitted to the upper computer via a serial communication protocol. Then, different regression models set in the upper computer train and process the data. The model with the best performance index was integrated into the microcontroller, and the predicted grain moisture content was displayed on the liquid crystal display (LCD) screen in real time.

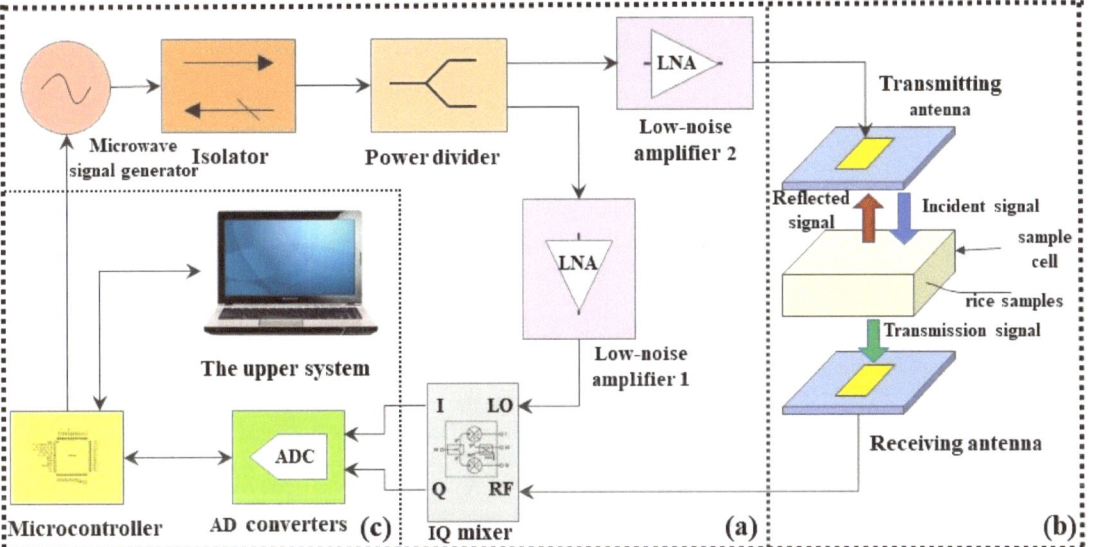

Figure 2. Schematic diagram of detection device: (**a**) a microwave signal transmission-generation module; (**b**) a real-time measurement module; (**c**) a signal receiving-processing module.

2.2.2. The Measurement of Attenuation and Phase Shift

The attenuation and phase shift of the microwave signal were derived from the two output signals of the IQ mixer. As shown in Figure 2a, the IQ mixer had a total of four ports: two input ports (RF and LO), and two output ports (I and Q). The microwave signal that did not penetrate the sample to the input port LO was treated as a reference signal; the microwave transmission signal that did penetrate the sample to the input port RF was considered a radio-frequency signal. The IQ mixer generated the identity signal of the radio-frequency signal and the reference signal in the output port I. An orthogonal signal was generated by mixing the radio-frequency signal from the output port Q and the reference signal that has undergone a 90-degree phase-shift transformation. The attenuation and phase shift of the microwave were calculated from the reference signal LO, and the radio-frequency signal RF was calculated from the output signal at the port I end and the output signal at the port Q [21].

The output signals from port I and port Q were direct currents in this study, and the signal processing procedure was same as the work in our preview work [22].

2.3. The Hardware Part of the Device

2.3.1. Circuit Module

The circuit module is shown in Figure 3. The whole device was divided into two layers: Figure 3a shows the superstructure, which was the high-frequency microwave circuit module. Figure 3b shows the lower structure, which was the low-frequency circuit and power module. Figure 3c shows the 3D simulation of the overall structure. The two parts were supported by four copper columns with fixed bolts at each corner. The height distance between the two parts was slightly greater than that of the highest circuit components. This design scheme made the overall structure compact and stable.

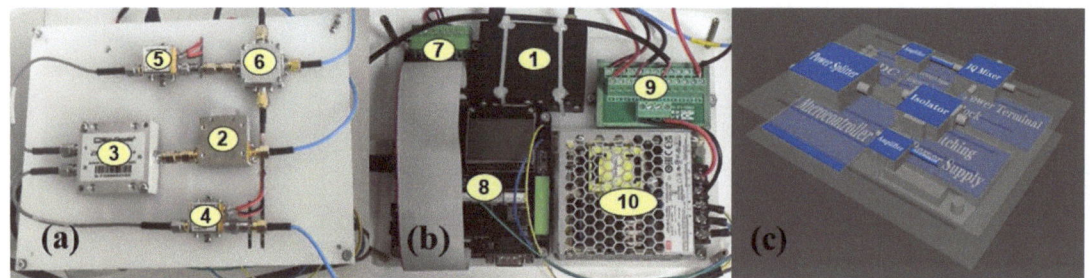

Figure 3. Circuit module: (**a**) high-frequency microwave circuit module; (**b**) low-frequency circuit power module; (**c**) 3D simulation of the whole module; ①. Microwave signal generator; ②. Isolator; ③. Power splitter; ④. Low-noise amplifier 2; ⑤. Low-noise amplifier 1; ⑥. IQ mixer; ⑦. AD converter; ⑧. Microcontroller; ⑨. Power terminal block; ⑩. Switching power supply.

In Figure 3a,c, the high-frequency microwave circuit module consisted of an isolator, a power divider, two low-noise amplifiers, and an IQ mixer. The main functions of this part were the transmission of the high-frequency microwave signal, isolation of the unwanted reflected wave signals received from the microstrip transmitting antenna, amplification of the microwave signal, and conversion of the transmission signal received by the microstrip receiving antenna into a low-frequency signal. In Figure 3b,c, the low-frequency circuit and power module in the lower structure consisted of an AD converter, an STM32 module, a power terminal block, and a switching power supply. The main functions of this part were the power supply of the whole device, reception/conversion/processing of low-frequency signals, and conversion of low-frequency signals into attenuation and phase-shift data. This structure effectively separated high- and low-frequency signals from each other and prevented interference, while greatly saving installation space and making the overall device smaller.

The microwave signal generator was DSINSTRUMENTS SG6000L (Suzhou Ruibeis Electronic Technology Co., Ltd., Suzhou, China), whose output frequency ranged from 25 Mhz to 6000 Mhz, and the phase noise is less than −72 dbc. The signal generator's maximum output power level was over +10 dBm. The size of this part was 7.0 cm × 3.2 cm × 5.5 cm, which is easy to integrate and carry. Marki's IQ-0205 model was selected as the IQ mixer. The IQ mixer operated in the frequency range of 2000–5000 Mhz and had a maximum conversion loss of 8 dB. Designed with a small size, IQ-0205 is quite convenient for portable device design.

2.3.2. Real-Time Measurement Module

As shown in Figure 4, the microstrip transmitting antenna and the microstrip receiving antenna were the same sizes in the real-time measurement module. Figure 4b shows one sample of a sample cell, and there were six sample containers of the same material, same height, same width, but different lengths, in this experiment, and they were all made by 3D printing.

Figure 4c shows that the real-time measurement module consists of two microstrip antennae, a sample cell, a fixed base, two guide rails, and a four-T antenna fixed frame. Both the fixed base and the T-shaped fixed antenna stand are made by 3D printing.

The two-microstrip antenna sensor GAUA3000M-40M-A (Beijing Gwave Technology Co., Ltd., Beijing, China) operate at 3 Ghz and has a bandwidth of 20 Mhz and a gain of 7 dBi with A coaxial SMA interface. Compared with the common horn antenna, the microstrip antenna has a narrow working bandwidth with a small volume advantage. The size of the microstrip antenna sensor is 80 mm × 80 mm × 1.5 mm, which is very suitable for integration in small-scale equipment.

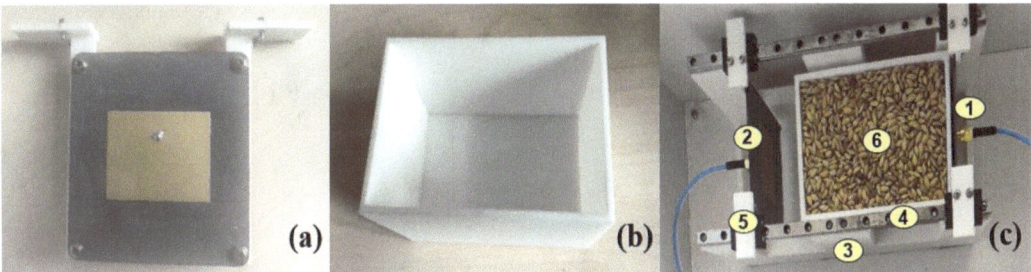

Figure 4. Real-time measurement module: (**a**) microstrip antenna; (**b**) sample cell; (**c**) real-time measurement module; ①, transmitting antenna; ②, receiving antenna; ③, fixed base; ④, guide rail; ⑤, antenna fixing frame; ⑥, measuring sample.

2.3.3. Integral Structure of the Portable Moisture Detecting Device

As shown in Figure 5, the circuit module and the real-time measurement module are installed in the same cuboid white plastic box. The two modules communicate with each other through the coaxial line. The size of the entire shell is 42 cm × 32 cm × 16 cm. The device is appropriate, lightweight, and easy to carry, thus offering the possibility to measure the moisture content of grains in situ.

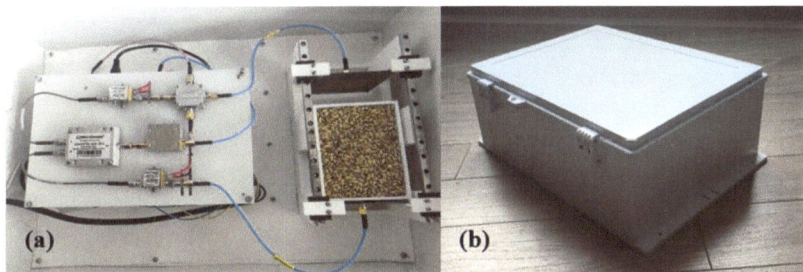

Figure 5. Integral measuring device: (**a**) inner structure; (**b**) external structure.

2.4. The Software Parts

An STM32F103RBT6 chip (Sichuan Kangwei Technology Co., Ltd., Sichuan, China) was installed as the microcontroller for signal receiving and processing and to realize the functions of the signal AD conversion, attenuation, and phase-shift data calculation, grain moisture content calculation, LCD control, temperature sensor control, and serial communication. The attenuation and phase-shift data were converted into code calculation by Formulas (1)–(6), and the moisture content was calculated by the best regression model selected by the upper system. The AD7606 module (Sichuan Kangwei Technology Co., Ltd., Sichuan, China) was applied to the AD conversion. An SPI communication protocol, a 16-bit analog-to-digital converter (ADC), and an eight-channel synchronous sampling were adopted to obtain a more stable signal compared to the ADC of the STM32, and the frequency of the obtained signal could reach up to 200 kSPS. The real-time ambient temperature data were collected by the DS18B20 (Sichuan Kangwei Technology Co., Ltd., Sichuan, China), which has a small volume and light weight, by separately calculating the integer and decimal digits of the parameters. The organic light-emitting diode (OLED) with the resolution of 128 × 64 was chose as the device screen. The result was displayed on the OLED screen.

The whole program adopted a modular design and was divided into different program modules according to the different functions. The defined peripheral header files only needed to import to the main program, and the whole set worked immediately once the main program was started. If the relevant functions needed modification, the corresponding header files were changed (or replaced) directly, which was quite convenient. The whole program was written in C language, and the code was compiled by the Keil software (Version 5, ARM Germany GmbH, Grasbrunn, Germany). The program flow chart is shown in Figure 6.

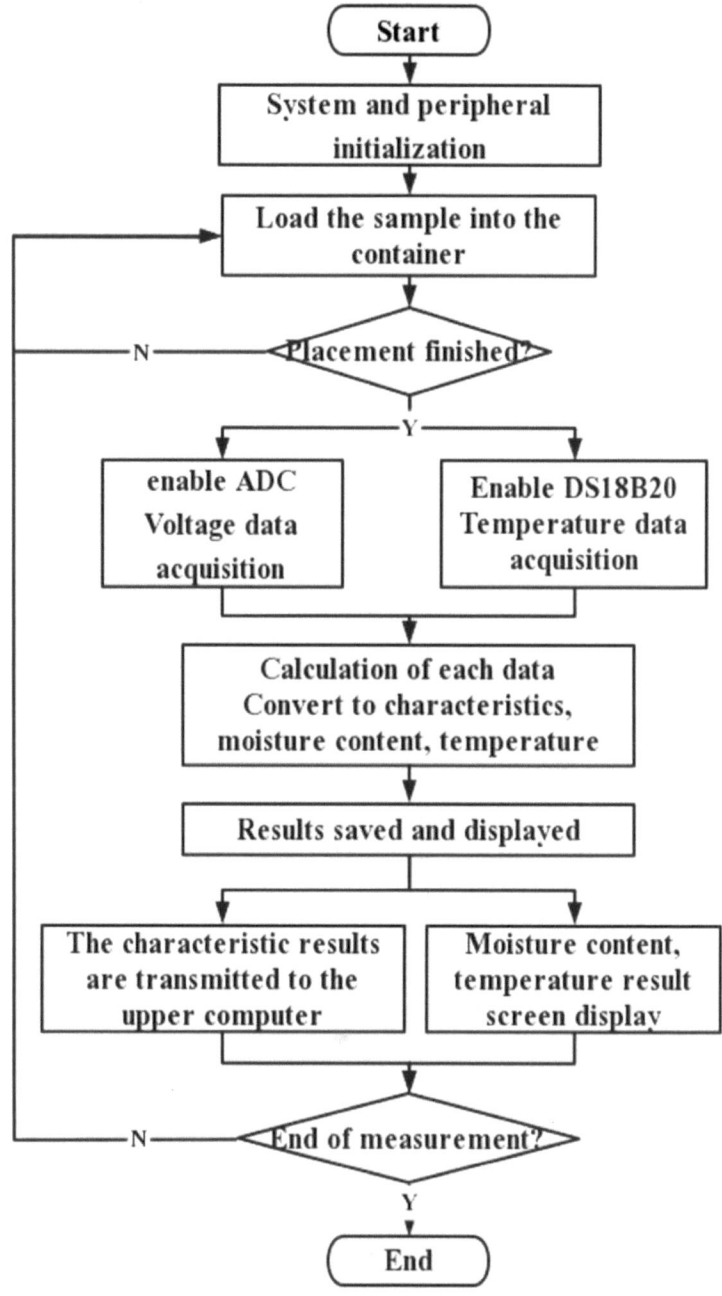

Figure 6. Program flow chart.

2.5. Sample Preparation

The fresh paddy rice (*Oryza sativa* L.) samples were purchased from a local supermarket (Hangzhou, China), and the initial moisture content of the samples was 13.66%. Before the experiment, all paddy rice was filtered with mesh to remove particles and other matters, and the rice with minimal damage and uniform shape were selected as the final experimental samples. An amount of 500.00 g of the paddy rice sample was weighed for each group, and eight groups were selected in this study. According to the initial moisture content of paddy rice, deionized water (0, 10.88, 29.69, 43.02, 49.94, 64.31, 79.46, and 95.45 g) was added to the eight groups. Then, the samples were sealed in plastic bags and stored in a dry, cool, unventilated environment for 2–3 days. During storage, the paddy rice was fully agitated to absorb as much water as possible for the sample. A standard oven-drying method [23] was applied to detect the moisture content of processed samples (5 g paddy rice was used here), and the results were 13.66, 15.65, 18.20, 20.17, 21.22, 23.21, 25.45, and 27.02% for each group of paddy rice. The eight groups with different moisture contents are shown in Figure 7. To verify the results, two groups of additional samples were prepared with random moisture contents (15.01 and 21.90%); the procedure was the same as for the eight groups of paddy rice above.

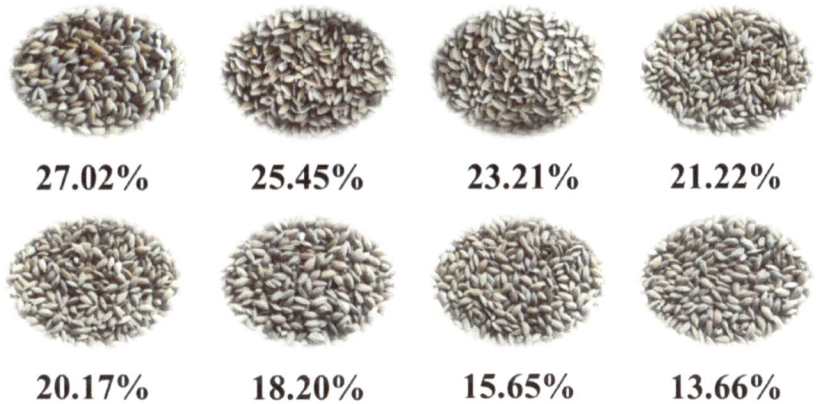

Figure 7. Paddy rice samples with different moisture contents.

2.6. Measurement Procedure and Data Acquisition

The effects of temperature and sample thickness on the signals were considered. The tests under the same temperature with different thicknesses were carried out, as well as the tests under the same thickness with different temperatures. The samples were filled with sample cells in the way of free fall. Before the moisture detection experiment, the conductivity of an empty sample container was tested three times, and the average value was taken as the reference data, which was used to obtain the real attenuation and phase-shift data. The results of the three measurements were averaged. The measured distance between the transmitting antenna and the receiving antenna depended on the thickness of the sample container, and the two antennae were closely connected with the sample container.

The thicknesses of the samples were 8–13 cm, respectively. The sample containers with different thicknesses are shown in Figure 8. All sample containers were made by 3D printing (WESTI Science Park Incubator Co., Ltd., Xi'an, China) with nylon (polyamide material) (WESTI Science Park Incubator Co., Ltd., Xi'an China). During measurement of the samples with different thicknesses, the ambient temperature was 21 °C. During the measurement of the samples with the same thickness, the ambient set temperatures were 15 °C, 20 °C, 25 °C, and 30 °C, respectively. For each measurement, three replicate trials were taken, and the detecting time was set to 5 s. After each measurement, the true values of the samples and the no-load values with on samples and were detected at different temperatures. All the obtained data were translated and transmitted through serial communication, and stored at the upper computer.

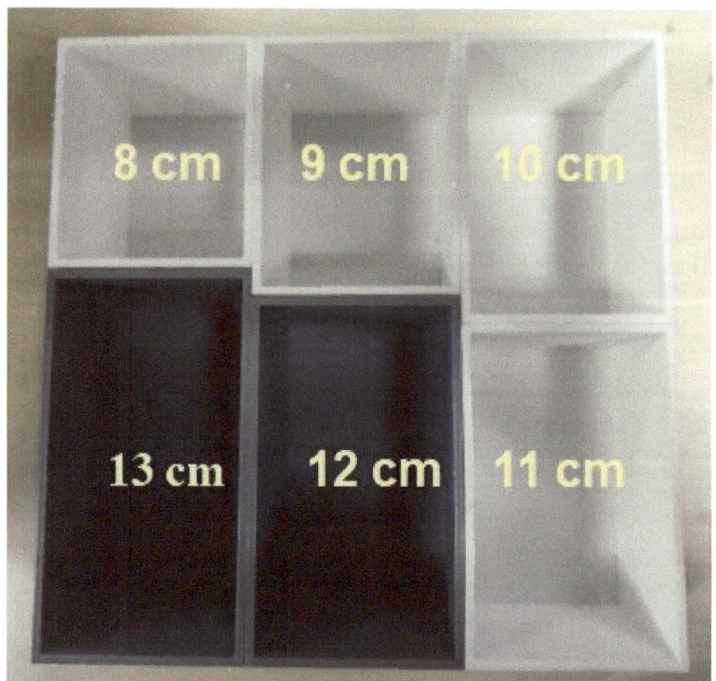

Figure 8. Sample containers with different thicknesses.

3. Results and Discussion

3.1. Eigenvalue Selection

The data obtained from the experiment included attenuation data, phase-shift data, and temperature data. For experiments with different thicknesses, the temperature was constant, and the feature values were attenuation and phase shift. For experiments at different temperatures, the sample thickness was constant, and the temperature was variable. To compensate for the temperature of the model, it was set as the characteristic value in the regression models. Our previous studies have shown that the microwave attenuation values at different frequencies were different for corn samples with different moisture contents [22]. When the microwave frequency was 2–10 Ghz, the attenuation value of samples with a moisture content lower than 29% did not change much, and when moisture content was higher than 29%, only the phase shift signals could distinguish the samples [21,24].

In this study, the microwave signal frequency was 3 Ghz, and the frequency selection source was from our previous work [24]. For the quantitative analysis of the importance of the attenuation and phase shift at 3 Ghz, a random forest was used to analyze the data with different thicknesses, and the final features were selected according to the importance of the feature values given by the random forest regression model. By calculating the contribution value of each feature, the importance of each feature is sorted according to its contribution value when constructing each tree within a random forest model. Usually, the Gini index or OOB error rate can be used to calculate the contribution value. The results are shown in Table 1.

The importance of the phase shift was much greater than that of the attenuation under different thicknesses. The average value of the attenuation under six thicknesses is 10.27%, and the average value of the phase shift is 89.73%. To verify the accuracy of the characteristic importance results given by a random forest, the cross-validation results of the random forest were applied. The input eigenvalues were divided into three cases: attenuation, phase shift, and attenuation + phase shift. The evaluation index took the determination coefficient R^2, the mean absolute error (MAE), and the root mean square error RMSE (RMSE), and the results are shown in Figure 9.

Table 1. Feature importance.

Thickness (cm)	The Importance of Feature Values (%)	
	Attenuation (dB)	Phase Shift (m)
8	16.84	83.16
9	6.55	93.45
10	7.43	92.57
11	9.21	90.79
12	7.94	92.06
13	8.81	91.19

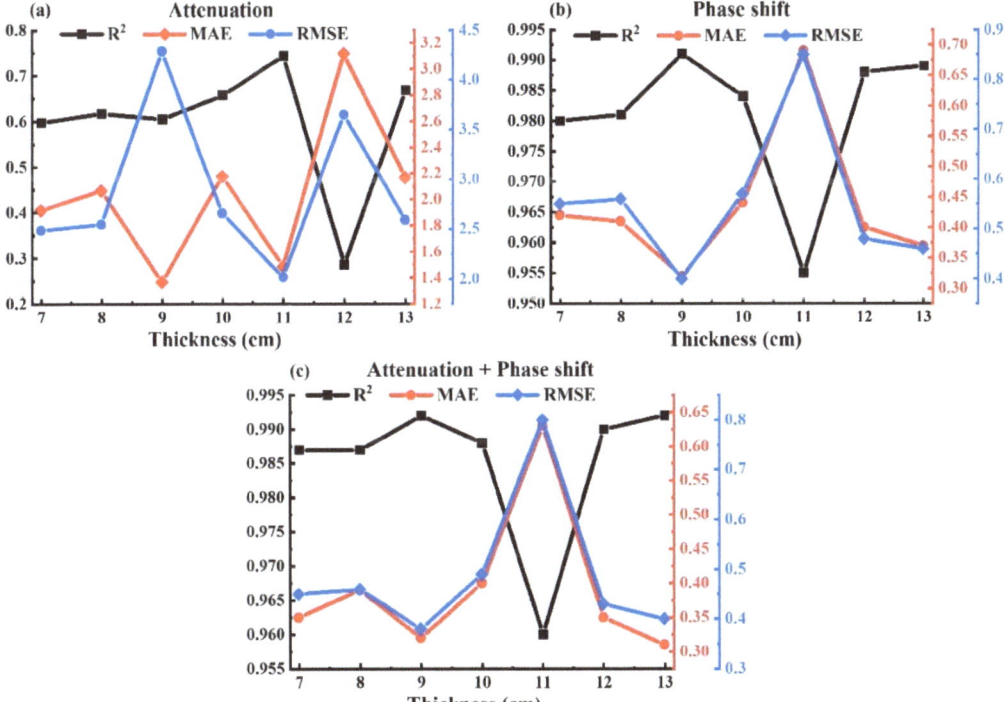

Figure 9. Feature importance verification: (**a**) the results based on attenuation; (**b**) the results based on phase shift; (**c**) the results based on attenuation and phase shift.

Figure 9 demonstrates: When the attenuation data were used as the only input eigenvalue, the average value of R^2 under six thicknesses was about 0.6; the average value of the MAE was about 2%, and the average value of the RMSE was about 3%. When the phase-shift data were used as the only input eigenvalue, the average value of R^2 was about 0.981, the average value of the MAE was about 0.4%, and the average value of the RMSE was about 0.6%. The model based on phase shift was obviously better than the model based on the attenuation characteristics, which may indicate that the influence of the phase shift on sample moisture content was dominant in this study. When the random forest model is combined with the attenuation and phase shift, the average values of the three indexes performed better than those of the models based on attenuation or phase shift alone. Therefore, the combination of attenuation and phase shift was applied as the input eigenvalue for the final regression model.

3.2. Sample Thickness Selection

The distance between the two antenna sensors was changed by changing the thicknesses of different sample cells, and the optimal sample thickness was decided by the prediction results. The sample thickness was divided into six groups, from 8 to 13 cm, with an interval of 1 cm. The random forest and decision tree have too many parameters to be adjusted and are easy to fall into over-fitting, and the k-nearest neighbor algorithm has the problems of difficult neighbor selection and easy classification error. The multiple linear regression (MLR) was selected to choose the best thickness.

The signal data of 240 samples (five samples for each group × eight groups × six levels of thickness) were used for the MLR model training. As mentioned in Section 2.5, two additional samples were made with random moisture contents to test the trained model; thus, 10 groups of samples were prepared for testing the established MLR model. The moisture content predictions of 180 samples (three samples for each group × 10 groups × six levels of thickness) under the thickness of 8 cm to 13 cm are displayed in Figure 10. Figure 10 shows the true and predicted values of the moisture contents for each sample with different colors.

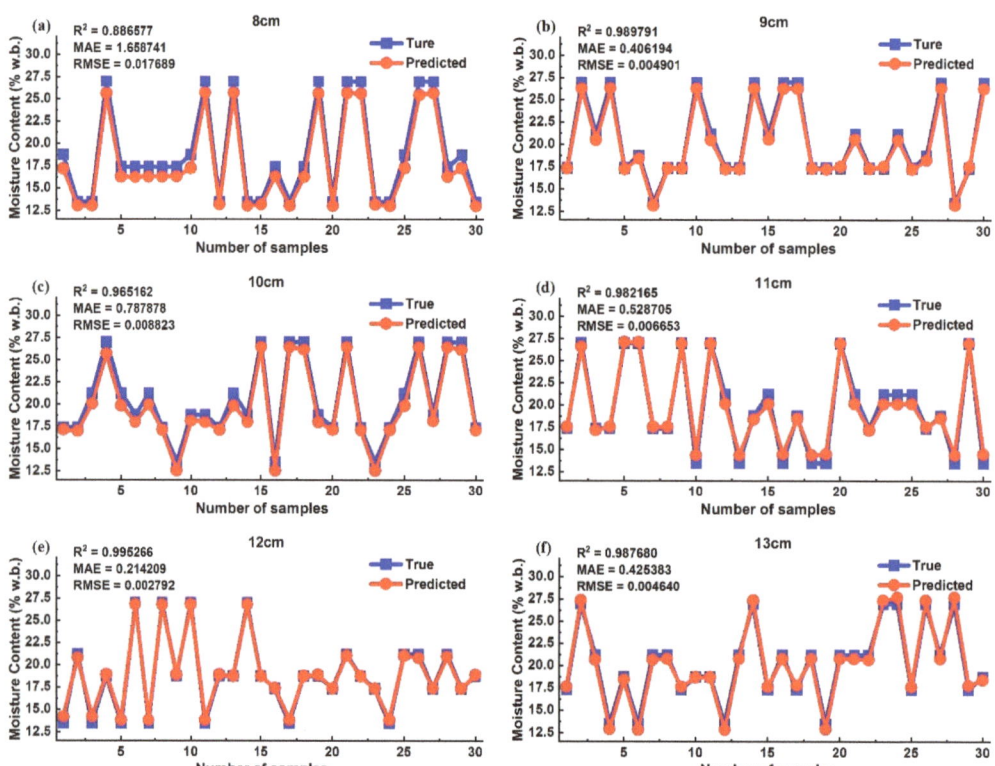

Figure 10. Prediction results under different sample thicknesses: (**a**) with the thickness of 8 cm; (**b**) with the thickness of 9 cm; (**c**) with the thickness of 10 cm; (**d**) with the thickness of 11 cm; (**e**) with the thickness of 12 cm; (**f**) with the thickness of 13 cm.

As shown in Figure 10, when the sample thickness was 8 cm: The R^2 value was only 0.887, while the R^2 values of other thicknesses were higher than 0.960. The MAE value was higher than 1.5%, while the MAE values of other thicknesses were lower than 0.8%. The

RMSE value was higher than 1.6%, while the RMSE values of other thicknesses were lower than 0.9%. It can be concluded that under the thickness of 8 cm, the prediction of moisture contents in paddy rice was the worst. According to the values of R^2, MAE, and RMSE, the prediction performance with a 12 cm sample thickness was best among those of 8–13 cm thickness. Thus, the final sample thickness of 12 cm was selected as the best test thickness.

3.3. Temperature Compensation

The ambient temperature has a certain impact on the accuracy of moisture content measurement. Based on the input characteristics determined above and the test thickness (12 cm), tests on four groups with different temperatures (15 °C, 20 °C, 25 °C, and 30 °C, respectively) were carried out respectively. The results are shown in Figure 11.

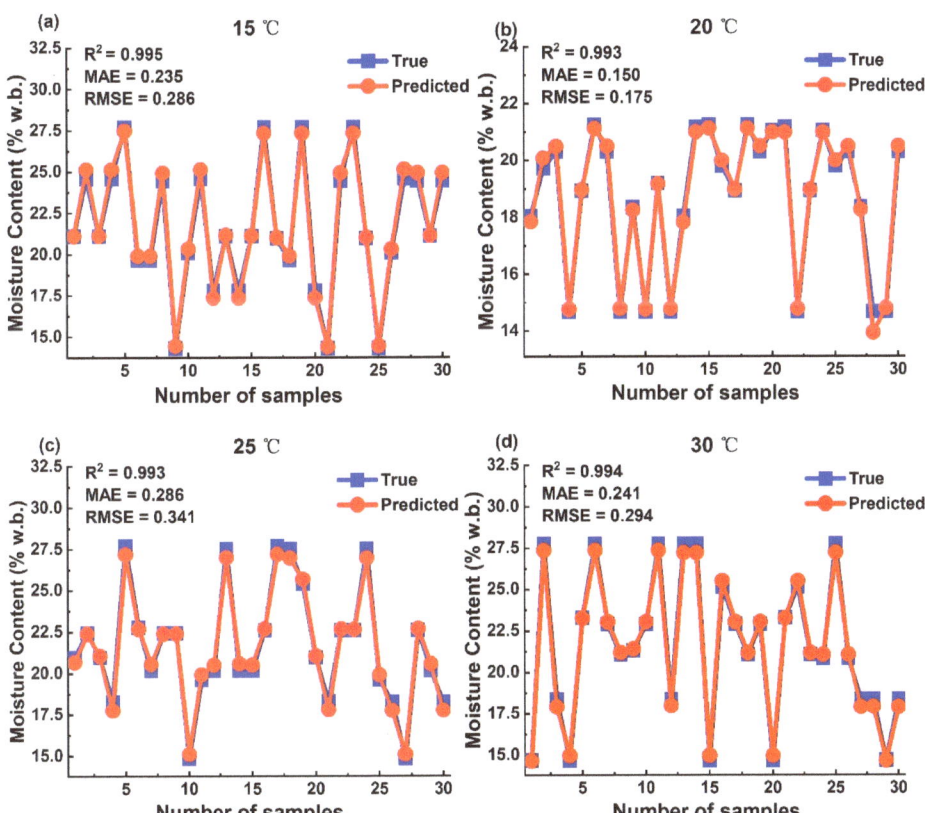

Figure 11. Performance results of different temperature test sets: (**a**) results at 15 °C ambient temperature; (**b**) results at 20 °C ambient temperature; (**c**) results at 25 °C ambient temperature; (**d**) results at 40 °C ambient temperature.

To reduce the influence that the environment temperature might cause during in situ testing, the temperature data were put into the training model as compensation data. As shown in Figure 11, the R^2 values at the four temperatures showed no obvious difference. When the ambient temperature was 20 °C, the MAE and RMSE values were the smallest, indicating that 20 °C might be the best ambient temperature for moisture content detection. Then, the model was verified with a random temperature (18 °C in this work). The results are shown in Figure 12.

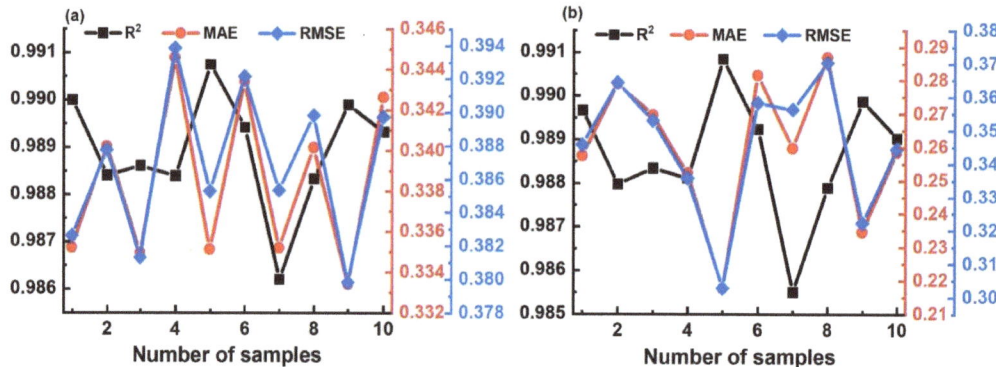

Figure 12. Performance index results before and after adding temperature compensation: (**a**) results before temperature compensation, (**b**) results after temperature compensation.

Figure 12 shows that before and after the addition of temperature compensation, the improvement in the R^2 value was not significant. However, the MAE and RMSE values after adding temperature compensation were lowered obviously. Therefore, it is necessary to compensate the temperature as the characteristic data in the model.

3.4. The Prediction of Moisture Content Based on Four Models

After the eigenvalues, sample thickness, and temperature compensation are all determined, it is necessary to find an algorithm with the best performance to return to the microcontroller as the final model to predict the unknown samples. A random forest, decision tree, k-nearest neighbor, and support vector machine were used to train the data, and the trained models were verified by predicting the new verification samples. The prediction results of the four models are shown in Figure 13.

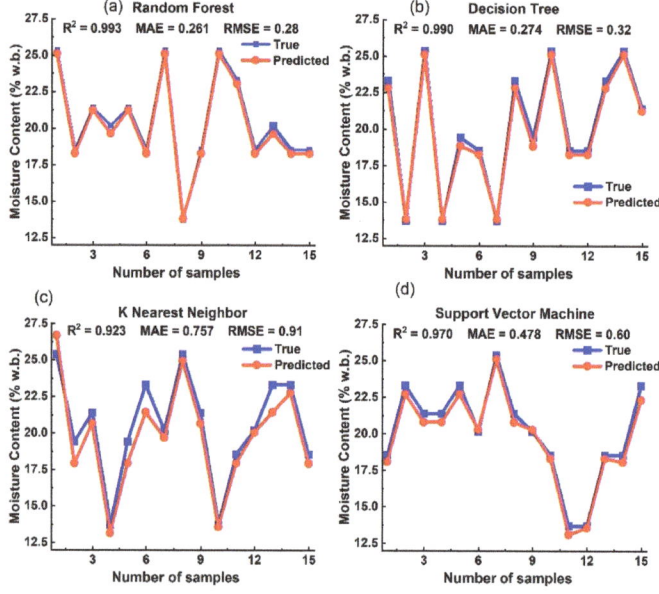

Figure 13. Prediction results of four algorithms: (**a**) results based on random forest; (**b**) results based on decision tree; (**c**) results based on k-nearest neighbor; (**d**) results based on support vector machine.

Figure 13 shows that the results of the three performance indexes of the random forest were better than the other three models, and the performance of the k-nearest neighbor algorithm was the worst (There was a large deviation between the predicted value and the real value of the moisture content in paddy rice.). Therefore, the random forest was chosen as the final model returned to the microprogrammed control unit (MCU) to predict the moisture content in the paddy rice samples.

3.5. The Final Display

The selected model parameters (random forest) were written into the MCU code, and portable moisture detection was applied to detect the moisture content of the rice. Some actual display effects are shown in Figure 14, where "Temp" refers to the ambient temperature measured in real time, "MC" refers to the measured moisture content value, and the measurement is repeated three times.

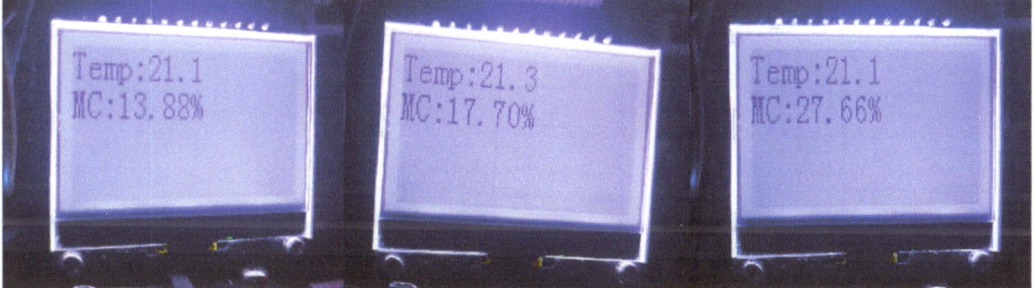

Figure 14. Display of actual moisture content measurement results.

As shown in Table 2, the maximum average absolute error of the equipment was 0.55%, and the minimum absolute error was 0.17% when compared to the results obtained from the standard drying method; thus, the measurement results from the portable moisture detection device are acceptable. The mean standard deviation was 0.18%, the maximum standard deviation was 0.41%, and the minimum standard deviation was 0.08%, indicating that the measurement results are relatively stable. In practical moisture detection, the content range of rice is generally 13–30%, and content outside the range is of little significance. The prediction result clearly follows the normal distribution, and the accuracy of the moisture measurement is acceptable [25,26]. Therefore, the device can be applied to the detection of grain moisture content.

Table 2. Comparison of measurement results of samples with different moisture content.

No.	The Moisture Content Measured by the Standard Drying Method ± SD(%)	The Measured Content Measured by the Portable Moisture Detection ± SD(%)	MAE (%)
1	27.02 ± 0.10	27.57 ± 0.10	0.55
2	25.58 ± 0.23	25.32 ± 0.10	0.26
3	23.21 ± 0.10	22.94 ± 0.08	0.27
4	21.45 ± 0.26	21.47 ± 0.41	0.28
5	20.17 ± 0.14	20.22 ± 0.25	0.17
6	19.19 ± 0.12	18.74 ± 0.11	0.45
7	18.20 ± 0.11	17.77 ± 0.10	0.43
8	15.65 ± 0.15	15.39 ± 0.15	0.26
9	13.66 ± 0.20	13.64 ± 0.23	0.17

4. Conclusions

In this study, a novel portable real-time detection device for grain moisture measurement based on microwave microstrip antenna sensors is developed, which can operate in situ with high accuracy. The main conclusions are as follows:

1. A real-time detection device for paddy rice moisture content was designed, and a microwave measurement module was built to carry out non-destructive measurements. The STM32F103RBT6 chip was designed as the signal receiving and processing module to realize the functions of AD conversion, microwave attenuation and phase-shift calculation, temperature acquisition and display, and screen control and serial port communication. The final moisture content was displayed in real time on the LCD screen.
2. The attenuation, phase-shift, and temperature data were obtained. The characteristic value of the best combined form was determined using a random forest, after comparison with the performances of decision tree model, k-nearest neighbor model, and support vector machine model. The temperature compensation was added to optimize the model. Finally, the model parameters with the best performance index in the above four algorithms were selected and returned to the single-chip microcomputer to display the measured value of the grain moisture content in real time.
3. Nine groups of grains with different moisture contents were measured by the detection device. The maximum and minimum average absolute errors of the measurement results were 0.55% and 0.17%, respectively. The maximum standard deviation was 0.41%, and the minimum was 0.08%. The accuracy and stability of the measurement results were within the acceptable range.

In this study, the portable moisture detection device designed for grain moisture content is small, light, highly expansive, and accurate. This work is expected to provide an important reference for the development of real-time measurement of other agricultural products, and to be of great significance for the development of intelligent agricultural equipment and industrial applications. In our future studies, the adaptability of the microwave moisture measurement system for moisture measurement of agricultural products with complex surfaces and shapes, such as tea leaf, soybean, and crop straw, will be investigated.

Author Contributions: Conceptualization, Z.W. and S.Q.; methodology, J.L.; software, J.L.; validation, J.L., S.Q. and Z.W.; formal analysis, S.Q.; investigation, S.Q. and J.L.; resources, Z.W.; data curation, J.L.; writing—original draft preparation, S.Q. and J.L.; writing—review and editing, S.Q. and J.L.; visualization, S.Q. and J.L.; supervision, S.Q. and J.L.; funding acquisition, S.Q. All authors have read and agreed to the published version of the manuscript.

Funding: This research was funded by the National Key R&D Program of China [2019YFE0124600].

Acknowledgments: The authors acknowledge the financial support of the National Key R&D Program of China (2019YFE0124600).

Conflicts of Interest: The authors declare no conflict of interest.

References

1. Besharati, B.; Lak, A.; Ghaffari, H.; Karimi, H.; Fattahzadeh, M. Development of a model to estimate moisture contents based on physical properties and capacitance of seeds. *Sens. Actuators A Phys.* **2021**, *318*, 112513. [CrossRef]
2. Du, J.; Lin, Y.; Gao, Y.; Tian, Y.; Zhang, J.; Fang, G. Nutritional Changes and Early Warning of Moldy Rice under Different Relative Humidity and Storage Temperature. *Foods* **2022**, *11*, 185. [CrossRef] [PubMed]
3. Risius, H.; Prochnow, A.; Ammon, C.; Mellmann, J.; Hoffmann, T. Appropriateness of on-combine moisture measurement for the management of harvesting and postharvest operations and capacity planning in grain harvest. *Biosyst. Eng.* **2017**, *156*, 120–135. [CrossRef]
4. Ramli, N.A.M.; Rahiman, M.H.F.; Kamarudin, L.M.; Zakaria, A.; Mohamed, L. A Review on Frequency Selection in Grain Moisture Content Detection. In Proceedings of the 5th International Conference on Man Machine Systems, Pulau Pinang, Malaysia, 26–27 August 2019; p. 012002.

5. Klomklao, P.; Kuntinugunetanon, S.; Wongkokua, W. Moisture content measurement in paddy. *J. Phys. Conf. Ser.* **2017**, *901*, 012068. [CrossRef]
6. Wang, W.C.; Wang, L. Design of Moisture Content Detection System. *Phys. Procedia* **2012**, *33*, 1408–1411. [CrossRef]
7. Liu, D.; Zeng, X.-A.; Sun, D.-W. Recent Developments and Applications of Hyperspectral Imaging for Quality Evaluation of Agricultural Products: A Review. *Crit. Rev. Food Sci. Nutr.* **2015**, *55*, 1744–1757. [CrossRef]
8. Li, C.; Li, B.; Huang, J.; Li, C. Developing an Online Measurement Device Based on Resistance Sensor for Measurement of Single Grain Moisture Content in Drying Process. *Sensors* **2020**, *20*, 4102. [CrossRef]
9. Tinna, A.; Parmar, N.; Bagla, S.; Goyal, D.; Senthil, V. Design and development of capacitance based moisture measurement for grains. *Mater. Today Proc.* **2021**, *43*, 263–267. [CrossRef]
10. Peiris, K.H.S.; Bean, S.R.; Chiluwal, A.; Perumal, R.; Jagadish, S.V.K. Moisture effects on robustness of sorghum grain protein near-infrared spectroscopy calibration. *Cereal Chem.* **2019**, *96*, 678–688. [CrossRef]
11. Nirmaan, A.M.C.; Rohitha Prasantha, B.D.; Peiris, B.L. Comparison of microwave drying and oven-drying techniques for moisture determination of three paddy (*Oryza sativa* L.) varieties. *Chem. Biol. Technol. Agric.* **2020**, *7*, 1. [CrossRef]
12. Heřmanská, A.; Středa, T.; Chloupek, O. Improved wheat grain yield by a new method of root selection. *Agron. Sustain. Dev.* **2015**, *35*, 195–202. [CrossRef]
13. Clarys, P.; Clijsen, R.; Taeymans, J.; Barel, A.O. Hydration measurements of the stratum corneum: Comparison between the capacitance method (digital version of the Corneometer CM 825®) and the impedance method (Skicon-200EX®). *Int. Soc. Ski. Imaging (ISSI)* **2012**, *18*, 316–323. [CrossRef] [PubMed]
14. Liu, H.; Liu, H.; Liu, H.; Zhang, X.; Hong, Q.; Chen, W.; Zeng, X. Microwave Drying Characteristics and Drying Quality Analysis of Corn in China. *Processes* **2021**, *9*, 1511. [CrossRef]
15. Chen, Z.; Wu, W.; Dou, J.; Liu, Z.; Chen, K.; Xu, Y. Design and Analysis of a Radio-Frequency Moisture Sensor for Grain Based on the Difference Method. *Micromachines* **2021**, *12*, 708. [CrossRef]
16. Zhang, W.; Yang, G.; Lei, J.; Liu, C.; Tao, J.; Qin, C. Development of on-line detection device for grain moisture content using microwave reflection method. *Trans. Chin. Society Agric. Eng.* **2019**, *35*, 21–28.
17. Taheri, S.; Brodie, G.; Jacob, M.V.; Antunes, E. Dielectric properties of chickpea, red and green lentil in the microwave frequency range as a function of temperature and moisture content. *J. Microw. Power Electromagn. Energy* **2018**, *52*, 198–214. [CrossRef]
18. Julrat, S.; Trabelsi, S. Measuring Dielectric Properties for Sensing Foreign Material in Peanuts. *IEEE Sens. J.* **2019**, *19*, 1756–1766. [CrossRef]
19. Yigit, E.; Duysak, H. Determination of Flowing Grain Moisture Contents by Machine Learning Algorithms Using Free Space Measurement Data. *IEEE Trans. Instrum. Meas.* **2022**, *71*, 1–8. [CrossRef]
20. Gundewar, P.P.; Patel, V.U.; Chaware, T.S.; Askhedkar, A.R.; Raje, R.S.; Subhedar, M.M.; Udgire, V.N. Design of a microstrip patch antenna as a moisture sensor. In Proceedings of the 2019 IEEE Pune Section International Conference, Pune, India, 18–20 December 2019; pp. 1–5.
21. Zhang, J.; Du, D.; Bao, Y.; Wang, J.; Wei, Z. Development of Multifrequency-Swept Microwave Sensing System for Moisture Measurement of Sweet Corn With Deep Neural Network. *IEEE Trans. Instrum. Meas.* **2020**, *69*, 6446–6454. [CrossRef]
22. Zhang, J.; Wu, C.; Shao, W.; Yao, F.; Wang, J.; Wei, Z.; Du, D. Thickness-Independent Measurement of Grain Moisture Content by Attenuation and Corrected Phase Shift of Microwave Signals at Multiple Optimized Frequencies. *IEEE Trans. Ind. Electron.* **2022**, *69*, 11785–11795. [CrossRef]
23. *GB/T21305*; Cereal and Cereal Products-Determination of Moisture Content-Routine Reference Method. The China National Standardization Management Committee: Beijing, China, 2007.
24. Zhang, J.; Bao, Y.; Du, D.; Wang, J.; Wei, Z. OM2S2: On-Line Moisture-Sensing System Using Multifrequency Microwave Signals Optimized by a Two-Stage Frequency Selection Framework. *IEEE Trans. Ind. Electron.* **2021**, *68*, 11501–11510. [CrossRef]
25. Tian, L.; Cai, T.; Goetghebeur, E.; Wei, L.J. Model evaluation based on the sampling distribution of estimated absolute prediction error. *Biometrika* **2007**, *94*, 297–311. [CrossRef]
26. Vera Zambrano, M.; Dutta, B.; Mercer, D.G.; MacLean, H.L.; Touchie, M.F. Assessment of moisture content measurement methods of dried food products in small-scale operations in developing countries: A review. *Trends Food Sci. Technol.* **2019**, *88*, 484–496. [CrossRef]

Article

Design and Optimization of Electronic Nose Sensor Array for Real-Time and Rapid Detection of Vehicle Exhaust Pollutants

Jin Tong [1,2], Chengxin Song [1,2], Tianjian Tong [3], Xuanjie Zong [4], Zhaoyang Liu [5], Songyang Wang [5], Lidong Tan [6], Yinwu Li [1,2,*] and Zhiyong Chang [1,2,7,*]

1. College of Biological and Agricultural Engineering, Jilin University, Changchun 130022, China
2. Key Laboratory of Bionic Engineering, Ministry of Education, Jilin University, Changchun 130022, China
3. Agricultural and BioSystems Engineering, Iowa State University, Ames, IA 50010, USA
4. Zibo Municipal Transport Management Service Center, Zibo 255000, China
5. Digital Intelligent Cockpit Department, Intelligent Connected Vehicle Development Institute, China FAW Group CO., Ltd., Changchun 130013, China
6. College of Transportation, Jilin University, Changchun 130022, China
7. Weihai Institute for Bionics, Jilin University, Weihai 264401, China
* Correspondence: lyw@jlu.edu.cn (Y.L.); zychang@jlu.edu.cn (Z.C.)

Abstract: Traditional vehicle exhaust pollutant detection methods, such as bench test and remote sensing detection, have problems such as large volume, high cost, complex process, long waiting time, etc. In this paper, according to the main components of vehicle exhaust pollutants, an electronic nose with 12 gas sensors was designed independently for real-time and rapid detection of vehicle exhaust pollutants. In order to verify that the designed electronic nose based on machine learning classification method can accurately identify the exhaust pollutants from different engines or different concentration levels from the same engine. After feature extraction of the collected data, Random Forest (RF) was used as the classifier, and the average classification accuracy reached 99.92%. This result proved that the designed electronic nose combined with RF method can accurately and sensitively judge the concentration level of vehicle exhaust pollutants.. Then, in order to enable the electronic nose to be vehicle-mounted and to achieve real-time and rapid detection of vehicle exhaust pollutants. We used Recursive Feature Elimination with Cross Validation (RFECV), Random Forest Feature Selector (RFFS) and Principal Component Analysis (PCA) to optimize the sensor array. The results showed that these methods can effectively simplify the sensor array while ensuring the RF classifier's classification recognition rate. After using RFECV and RFFS to optimize the sensor array, the RF classifier's classification recognition rate of the optimized sensor arrays for vehicle exhaust pollutants reached 99.77% and 99.44%, respectively. The numbers of sensors in the optimized sensor arrays were six and eight respectively, which achieved the miniaturization and low-cost of the electronic nose. With the limitation of six sensors, RFECV is the best sensor array optimization method among the three methods.

Keywords: vehicle exhaust pollutants; electronic nose; sensor array optimization; feature extraction; feature selection

1. Introduction

Vehicle exhaust contains hundreds of harmful substances and a large number of greenhouse gases [1]. It is of great significance to strengthen the monitoring of vehicle exhaust pollutants for environmental protection. The main components of vehicle exhaust pollutants include carbon monoxide (CO), nitrogen oxides (NO_x), hydrocarbons (HC) and sulfur dioxide (SO_2) [2].

At present, the main vehicle exhaust pollutant detection methods include remote sensing detection [3], vehicle equipment detection [4] and traditional bench test [5]. However, these methods have many problems, such as large volume, high cost, complex process, long waiting time, etc.

An electronic nose is an odor recognition system [6–8] composed of sensor arrays, which simulates the working principle of mammalian olfactory organs. Electronic nose has been widely used in various fields, such as the food industry, chemical industry, medical field, etc. [9–13]. In previous work, it has been proved that the use of gas sensors can identify CO, NO_x, HC, SO_2 and other exhaust pollutants and judge their concentration levels [14–17]. Therefore, the electronic nose with 12 gas sensors designed independently was used in this paper to carry out real-time and rapid detection of vehicle exhaust pollutants.

In order to make it possible for the vehicle-mounted electronic nose to rapidly detect vehicle exhaust pollutants in real time, it is necessary to simplify the sensor array to achieve the miniaturization and low-cost of the electronic nose. At the same time, because the sensor array in the electronic nose has cross sensitivity, there will be redundancy in the sensor array. Optimizing the sensor array can not only reduce the volume and cost [18–20], but also remove redundant information and improve the recognition rate of the electronic nose in identifying pollutants.

At present, many feature selection methods have been used to optimize the sensor array of electronic nose, and have achieved good results. Recursive Feature Elimination (RFE) is a feature selection algorithm that searches for the optimal feature subset by repeatedly constructing models, and has been widely used in the optimization of electronic nose's sensor array [21,22]. Genetic Algorithm (GA) is a randomized search method with global optimization capability, and has also been used in sensor array optimization of electronic nose [23,24]. While Random Forest (RF) can not only solve classification and regression problems, it also has certain applications in the optimization of an electronic nose's sensor array [25]. This paper used two popular feature selection methods: Recursive Feature Elimination with Cross Validation (RFECV) and Random Forest Feature Selector (RFFS). As a contrast, this paper also used the traditional Principal Component Analysis method to optimize the sensor array, and compares the optimization results of the popular feature selection methods and the traditional method.

2. Materials and Methods

2.1. Structure of the Electronic Nose

The electronic nose system designed in this paper mainly includes a sampling unit and a detection unit. The sampling unit is mainly composed of a sampling pipe, three-way valve, flowmeter and chamber. The detection unit mainly includes a sensor array, regulating circuit board, analog digital converter (16 channel 12 bits of Beijing Pop WS-5921/U60216), computer, several connecting wires and a power supply (5V DC). The sensor converts the odor information into electrical signals through the change of the internal resistance value, then transmits it to the regulating circuit board in turn, and converts the electrical signals into digital signals through the analog digital converter, finally transmits the digital signals to the Vib'SYS signal acquisition, processing and analysis software at the computer terminal.

As the main components of vehicle exhaust pollutants are carbon monoxide (CO), nitrogen oxides (NO_x), hydrocarbons (HC), Pb compounds (Pb), carbon dioxide (CO_2), particulate emissions (PM), sulfur dioxide (SO_2) and other gases [2], this paper used 12 gas sensors that are sensitive to the above gases. The details of the selected gas sensors are shown in Table 1.

The sensor array chamber is composed of an external cavity column and an octagonal plate, respectively. The gas sensor is fixed through the hole on the octagonal plate in the same direction, and the detection surface of the gas sensor is in the inner plane of the chamber. The 3D schematic diagram of the sensor cavity structure is shown in Figure 1.

Table 1. Information of the 12 gas sensors.

Serial Number	Sensor	Response Characteristics	Producer (Country)
A	GSBT11	VOCs, toluene, benzene, formaldehyde	Ogam (Korea)
B	MP135	Ethanol, cigarette smoke, air pollutants	Winsen (China)
C	MP901	Alcohol, smoke, formaldehyde	Winsen (China)
D	WSP1110	Carbon dioxide	Winsen (China)
E	WSP2110	Toluene, benzene, alcohol, acetone	Winsen (China)
F	TGS2600	Air quality gas (hydrogen sulfide, etc.)	Figaro (Japan)
G	TGS2602	Formaldehyde and VOCs	Figaro (Japan)
H	TGS2603	Sulfur odor gas	Figaro (Japan)
I	TGS2610	Butane, liquefied gas	Figaro (Japan)
J	TGS2611	Natural gas	Figaro (Japan)
K	TGS2612	Combustible gas	Figaro (Japan)
L	TGS2620	Liquor	Figaro (Japan)

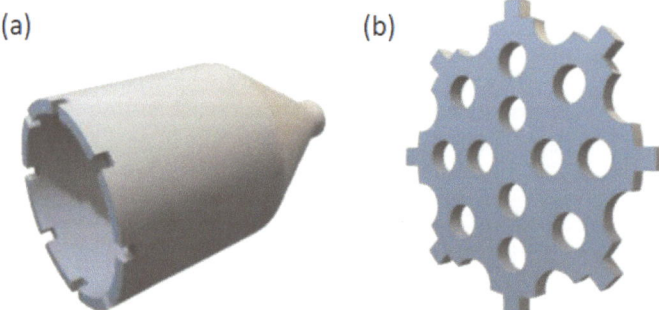

Figure 1. 3D schematic diagram of sensor cavity structure. (**a**) The schematic diagram of external cavity column. (**b**) The schematic diagram of octagonal plate.

2.2. Engine Bench Test

In order to summarize the concentration variation law of diesel engine's exhaust pollutants, we conducted the engine bench test at first. The CA4D28C5 diesel engine and the G01 gasoline engine were used in the test. The bench was set at the same rotary speed and different torques, and the exhaust pollutants concentration was measured in a continuous time. The engine's testing bench and device are shown in Figure 2.

Figure 2. Engine bench test device diagram. (**a**) Engine's testing bench. (**b**) The CA4D28C5 diesel engine testing device.

An AVL DICOM 4000 pollutant analyzer and a HORIBA MEXA-7100DEGR pollutant analyzer were used for vehicle exhaust pollutant detection, respectively. The former was used for diesel engine pollutant detection, and the latter was used for gasoline engine pollutant detection. These two pollutant analyzers not only meet the standards of vehicle exhaust pollutants detection, but also have the advantages of moderate measurement range, high measurement accuracy, stable reading and good anti-interference performance. The two pollutant analyzers are shown in Figure 3. The changes diagrams of exhaust pollutants concentration of CA4D28C5 diesel engine measured by AVL DICOM 4000 pollutant analyzer under the operating conditions of 1600 r/min and 2200 r/min, respectively with different torques (Nm) are shown in Figure 4.

Figure 3. Physical diagram of the two pollutant analyzers. (**a**) AVL DICOM 4000 pollutant analyzer. (**b**) HORIBA MEXA-7100DEGR pollutant analyzer.

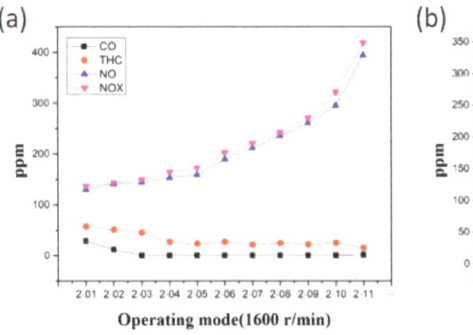
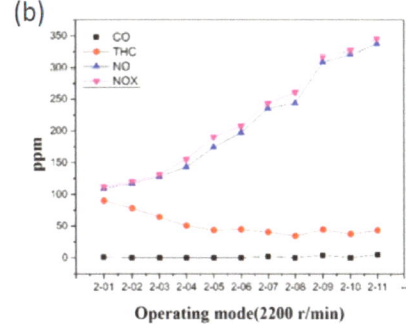

Figure 4. Exhaust pollutants concentration diagrams of CA4D28C5 diesel engine under different torques (Nm) with different rotary speeds conditions. (**a**) Variation in exhaust pollutants concentration under different torques (Nm) with 1600 r/min. (**b**) Variation in exhaust pollutants concentration under different torques (Nm) with 2200 r/min.

It can be seen from the changes in exhaust pollutants concentrations under different working conditions in Figure 4 that, at the same rotary speed, with the increase in torque, the concentrations of carbon monoxide (CO) and total hydrocarbons (THC) gradually decrease, however the concentrations of nitric oxide (NO) and nitrogen oxides (NOx) gradually increase, and their change ranges are similar; at the same torque with different rotary speeds, the change trends of CO, THC, NO and NOx are similar, but the concentrations are different.

According to the above concentration variation law of the diesel engine's exhaust pollutants, we believe that the sensor array in the electronic nose needs to have the ability to accurately identify the exhaust pollutants from different engines or different concentration levels from the same engine. The test scheme developed according to the concentration variation law of the diesel engine's exhaust pollutants obtained from the above engine bench test is shown in Section 2.3.

2.3. Experimental Setups

According to the concentration variation law of the diesel engine's exhaust pollutants in Section 2.2, we designed five groups of experiments using electronic nose to detect vehicle exhaust pollutants. The first four groups used a CA4D28C5 diesel engine, which was tested under the conditions of the same rotary speed with different torques. That is, the experiments were conducted under the conditions that the types of exhaust pollutants gases were the same, but the concentration levels of each pollutant were only slightly different. The fifth group used G01 gasoline engine, which means that the experimental conditions of different exhaust pollutants gas types and different concentration levels were taken as the control group.

The total test time of each sample was 300 s, of which 90 s is the data acquisition time, and the data acquisition frequency was 50 Hz, so that each sample contains $12 \times 90 \times 50 = 54{,}000$ data points. The other 210 s is the cleaning time of the electronic nose chamber and the zero-setting time of the resistance.

The actual experimental steps are as follows:

(1) Check the connection tightness and safety of each component.
(2) Power on, start the electronic nose detection system, warm up for 30 min, and expose the sensor to clean air.
(3) The sample gas is introduced into the electronic nose's chamber through the catheter connected with the three-way valve.
(4) Collect and store the signal of the sample gas.
(5) After the data collection of one group of samples is completed, clean the electronic nose with clean air for about 210 s.
(6) Perform the next set of experiments and repeat steps (2–5).

A total of 225 samples were obtained in the experiments. The concentration levels of the main exhaust pollutants under the test conditions and corresponding operating conditions of the samples are shown in Table 2.

Table 2. Composition of experimental samples.

Experimental Number	Engine	Rotary Speed (r/min)	Torque (Nm)	CO (%vol)	CO_2 (%vol)	THC (ppm)	Number of Samples
1	Diesel engine	2000	29	0.03	9.6	96	45
2		2000	41.5	0.08	6.1	90	45
3		2000	103.9	0.02	9.0	54	45
4		2000	153	0.01	9.7	42	45
5	Gasoline engine	1200	67	1200	926	1.38	45

2.4. Feature Extraction

After completing the data acquisition steps in Section 2.3, in order to reduce the data dimension and ensure the effectiveness of the subsequent pattern recognition algorithm [26], we extracted four features from each data sample obtained: Maximum Value (MAX), Average Value (Mean), Integral Value (IV) and Wavelet Transform (WT). MAX reflects the steady state information of the whole gas sensor response curve. Mean and IV combine all the information of the whole gas sensor response curve. WT can better reflect the transient information of the whole gas sensor response curve. MAX, Mean, IV and WT were extracted from each data sample obtained from 12 sensors. After feature extraction, each data sample changes from a data sample containing 54,000 data points to a feature sample containing only 12 data points. The feature samples extracted from 225 samples were spliced together to obtain a feature matrix containing 12 feature vectors and each feature vector contains 225 feature values.

2.5. Sensor Array Optimization

In order to enable the electronic nose to realize real-time and rapid detection of vehicle exhaust pollutants, it is necessary to make the electronic nose more miniaturized and low-cost. On the one hand, the sensor array optimization method can simultaneously realize the miniaturization and low-cost of the electronic nose. On the other hand, due to the cross sensitivity of gas sensors, sensor array optimization can reduce the training time of classification models, improve the recognition rate of classification models, and avoid the occurrence of over fitting problems. Three sensor array optimization (i.e., feature selection) methods based on different principles and optimization strategies were used in this paper: Recursive Feature Elimination with Cross Validation (RFECV) based on the packaging method, Random Forest Feature Selector (RFFS) based on an embedding method, and traditional Principal Component Analysis (PCA) as the comparison. They are briefly introduced below.

2.5.1. Sensor Array Optimization Based on RFECV

Recursive Feature Elimination (RFE) was proposed by Guyon [27] and has been widely used in solving feature selection problems. RFECV is a feature selection process for recursive feature elimination in the cross-validation cycle [28], which can automatically find the feature subset with the optimal number of features to obtain feature selection results. In this paper, Random Forest (RF) was used as the classification model in the process of recursive feature elimination. In the recursive step of each iteration, remove the last feature according to feature ranking, retrain the RF model with the retained features, and cross verify the performance of the RF model until there is only one feature left. Finally, according to the performance of RF model in different feature numbers and feature subsets composed of different features, the optimal number of features and the optimal feature subset can be obtained.

2.5.2. Sensor Array Optimization Based on RFFS

Random forests can not only deal with classification and regression problems, but also can be used to evaluate and select features because they can estimate the importance of features [29]. The principle of using random forests to evaluate and select features is based on the difference between the classification performance of random forest on the original dataset and the randomly extracted dataset. By calculating the classification performance difference of each decision tree in the random forest on different randomly extracted datasets, the importance of features can be estimated and the feature ranking can be obtained. The importance of the features is estimated by Equations (1) and (2):

The importance of feature A_j is estimated as:

$$I(A_j) = \sum \frac{d_i}{n \times SEd_i}, \qquad (1)$$

where d_i represents the performance difference of decision tree i and SEd_i represents the standard error of all decision trees:

$$SEd_i = \frac{SDd_i}{\sqrt{n}}, \qquad (2)$$

where SDd_i is the standard deviation of d_i and n is the number of elements in the dataset.

2.5.3. Sensor Array Optimization Based on PCA

Principal Component Analysis is a dimension reduction method, which transforms the original multivariable in high-dimensional space into a set of linear independent comprehensive indexes in low-dimensional space through orthogonal transformation [30]. The eigenvalues obtained through Principal Component Analysis are sorted from large to small to measure the importance of features. Finally, features are selected according to the importance of features [18].

Assuming that $X = (x_1, x_2, \ldots, x_{12})$ is the original variable (the feature extracted from 12 gas sensors), $Z = (z_1, z_2, \ldots, z_p)$, $(p \leq 12)$ as the comprehensive indexes in the low dimensional space. The transformation process from the original variable matrix X to the comprehensive index matrix Z can be expressed as:

$$\begin{aligned} z_1 &= a_{1,1}x_1 + a_{1,2}x_2 + \ldots + a_{1,12}x_{12} \\ z_2 &= a_{2,1}x_1 + a_{2,2}x_2 + \ldots + a_{2,12}x_{12} \\ &\vdots \\ z_p &= a_{p,1}x_1 + a_{p,2}x_2 + \ldots + a_{p,12}x_{12} \end{aligned} \quad (3)$$

where $a_{p,12}$ represents the 12th coefficient in the p-th comprehensive index.

Calculate the absolute value of the sum of the coefficients of the original variable corresponding to each feature in all comprehensive indexes, such as the absolute value of the sum of coefficients of the feature values extracted from the 12th sensor in all comprehensive indexes A_{12}:

$$z_2 = |a_{1,12}| + |a_{2,12}| + \ldots + |a_{p,12}|. \quad (4)$$

The absolute value of the sum of the coefficients of each feature is used to represent the contribution degree of each feature in the comprehensive indexes, and then the feature selection can be obtained by descending order.

3. Results and Discussion

Different sensor array optimization methods will result in different sensor array optimization results, in which the number and combination of sensors in the sensor array will be different. The experimental results of this paper were obtained by taking the original data and the data after sensor array optimization as the input data of the Random Forest (RF) classifier. The classification recognition rate of the test set is the main index to evaluate the effectiveness of the original sensor array and the optimized sensor array. Therefore, this paper used the classification recognition rate of the test set to evaluate the results of sensor array optimization. In order to make the classification results more reliable and credible, the stratified sampling strategy was used to conduct 3-fold cross-validation 100 times, and the average of the 300-test set classification recognition rate was calculated as the final classification recognition rate of the test set. In the sensor array optimization stage, the original data set was divided into 2/3 training set and 1/3 test set; the test set was not used in the sensor array optimization stage.

The classification recognition rate obtained using RF without sensor array optimization is shown in Figure 5a. It can be seen that the four feature extraction methods have achieved high classification recognition rate when using RF as the classifier. The highest MAX has an average classification recognition rate of 99.92%, and the WT with the lowest average classification recognition rate has also reached 98.16%. This shows that the original sensor array is effective and has the ability to accurately identify exhaust pollutants from different engines or different concentration levels from the same engine.

The results of sensor array optimization of the four extracted eigenvalues based on RF model and RFECV method are shown in Table 3. The optimized sensor array includes six gas sensors: MP135, TGS2600, TGS2610, TGS2611, TGS2612, and TGS2620. The optimization of sensor array based on RFECV method has achieved good results. Figure 5b shows that Mean and IV, which have the lowest average classification recognition rate after the optimization of sensor array, still reached 97.94%, almost without any loss. This shows that the RFECV method is very effective for sensor array optimization.

The classification recognition rate of sensor array optimization using RFFS is shown in Figure 5c. Compared with the RFECV method, RFFS retains a total of eight gas sensors. In addition to the six same gas sensors selected in the RFECV method, it also retains two gas sensors, GSBT11 and TGS2602. It can be seen from the comparison of average classification recognition rate between Figure 5b,c that the classification recognition rate of RFFS with eight gas sensors is slightly improved in Mean and IV compared with RFECV with six gas sensors, but it is decreased in MAX with the highest classification recognition rate, and it also need to bear the cost of increasing the development of two gas sensors.

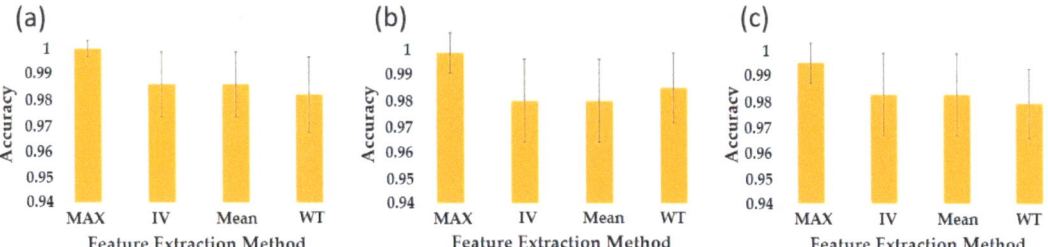

Figure 5. (a) The RF classifier's average classification recognition rate of the four feature extraction methods without sensor array optimization. (b) The RF classifier's average classification recognition rate of the four feature extraction methods after using RFECV as the sensor array optimization method. (c) The RF classifier's average classification recognition rate of the four feature extraction methods after using RFFS as the sensor array optimization method. (The error bars represent the standard deviations of the recognition rates, and the calculation formula is $\sigma = \sqrt{\left((x_1 - \bar{x})^2 + (x_2 - \bar{x})^2 + \ldots + (x_n - \bar{x})^2\right)/n}$, where σ represents the standard deviation, x_1, $x_2, \ldots, x_n$ represent recognition rates, $\bar{x}$ represents the average recognition rate, n is the number of recognition rates. In this paper, we used 3-fold cross validation 100 times, so the value of n is 300.)

Table 3. The sensor array optimization results of RFECV based on RF.

Feature Extraction Method	Number of Sensors after Optimization	Sensor Serial Number Retained after Optimization
MAX		
Mean		B, E, H
IV	6	I, K, L
WT		

Using PCA to optimize sensor arrays requires setting corresponding thresholds to limit the absolute value of the sum of coefficients of each feature. In this paper, in order to compare with RFECV and RFFS, the threshold values are set at the values required when six and eight gas sensors were reserved. The sensor array optimization results and the corresponding RF classification recognition rate is shown in Table 4. Obviously, when the number of optimized sensors is limited, the classification recognition rate of sensor array optimized by PCA is worse than that of other sensor array optimization methods. When eight sensors are retained, the classification recognition rate of MAX decreased by more than 2% compared with RFFS. When six sensors are retained, the classification recognition rate of MAX, Mean and IV declined compared with RFECV.

When PCA was used as the sensor array optimization method, TGS2600, TGS2603, TGS2610, TGS2611, TGS2612 and TGS2620 were selected to be retained in almost every feature extraction method, regardless of whether six or eight gas sensors were retained, which indicates that they have a good response to vehicle exhaust pollutants. However, MP135 and MP901, which were retained for many times when eight gas sensors were retained, were rarely selected when six gas sensors were retained. This may be because the target gases detected by them overlap with the six gas sensors frequently selected above. When different feature extraction methods retained the same number of gas sensors, the main reason why the gas sensors selected for retention were different and the main reason why the sensor array optimization using PCA method was not as effective as the other two sensor array optimization methods may be because it is an unsupervised dimension reduction method, and its realization method is to maximize the variance in the projection direction, so the category information is not fully utilized.

Table 4. The sensor array optimization results and the classification recognition rate based on RF after using PCA.

Feature Extraction	Number of Sensors after Optimization	Sensor Serial Number Retained after Optimization	Recognition Rate (%)
MAX	8	A, C, F, H, I, J, K, L	97.12
Mean	8	B, C, F, H, I, J, K, L	98.34
IV	8	B, C, F, H, I, J, K, L	98.34
WT	8	B, F, G, H, I, J, K, L	98.59
MAX	6	A, F, I, J, K, L	97.44
Mean	6	F, H, I, J, K, L	96.26
IV	6	F, H, I, J, K, L	96.26
WT	6	B, F, H, J, K, L	98.63

When the number of sensors retained after using different methods to optimize the sensor array is the same, the sensor array with higher classification recognition rate is better. When the classification recognition rate of sensor arrays is the same, the sensor array with fewer sensors is better. A good sensor array needs to achieve the highest recognition rate when the number of sensors is as small as possible. Considering that the recognition rate of the sensor array composed of six gas sensors is almost no lower than that of the original sensor array composed of 12 gas sensors, and the sensor array composed of six gas sensors can reduce volume and save the cost to make it possible for the vehicle-mounted electronic nose to rapidly detect vehicle exhaust pollutants in real time. We believe that the optimal number of sensors is to retain six gas sensors. After limiting the number of sensors in the sensor array to six, the recognition rate of each sensor array optimization method is shown in Table 5.

Table 5. The classification recognition rate based on RF after using different methods (limited the number of sensors in the sensor array to six).

Feature Extraction	Sensor Array Optimization Method	Recognition Rate (%)
MAX	RFECV	99.77
MAX	RFFS	99.77
MAX	PCA	97.44
Mean	RFECV	97.94
Mean	RFFS	97.94
Mean	PCA	96.26
IV	RFECV	97.94
IV	RFFS	97.94
IV	PCA	96.26
WT	RFECV	98.44
WT	RFFS	98.33
WT	PCA	98.63

In this case, the average RF classification recognition rate of MAX after using RFECV and RFFS for sensor array optimization has both reached 99.77%. The classification recognition rate of WT after using RFECV was higher than using RFFS. It means that RFECV is a better sensor array optimization method, and MAX is a better feature extraction method than the other three. In addition, the sensor array optimized by RFECV method and using MAX as the feature extraction method includes sensors: MP135, TGS2600, TGS2610, TGS2611, TGS2612 and TGS2620.

The cost of using only the above six gas sensors is $32, which can save about 56% of the cost compared with the original 12 sensor arrays. It can also greatly reduce the volume of the electronic nose to achieve the purpose of miniaturization. The average time of using MAX as the feature extraction after using RFECV to test a new real sample was 0.021 s (using Python 3.10.5 and Visual Studio Code 2022). The miniaturization and the rapid detection time make it possible for the vehicle-mounted electronic nose to rapidly detect vehicle exhaust pollutants in real time.

4. Conclusions

In this paper, a self-designed electronic nose composed of 12 gas sensors was used to detect vehicle exhaust pollutants from different engines or the same engine at different concentration levels. Firstly, we conducted an engine bench test to summarize the concentration variation law vehicle exhaust pollutants. After analyzing the experimental data and extracting the features, the highest RF classification recognition rate was up to 99.92% without optimizing the sensor array. In order to enable the vehicle-mounted electronic nose to quickly detect vehicle exhaust pollutants in real time, reduce the volume, save development cost, and save detection time, we used RFECV, RFFS, and PCA to optimize the sensor array. When the number of sensors was not fixed, the classification recognition rate of MAX after using RFECV and RFFS methods reached 99.77% and 99.44% respectively, while RFECV retained less sensors. When the number of sensors was limited to six, the classification recognition rate of MAX after using RFFS was up to 99.77%, as high as that of RFECV. The classification recognition rate of WT after using RFECV was higher than using RFFS, which means RFECV is a better sensor array optimization method in this case. The cost of the sensor array optimized by RFECV method and using MAX as the feature extraction method is only 32$ with almost no loss of recognition rate. In addition, its average detection time of a new sample was 0.021 s.

In summary, through the research in this paper, we found that it is feasible to use electronic nose for real-time and rapid detection of vehicle exhaust pollutants, and electronic nose has a good application prospect in this area. At the same time, combined with the sensor array optimization methods, the electronic nose can be miniaturized and low-cost, which makes it possible to detect vehicle exhaust pollutants in real time. In the future, we will use the vehicle-mounted electronic nose in combination with the developing edge computing and cloud computing technologies to develop corresponding cloud calculating platforms. We can monitor vehicle exhaust pollutants in real time, and make corresponding predictions regarding air quality.

Author Contributions: Conceptualization, Z.C.; methodology, J.T., Z.C., Y.L. and L.T.; software, C.S., X.Z., Z.L. and S.W.; validation, C.S., X.Z., Z.L. and S.W.; formal analysis, J.T., Y.L. and L.T.; investigation, Z.C., Y.L. and L.T.; resources, X.Z., Z.L. and S.W.; data curation, Y.L., L.T., X.Z., Z.L. and S.W.; writing—original draft preparation, C.S., T.T., X.Z., Z.L. and S.W.; writing—review and editing, J.T., Y.L. and L.T.; visualization, C.S., T.T. and X.Z.; supervision, J.T. and Z.C.; project administration, X.Z., Z.L. and S.W.; funding acquisition, X.Z., Z.L. and S.W. All authors have read and agreed to the published version of the manuscript.

Funding: This work was supported by the National Natural Science Foundation of China (51875245), the Science-Technology Development Plan Project of Jilin Province (20200501013GX, 20200403038SF, 20210203004SF, 20220401087YY), the Special Project of Industrial Technology Research and Development of Jilin Province (2022C045-6), the "13th Five-Year Plan" Scientific Research Foundation of the Education Department of Jilin Province (JJKH20220193KJ, JJKH20221011KJ).

Institutional Review Board Statement: Not applicable.

Informed Consent Statement: Not applicable.

Data Availability Statement: The datasets generated and/or analyzed during the current study are available from the corresponding author upon reasonable request.

Conflicts of Interest: The authors declare no conflict of interest.

References

1. Twigg, M. Progress and future challenges in controlling automotive exhaust gas emissions. *Appl. Catal. B* **2007**, *70*, 3523. [CrossRef]
2. Šarkan, B.; Kuranc, A.; Kučera, Ľ. Calculations of exhaust emissions produced by vehicle with petrol engine in urban area. In Proceedings of the 4th International Conference of Computational Methods in Engineering Science (CMES), Kazimierz Dolny, Poland, 21–23 November 2019.
3. Zhang, Y.; Stedman, D.H.; Bishop, G.A.; Guenther, P.L.; Beaton, S.P.; Peterson, J.E. On-road hydrocarbon remote sensing in the Denver area. *Environ. Sci. Technol.* **1993**, *27*, 1885–1891. [CrossRef]
4. Tutuianu, M.; Bonnel, P.; Ciuffo, B.; Haniu, T.; Ichikawa, N.; Marotta, A.; Pavlovic, J.; Steven, H. Development of the World-wide harmonized Light duty Test Cycle (WLTC) and a possible pathway for its introduction in the European legislation. *Transp. Res. D Transp. Environ.* **2015**, *40*, 61–75. [CrossRef]
5. Guo, W.; Qiong, W.; Zhao, M. Exhaust emissions of diesel engines with nano-copper additives. *Appl. Nanosci.* **2020**, *10*, 1045–1052. [CrossRef]
6. Jeong, S.Y.; Kim, J.S.; Li, J.H. Rational design of semiconductor-based chemiresistors and their libraries for next-generation artificial olfaction. *Adv. Mater.* **2020**, *32*, 2002075. [CrossRef]
7. Lee, J.M.; Devaraj, V.; Jeong, N.N.; Lee, Y.; Kim, Y.J.; Kim, T.; Yi, S.H.; Kim, W.G.; Choi, E.J.; Kim, H.M.; et al. Neural mechanism mimetic selective electronic nose based on programmed M13 bacteriophage. *Biosens. Bioelectron.* **2022**, *196*, 113693. [CrossRef] [PubMed]
8. Machungo, C.; Berna, A.Z.; McNevin, D.; Wang, R.; Trowell, S. Comparison of the performance of metal oxide and conducting polymer electronic noses for detection of aflatoxin using artificially contaminated maize. *Sens. Actuators B.* **2022**, *360*, 131681. [CrossRef]
9. Chang, Z.Y.; Li, J.H.; Qi, H.Y.; Ma, Y.H.; Chen, D.H.; Xie, J.; Sun, Y.H. Bacterial infection potato tuber soft rot disease detection based on electronic nose. *Open Life Sci.* **2017**, *12*, 379–385. [CrossRef]
10. Zhao, R.S.; Kong, C.; Ren, L.Q.; Sun, Y.H.; Chang, Z.Y. Real-time monitoring of the oil shale pyrolysis process using a bionic electronic nose. *Fuel* **2022**, *313*, 122672. [CrossRef]
11. Weng, X.H.; Sun, Y.H.; Xie, J.; Deng, S.H.; Chang, Z.Y. Bionic Layout Optimization of Sensor Array in Electronic Nose for Oil Shale Pyrolysis Process Detection. *J. Bionic Eng.* **2021**, *18*, 441–452. [CrossRef]
12. Tirzīte, M.; Bukovskis, M.; Strazda, G.; Jurka, N.; Taivans, I. Detection of lung cancer with electronic nose and logistic regression analysis. *J. Breath Res.* **2018**, *13*, 016006. [CrossRef] [PubMed]
13. Dragonieri, S.; Pennazza, G.; Carratu, P.; Resta, O. Electronic nose technology in respiratory diseases. *Lung* **2017**, *195*, 157–165. [CrossRef] [PubMed]
14. Kang, M.; Cho, I.; Park, J.; Jeong, J.; Lee, K.; Lee, B.; Henriquez, D.D.; Yoon, K.; Park, I. High Accuracy Real-Time Multi-Gas Identification by a Batch-Uniform Gas Sensor Array and Deep Learning Algorithm. *ACS Sens.* **2022**, *7*, 430–440. [CrossRef] [PubMed]
15. Arroyo, P.; Meléndez, F.; Suárez, J.I.; Herrero, J.L.; Rodríguez, S.; Lozano, J. Electronic nose with digital gas sensors connected via bluetooth to a smartphone for air quality measurements. *Sensors* **2020**, *20*, 786. [CrossRef] [PubMed]
16. Kun, C.; He, Y.; Li, Y.; Ng, A.; Yu, J. A Room Temperature Hydrocarbon Electronic Nose Gas Sensor Based on Schottky and Heterojunction Diode Structures. *IEEE Electron Device Lett.* **2019**, *41*, 163–166. [CrossRef]
17. Zhang, W.; Tian, F.; Song, A.; Hu, Y. Research on electronic nose system based on continuous wide spectral gas sensing. *Microchem. J.* **2018**, *140*, 1–7. [CrossRef]
18. Sun, H.; Tian, F.C.; Liang, Z.F.; Sun, T.; Yu, B.; Yang, S.X.; He, Q.H.; Zhang, L.L.; Liu, X.M. Sensor array optimization of electronic nose for detection of bacteria in wound infection. *IEEE Trans. Ind. Electron.* **2017**, *64*, 7350–7358. [CrossRef]
19. Yin, Y.; Yu, H.C.; Chu, B.; Xiao, Y.J. A sensor array optimization method of electronic nose based on elimination transform of Wilks statistic for discrimination of three kinds of vinegars. *J. Food Eng.* **2014**, *127*, 43–48. [CrossRef]
20. Bag, A.K.; Tudu, B.; Roy, J.; Bhattacharyya, N.; Bandyopadhyay, R. Optimization of sensor array in electronic nose: A rough set-based approach. *IEEE Sens. J.* **2011**, *11*, 3001–3008. [CrossRef]
21. Yan, K.; Zhang, D. Feature selection and analysis on correlated gas sensor data with recursive feature elimination. *Sens. Actuators B* **2015**, *212*, 353–363. [CrossRef]
22. Li, P.; Niu, Z.; Shao, K.; Wu, Z. Quantitative analysis of fish meal freshness using an electronic nose combined with chemometric methods. *Measurement* **2021**, *179*, 109484. [CrossRef]
23. Xu, Z.; Shi, X.; Lu, S. Integrated sensor array optimization with statistical evaluation. *Sens. Actuators B* **2010**, *149*, 239–244. [CrossRef]
24. Jia, P.; Tian, F.; Fan, S.; He, Q.; Feng, J.; Yang, S.X. A novel sensor array and classifier optimization method of electronic nose based on enhanced quantum-behaved particle swarm optimization. *Sens. Rev.* **2014**, *34*, 304–311. [CrossRef]

25. Wei, G.; Zhao, J.; Yu, Z.; Feng, Y.; Li, G.; Sun, X. An effective gas sensor array optimization method based on random forest. In Proceedings of the 17th IEEE SENSORS Conference, New Delhi, India, 28–31 October 2018.
26. Yan, J.; Guo, X.; Duan, S.; Jia, P.; Wang, L.; Peng, C.; Zhang, S. Electronic nose feature extraction methods: A review. *Sensors* **2015**, *15*, 27804–27831. [CrossRef] [PubMed]
27. Guyon, I.; Weston, J.; Barnhill, S.; Vapnik, V. Gene selection for cancer classification using support vector machines. *Mach. Learn.* **2002**, *46*, 389–422. [CrossRef]
28. Wang, S.; Chen, S. Insights to fracture stimulation design in unconventional reservoirs based on machine learning modeling. *J. Pet. Sci. Eng.* **2019**, *174*, 682–695. [CrossRef]
29. Čehovin, L.; Bosnić, Z. Empirical evaluation of feature selection methods in classification. *Intell. Data. Anal.* **2010**, *14*, 265–281. [CrossRef]
30. Jolliffe, I.T.; Cadima, J. Principal component analysis: A review and recent developments. *Philos. Trans. R. Soc. A* **2016**, *374*, 20150202. [CrossRef]

Article

Monitoring of MSW Incinerator Leachate Using Electronic Nose Combined with Manifold Learning and Ensemble Method

Zhongyuan Zhang [1], Shanshan Qiu [1,2,*], Jie Zhou [1,*] and Jingang Huang [1,2]

1. College of Materials and Environmental Engineering, Hangzhou Dianzi University, Hangzhou 310018, China
2. The Belt and Road Information Research Institute, Hangzhou Dianzi University, Hangzhou 310018, China
* Correspondence: qiuss@hdu.edu.cn (S.Q.); jane@hdu.edu.cn (J.Z.)

Abstract: Waste incineration is regarded as an ideal method for municipal solid waste disposal (MSW), with the advantages of waste-to-energy, lower secondary pollution, and greenhouse gas emission mitigation. For incineration leachate, the information from the headspace gas that varies at different processing processes and might be useful for chemical analysis, is ignored. The study applied a novel electronic nose (EN) to mine the information from leachate headspace gas. By combining manifold learnings (principal component analysis (PCA) and isometric feature mapping (ISOMAP), and uniform manifold approximation and projection (UMAP) and ensemble techniques (light gradient boosting machine (lightGBM) and extreme gradient boosting (XGBT)), EN based on the UMAP-XGBT model had the best classification performance with a 99.95% accuracy rate in the training set and a 95.83% accuracy rate in the testing set. The UMAP-XGBT model showed the best prediction ability for leachate chemical parameters (pH, chemical oxygen demand, biochemical oxygen demand, ammonia, and total phosphorus), with R^2 higher than 0.99 both in the training and testing sets. This is the first study of the EN application for leachate monitoring, offering an easier and quicker detection method than traditional instrumental measurements for the enforcement and implementation of effective monitoring programs.

Keywords: electronic nose; incinerator leachate; data mining; prediction; classification

1. Introduction

The world generates 2.01 billion tons of municipal solid waste (MSW) annually, and waste generated per person per day averages at 0.74 kg. When looking forward, global waste is expected to grow to 3.40 billion tons by 2050 [1]. Comprised of physicochemical and biological characteristics that are aggressive to the soil, water resources, fauna and flora, and MSW is difficult to handle for most countries and regions [2]. To date, the main disposal methods for MSW are landfill and incineration. MSW landfill causes some issues to the environment, including: (1) high greenhouse gas (GHG) emissions if landfill gas is not properly collected, (2) leachate produced damages the ecosystem, (3) a larger space is needed for the project set up [3]. Therefore, waste incineration is regarded as an ideal method for MSW disposal [4], with the advantages of waste-to-energy, lower secondary pollution, greenhouse gas emission mitigation, and so on.

However, for MSW incineration, a considerable number of challenges are still generated at different points, including but not limited to leachate processing. For incineration leachate, research has mainly focused on the characteristics of leachate concentrate [5], the organic matter molecular transformation in leachate [6], and the degradation of refractory organics [7]. All these direct or indirect studies relate to the leachate headspace gas, which hints that the information in the leachate headspace gas can be mined for leachate processing or monitoring. Until now, few in-depth studies have been conducted to fetch information from the vast amounts of original data about the varieties, concentrates, and changes of those materials.

The electronic nose (EN) appears to be a promising candidate for headspace gas detecting, mimicking the human nose, with a range of applications, including the food industry, disease diagnosis, and environment monitoring [8]. Different from instrumental methods, such as gas chromatography (GC) with a flame photometric detector (FID), photoionization (PID), or a mass spectrometer (MS), EN offers the whole information that is unique for each sample headspace gas instead of specific materials or their concentrations. EN has been predominately used for indoor air monitoring [9], soil contamination detection [10], and water quality monitoring [11]. However, studies on leachate detection based on EN technology are rare, according to our best knowledge.

Novel EN devices, based on machine learning algorithms capable of real-time detection of industrial and municipal pollutants, have been developed to monitor specific environmental-pollutant levels for enforcement and implementing effective pollution-abatement programs [12]. Manifold learning, such as isometric feature maps (ISOMAPs) and uniform manifold approximation and projection (UMAP), uncover a low-dimensional manifold embedded in a high-dimensional space while respecting the intrinsic geometry [13]. Manifold learning, as a novel data pre-processing technology, can improve the performance of EN detecting.

Ensemble methods are designed to overcome problems with weak predictors and meet the fast, high-performance requirement [14]. Combining manifold learning and ensemble methods techniques makes it possible to identify and differentiate between gases of different leachate samples based on EN and mine useful information for leachate monitoring. A novel EN based on manifold learning and ensemble methods was applied to monitor the changes in leachate headspace gas. The main objectives are: (1) to study the variation of leachate headspace gas based on EN with different processing procedures; (2) to investigate the manifold structure of EN original data based on principal component analysis, ISOMAP, and UMAP; and (3) to mine the relationship between leachate headspace gas and chemical parameter changes based on ensemble methods (extreme gradient boosting and light gradient boosting machines). The study provides insights into the relationship between leachate gas emission and chemical parameters based on EN combined with manifold learning and ensemble methods and offers an easier and quicker monitoring method than traditional instrumental measurements for the enforcement and implementation of effective monitoring programs.

2. Materials and Methods

2.1. Sample Collection

The incineration leachate samples collected were from Wenling Green New Energy Co., Ltd. (Wenling, China), which was invested in by Zheneng Jinjiang Environment Holding Co., Ltd. (Hangzhou, China), who is a forerunner and leading waste-to-energy operator in China's waste-to-energy (WTE) industry. The Wenling incineration power generation plant is located in the northern part of the Eastern New District of Wenling City, next to the East China Sea, with a total area of 7.3×10^5 sqm. The leachate treatment scale of the incineration plant is 1600 tons/day in two phases, and for now, the treatment scale of the first phase is 800 tons/day (600 tons/day of domestic waste and 200 tons/day of dry sludge). The leachate treatment process of the incineration power plants is shown in Figure 1.

Leachate samples from six water outlets were collected on the 15 July 2022. In Figure 2, the samples from six water outlets were named leachate raw water (LRW), leachate effluent (LE), internal circulation reactor effluent, aerobic effluent (AeroE), anaerobic effluent (ANE), and MBR effluent (MBRE). The samples were preserved in a refrigerator at a temperature lower than 4 °C and were forwarded to the laboratory.

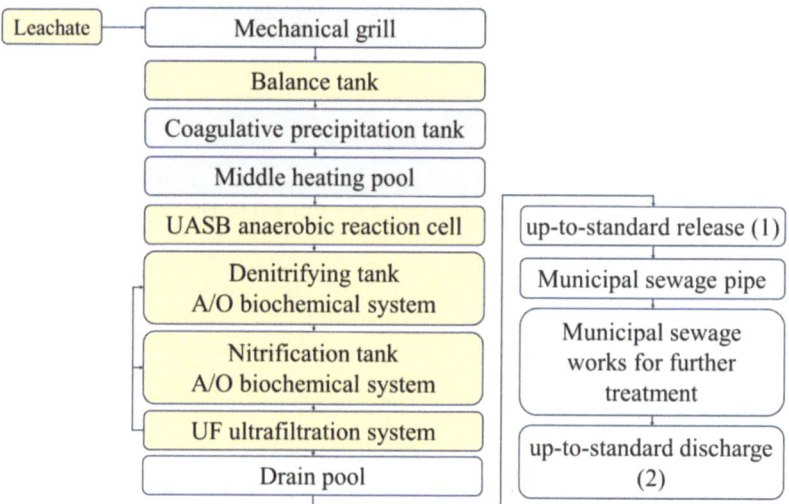

Figure 1. Process flow diagram of incineration leachate treatment in the study.

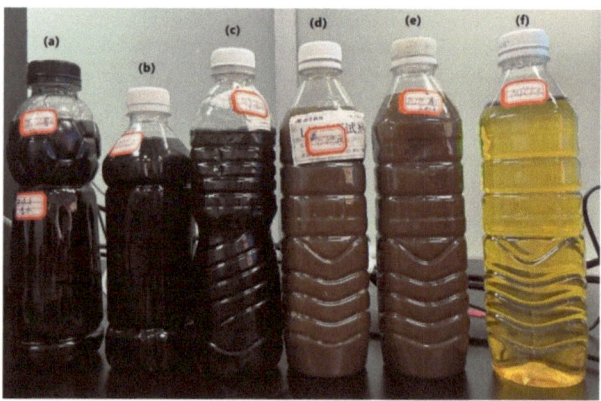

Figure 2. Leachate sample collection: (**a**) leachate raw water (LRW), (**b**) leachate effluent (LE), (**c**) internal circulation reactor effluent (ICRE), (**d**) aerobic effluent (AeroE), (**e**) Anaerobic effluent (ANE), and (**f**) MBR effluent (MBRE).

2.2. Chemical Parameters Detection for Incinerator Leachate

In general, incinerator leachate is tested by conventional parameters, including pH, chemical oxygen demand (COD), biochemical oxygen demand after 5 days (BOD_5), ammonia (NH_4^+-N), and total phosphorus (TP). The values of those conventional parameters exhibit considerable differences due to variations in composition and moisture content, as well as seasonal factors and incinerator location. The value of pH was tested by the electrode method [15]. For COD detection, the dichromate method is not suitable for the water samples in which the chloride ion concentration is higher than 1000 mg/L. The chlorine emendation method was applied to detect the contents of COD in the incinerator leachate samples [16]. The concentration of ammonia nitrogen was measured according to Nessler's reagent spectrophotometry [17]. The alkaline potassium persulfate digestion UV spectrophotometric method [18] was used to detect total nitrogen content. The ammonium molybdate spectrophotometric method [19] was applied to detect the content of TP.

2.3. E-Nose Detection

The EN mainly consists of two parts: the sensor array, which is applied to sense the information in the sample's headspace, and the software part, which handles the information received from the sensors. According to the sensing materials, metal-oxide-semiconductor (MOS), quartz crystal microbalance, and surface acoustic wave sensors are most applied in the EN system [20]. MOS gas sensors are most sensitive to hydrogen and unsaturated hydrocarbons or solvent vapors containing hydrogen atoms [21]. The headspace gas from incinerator leachate contents is mainly composed of volatile organic compounds (VOCs), including hydrogen sulphide, methyl mercaptan, acetylene, propylene, and ethylene, and also varies according to source, season, the incinerator site, and so on [22].

A commercial PEN2 electronic nose (Airsense Analytics, GmBH, Schwerin, Germany) was applied to detect the headspace gas from incinerator leachate samples at different processing procedures. For PEN2, MOS sensors are the core part, and the details of MOS sensors are presented in Table 1. The MOS sensors convert the information about gas types and concentrations into an electrochemical signal. The EN signal was expressed as G/G0, where G and G0 represent the resistance of a sensor in sample headspace gas and clean air. As the sensors are cross-sensitive to a class of gas compounds, EN does not give the specific information of one material but offers the headspace gas complementary information.

Table 1. Sensors used and their main applications in the EN.

No.	Sensor Name	General Description	Reference
S1	W1C	Aromatic compounds	Toluene, 0.1 g/kg
S2	W5S	Very sensitive with negative signal, broad range sensitivity, react on nitrogen oxides	NO_2, 1×10^{-3} g/kg
S3	W3C	Very sensitive with aromatic compounds	Benzene, 1×10^{-2} g/kg
S4	W6S	Mainly hydrogen, selectively, (breath gases)	H_2, 0.1 g/kg
S5	W5C	Alkanes, aromatic compounds, less polar compounds	Propane, 1×10^{-3} g/kg
S6	W1S	Sensitive to methane (environment). Broad range, similar to S8;	CH_4, 0.1 g/kg
S7	W1W	Reacts on sulfur compounds, or sensitive to many terpenes and sulfur organic compounds;	H_2S, 1×10^{-4} g/kg
S8	W2S	Detects alcohol's, partially aromatic compounds, broad range	CO, 0.1 g/kg
S9	W2W	Aromatics compounds, sulfur organic compounds	H_2S, 1×10^{-3} g/kg
S10	W3S	Reacts on high concentrations > 0.1 g/kg, sometime very selective (methane)	CH_4, 0.1 g/kg

During the EN detection, incinerator leachate liquid with a 5 mL volume was placed into a 500 mL beaker. The beaker was sealed by plastic wrap and was kept still for 30 min to balance the headspace gas generated from the incinerator leachate. Two holes were made, one for EN detection and the other for a steady stream of gas while EN detecting. The EN detection time was set to 80 s, and then the gas path and sensor chamber were cleaned with clean air. The gas flow rate was 200 mL/min, and one signal per second was collected. Landfill leachate was collected from six water outlets with 24 samples; thus, 144 samples (24 samples × 6 procedures) were prepared. All the detection was accomplished on the sample collection day.

2.4. Data Reduction Based on Manifold Learning

2.4.1. Principal Component Analysis

As a multivariate technique, Principal Component Analysis (PCA) was applied to analyze a data set consisting of several inter-correlated quantitative dependent variables [23]. By calculating eigenvalue and eigenvector from the covariance matrix of the original data set, the new orthogonal variables will be derived and usually called principal components (PC). The cumulative contribution rate of PCs should reach more than 85% of the total variance, then the PCA will be considered to have extracted the main information of the original data.

2.4.2. Isometric Feature Mapping

As a nonlinear dimensionality reduction technique, Isometric Feature Mapping (ISOMAP) maintains the essential geometric structure of nonlinear data [24]. ISOMAP is multidimensional scaling combined with geodesic distance for reducing the dimensionality of data sampled from a smooth manifold. ISOMAP tries to solve the shortest path to obtain the geodesic distance that preserves the characteristics of high-dimensional data structures as much as possible. Multidimensional scaling is used to calculate the coordinates of each data point in the low-dimensional space, and the original data is embedded in the high-dimensional set.

2.4.3. Uniform Manifold Approximation and Projection

As a nonlinear dimensionality reduction technique, uniform manifold approximation and projection (UMAP) was developed for the analysis of any type of high-dimensional data [25]. From a theoretical framework based in Riemannian geometry and algebraic topology, UMAP learns the data representation between points in high-dimensional space and maps to low dimensions by calculating the joint probability density between high-dimensional sample points. Spectral clustering analysis is used to initialize the low-dimensional data and then project it into the low-dimensional space. Adjustable parameters are used in joint probability density to control the change of conditional probability and ensure the symmetry of the data. Low-dimensional data also provides two parameters to adjust the aggregation of mapped data so that low-dimensional data can better fit high-dimensional spatial data.

2.5. Classification and Prediction

2.5.1. Classification and Regression Tree

Classification and Regression Tree (CART) selects features based on the minimization of the Gini coefficient to generate a binary tree. By pre-pruning through empirical judgment, the useless attributes can be removed. After the construction is completed, the algorithm can resist overfitting and has better generalization ability by cutting off a part of the information with less proportion.

In addition, the shortcomings that CART cannot handle large amounts of data, underfitting, and overfitting, can be overcome by integrating multiple CART classifiers into a single ensemble model with the ideas of bagging and boosting. The study selects a boosting algorithm with relatively stable generalization performance. Boosting is a kind of optimization algorithm based on the greedy strategy of selecting the fixed loss function (optimization function and objective function) based on a greedy strategy for the optimization of the loss function, committed to obtaining the minimum loss optimization function, such as eXtreme Gradient Boosting (XGBT) and Light Gradient Boosting Machine (lightGBM).

2.5.2. eXtreme Gradient Boosting

eXtreme Gradient Boosting (XGBT) uses the first and second partial derivatives, and the second derivatives help the gradient descend faster and more accurately. Using Taylor expansion to obtain the function as the second derivative form of the independent variable, the leaf splitting optimization calculation can be carried out only by relying on the value of the input data without selecting the specific form of the loss function, essentially separating the selection of the loss function from the optimization of the model algorithm/parameter selection [26]. The algorithm goes:

1. Initialize model with a constant value:

$$\hat{f}_0(x) = \mathrm{argmin} \sum_i^n L(y_i, \theta)$$

For m = 1 to M:
a. Compute so-called pseudo-residuals:

$$\hat{g}_m(x) = -\left[\frac{\partial L(y_i, f(x_i))}{\partial f(x_i)}\right]_{f(x)=\hat{f}_{m-1}(x)}$$

$$\hat{h}_m(x) = -\left[\frac{\partial^2 L(y_i, f(x_i))}{\partial f(x_i)^2}\right]_{f(x)=\hat{f}_{m-1}(x)}$$

b. Fit a base learner using the training set $\left\{x_i, \frac{\hat{g}_m(x_i)}{\hat{h}_m(x_i)}\right\}_{i=1}^n$ by solving the optimization problem below:

$$\hat{\varphi}_m = \operatorname{argmin} \sum_i^N \frac{1}{2}\hat{h}_m(x_i)\left[-\frac{\hat{g}_m(x_i)}{\hat{h}_m(x_i)} - \varphi(x_i)\right]^2$$

$$\hat{f}_m(x) = \alpha \hat{\varphi}_m(x)$$

c. Update the model:

$$\hat{f}_m(x) = \hat{f}_{m-1}(x) + \hat{f}_m(x)$$

2. output $F_M(x)$.

2.5.3. Light Gradient Boosting Machine

Light Gradient Boosting Machine (LightGBM) grows trees leaf-wise instead of level-wise, yielding the largest loss decrease. LightGBM implements a highly optimized histogram-based decision tree learning algorithm, which greatly improves efficiency and memory consumption [27]. The algorithm goes:

(1) The sample points are sorted in descending order according to the absolute value of their gradient;
(2) Select the first samples of the sorted results to generate a subset of large gradient sample points;
(3) For 100% samples of the remaining sample set (1 − a), randomly select b (1 − a) × 100% sample points to generate a set of small gradient sample points;
(4) Merge the large gradient samples with the sampled small gradient samples;
(5) Multiply the small gradient samples by a weight coefficient;
(6) Learn a new weak learner (CART) using the above-sampled samples;
(7) Continuously repeat steps (1)~(6) until the specified number of iterations or convergence is reached.

2.6. The Evaluation of Data Processing

To evaluate the accuracy and precision of the established models, 100 samples were set as the training data, and the rest, 44 samples, were set as the testing data.

The receiver operative curve (ROC) was deployed as a performance indicator for the classification models. True positive and negative rates are the most commonly used to evaluate the performance of classification tests. The higher the probability value of these two indicators, represents the better the judgment effect in the model [28].

The coefficient R^2 and RMSE were selected as the evaluation parameters for prediction models. The higher the R^2 was and the lower the RMSE was, the more accurate the prediction ability of the model would be.

3. Results and Discussion

3.1. The Chemical Parameter Changes of Leachate

The composition of leachate is highly variable and heterogeneous. In general, incinerator leachate is tested using conventional parameters, including pH, COD, BOD_5, ammonia, and TP. The changes in the chemical parameters for leachate samples are shown in Table 2. There were statistically significant differences (Turkey HSD, $p < 0.05$) in the contents of COD, BOD$_5$, ammonia, TN, and TP. It was noteworthy that the values of COD decreased significantly at each process procedure. The changes were also noticeable for BOD, ammonia, TN, and TP. Different from the other five chemical parameters, the value of pH changes a lot at the last processing procedure. All the chemical parameters were all up to standard, when the processed leachate was discharged to the municipal pipe network.

Table 2. Average values of leachate chemical parameters.

	pH	COD (mg/L)	BOD$_5$ (mg/L)	Ammonia (mg/L)	TN (mg/L)	TP (mg/L)
LRW	8b	4.23×10^3 f	1.10×10^3 c	1.92×10^3 d	2.18×10^3 c	15.4 b
RPE	8b	6.14×10^3 e	1.50×10^3 d	1.70×10^3 c	2.14×10^3 c	24.3 c
ICRE	8.1b	2.90×10^3 d	0.70×10^3 b	1.54×10^3 c	1.76×10^3 b	24.1 c
AnE	8.3b	2.00×10^3 c	0.52×10^3 b	0.71×10^3 b	1.32×10^3 a	15.2 b
AeroE	7.8b	1.5×10^3 b	0.10×10^3 a	0.12×10^3 a	1.20×10^3 a	10.3 a
MBRE	6a	0.33×10^3 a	0.09×10^3 a	0.04×10^3 a	1.13×10^3 a	4.61 a

The values are the average of the total score of the ten experts with respect to three replications of leachate samples. Mean in the same row followed by different inline letters (a, b, c, d, e, f) is statistically different as confirmed by Tukey's HSD test ($p < 0.05$).

3.2. The Result of EN Detection

The EN was used to analyze the headspace gas of the leachate samples at different process periods. A typical response of the EN sensors array during exposure to sample gas, which was randomly selected from the 144 samples, is depicted in Figure 3a. The procedure of extracting the sample gas from the beaker to the sensing chamber took 5 seconds, and then the sensors could react with the gas. The sensor signals changed significantly from 5 to 35 s, and then the signals achieved a dynamic equilibrium. The various signal values (maximum or minimum), the shifts, the response areas, and so on indicated that the sensors offered unique and abundant characteristics about the headspace gas of the leachate samples. To simplify data processing, sensor signals at 80th second were selected as the input data of the analysis models. To fully understand the sensor signals, the average values of 10 sensors were calculated and shown in Figure 3b. The overall signals (at the 80th second) varied a lot in the first three process periods (LRW, LE, and ICRE), and for ANE, AeroE, and MBRE samples, the signals changed not so remarkably. To further analyze the behaviors of those sensors, a radar fingerprint chart of EN signals is shown in Figure 3c: S2, the most sensitive sensor, showed the biggest variance, and S10 stayed almost still, and S10 stayed almost still. The impacts of leachate headspace gas on the responses of S8, S6, S9, and S5 were to different degrees, and those on S7 and S4 were not so obvious.

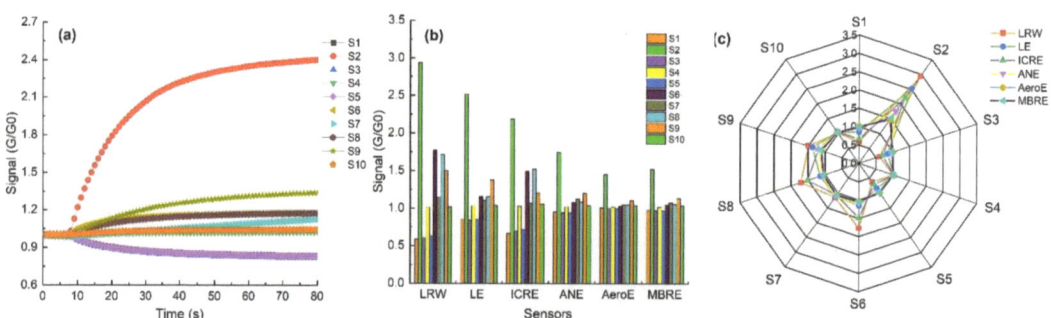

Figure 3. A typical response of EN sensor array during exposure to sample gas: (**a**) a typical response of EN during leachate detection, (**b**) the average values of EN signals at the 80th second, and (**c**) a radar fingerprint chart of EN signals.

3.3. Data Reduction Based on Manifold Learning

Data reduction helps transfer an abundant and disordered original data set into a simplified and ordered form. PCA is a popular technology in dimensionality reduction and is flexible, fast, and easily interpretable. PCA does not perform well when there are nonlinear relationships within the data. For high-dimensional data, it is difficult to affirm whether the EN data is linear or not linear. ISOMAP and UMAP, as two kinds of manifold learning, were applied and compared with PCA.

The best description of differences in the original data can be found by calculating the eigen-decomposition of positive semi-definite matrices and the singular value decomposition of rectangular matrices. The PCs are ordered by ranking according to their contribution (eigenvalue). In Figure 4a, the contributions of PCs are displayed, and the first three PCs have extracted the most information from the original EN data at the 80th second (more than 85%). The sample distributions of 144 samples based on the first three PCs (PC1, PC2, and PC3) are shown in Figure 4b. LPW, LE, and ICRE can be easily classified, but ANE, AeroE, and MBRE are overlapped in a three-dimensional space. The result is similar to those shown in Figure 3b in some ways.

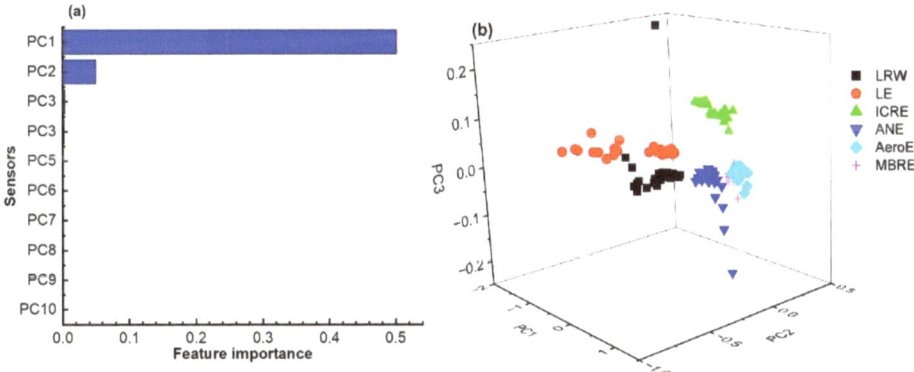

Figure 4. EN data reduction based on PCA: (**a**) PCs ordered according the eigenvalues, (**b**) sample distributions based on the first three PCs.

ISOMAP constructs the geodesic distance graph from the original EN data, and uses eigenvalue decomposition of MDS on the geodesic distance matrix to achieve low-dimensional embeddings. The ICs are ordered and displayed by ranking the contribution in Figure 5a. The first three ICs also extracted the most information of the original EN data at 80th second (more than 85%), which are very similar to that in Figure 4a. Because PCA and ISOMAP used eigen-decomposition and eigenvalue in this study, but there were minuscule differences in the data. The sample distributions based on the first three ICs are shown in Figure 5b. Similar as in Figure 4b, LPW, LE, and ICRE can be easily classified, but ANE, AeroE, MBRE overlapped.

UMAP preserves the local and global data structures and offers short run times based on Riemannian geometry and algebraic topology. Calculating the distance between embedding spaces is an approximate measure to determine how sensitive the canonical embedding space's topology is, which is the feature importance. Figure 6a provides a careful look at the feature importance of 10 UCs, which is quite different from Figures 4a and 5a. The first three UCs obviously did not extract more than 85% of the information from the original EN data, but not meaning that UMAP would do badly for later classification and prediction. In Figure 6b, most of the samples seem to be clustered narrowly. LPW, LE, and ICRE are classified clearly and can be easily distinguished in a three-dimensional space. ANE and AeroE are overlapped, along with two MERE samples. In general, in three-dimensional space, UMAP outperformed the PCA and ISOMAP in Figures 4b and 5b.

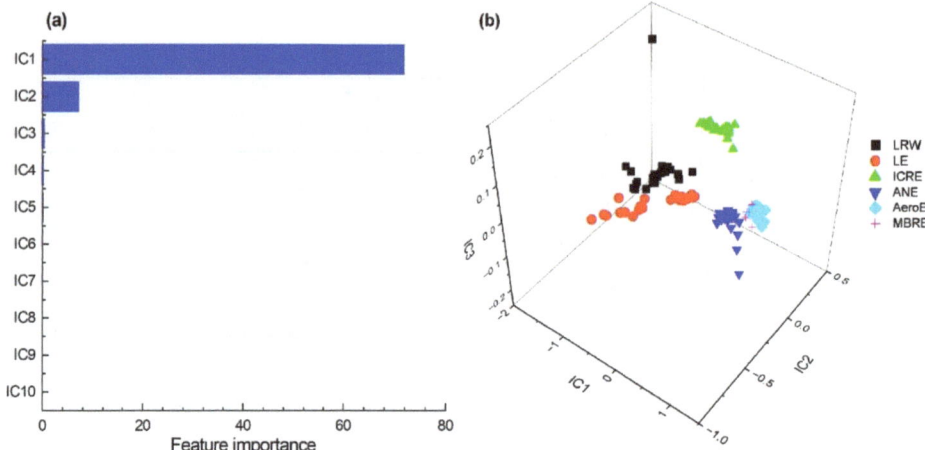

Figure 5. EN data reduction based on ISOMAP: (**a**) ICs ordered according the eigenvalues, (**b**) sample distributions based on the first three ICs.

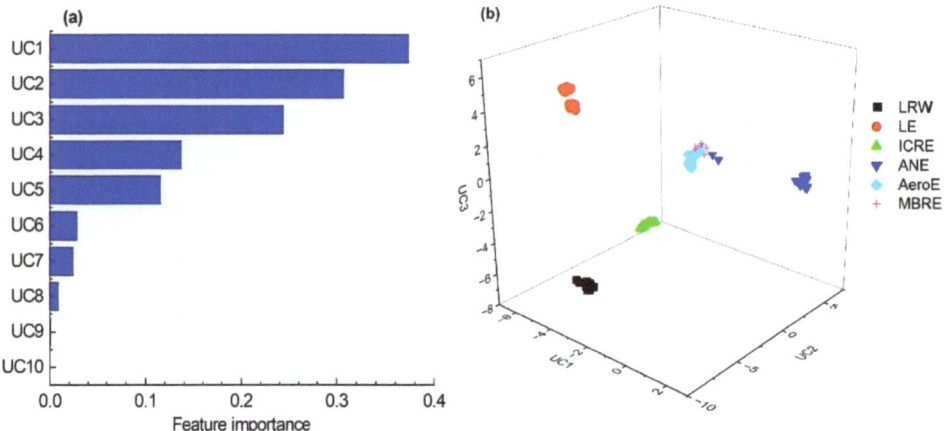

Figure 6. EN data reduction based on UMAP: (**a**) UCs ordered according the eigenvalues, (**b**) sample distributions based on the first three UCs.

3.4. Classification Based on EN Signals

3.4.1. Classification Result Based on LightGBM

ROC graphs are used to organize classifiers and visualize the results. As can be seen from the ROC curve of the lightGBM in Figure 7, the classification accuracy results of the original, PCA, ISOMAPa, and UMAP data are very different in the training set, respectively. In the view of the lightGBM models, the PCA better retains the majority of the information of the original data set according to the ROC curve, and the overall AUC area reaches 100%. From the ROC results, the performance of UMAP was better than that of ISOMAP, and only one category 6 classification showed partial errors. In the ISOMAP-lightGBM model, samples from classes 4 and 6 were misclassified.

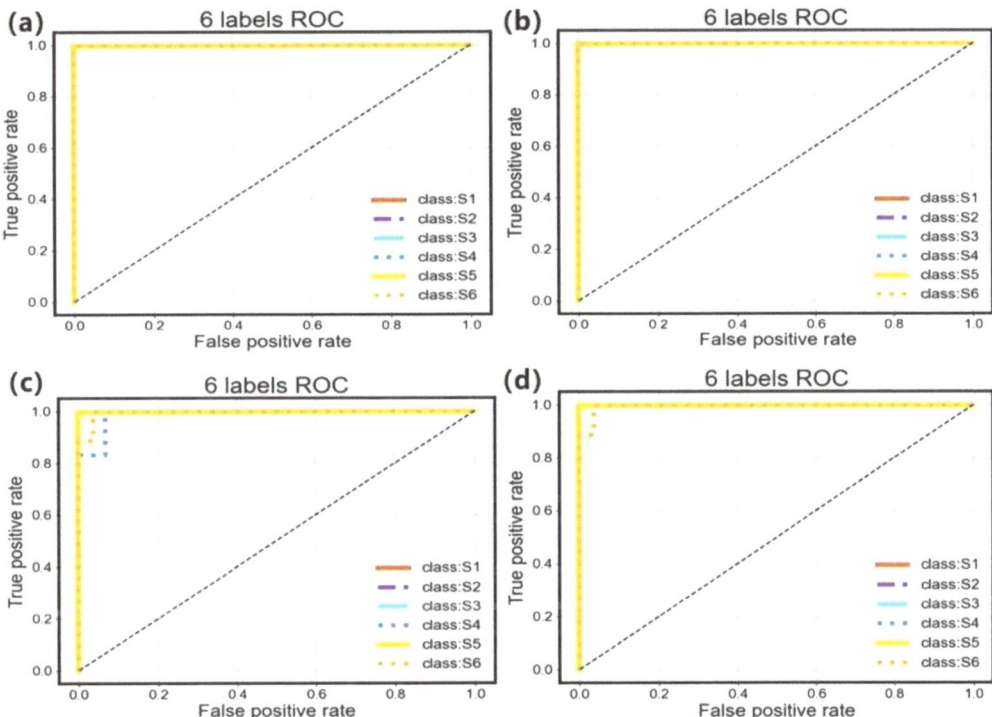

Figure 7. Overall receiver operating characteristics (ROC) curve of lightGBM showing the true positive and false-positive: (**a**) based on original data, (**b**) based on the PCA data, (**c**) based on the ISOMAP data, and (**d**) based on the UMAP data. Class S1 refers to LRW samples, class S2 refers to LE, class S3 refers to ICRE, class S4 refers to AeroE, class S5 refers to ANE, and class S6 refers to MBRE.

To explore further the models based on different data sets, testing data sets were applied to verify the classification result. Moreover, to reduce the volatility attributable, each model was run 20 times, and the average results are displayed in Table 3. The best classification performance is the UMAP-XGBT model, with a 99.95% accuracy rate in the training set and a 97.36% accuracy rate in the testing set, indicating that UMAP-XGBT has the most stable robustness. For PCA-lightGBM, the classification results were worse than those of the original-lightGBM model in the testing set. ISOMAP-lightGBM has a 100% average accuracy rate in the training set and a 96.81 average accuracy rate in the testing set.

Table 3. The classification results in the training set and testing set based on lightGBM.

Model	Accurate Rate in the Training Set (%)	Accurate Rate in the Testing Set (%)
Original-lightGBM	100	96.25
PCA-lightGBM	100	94.44
ISOMAP-lightGBM	100	96.81
UMAP-lightGBM	99.95	97.36

3.4.2. Classification Result Based on XGBT

According to the ROC results in Figure 8, the best performance of classification models is PCA-XGBT, with only two error classification cases. In the UMAP-XGBT model, samples of LE and AeroE were misclassified, but the overall performance was better than that of the original-XGBT. The ISOMAP-XGBT model, with the worst classification performance, has categories LE and AeroE misclassified. According to the training set, ISOMAP-XGBT has the worst performance.

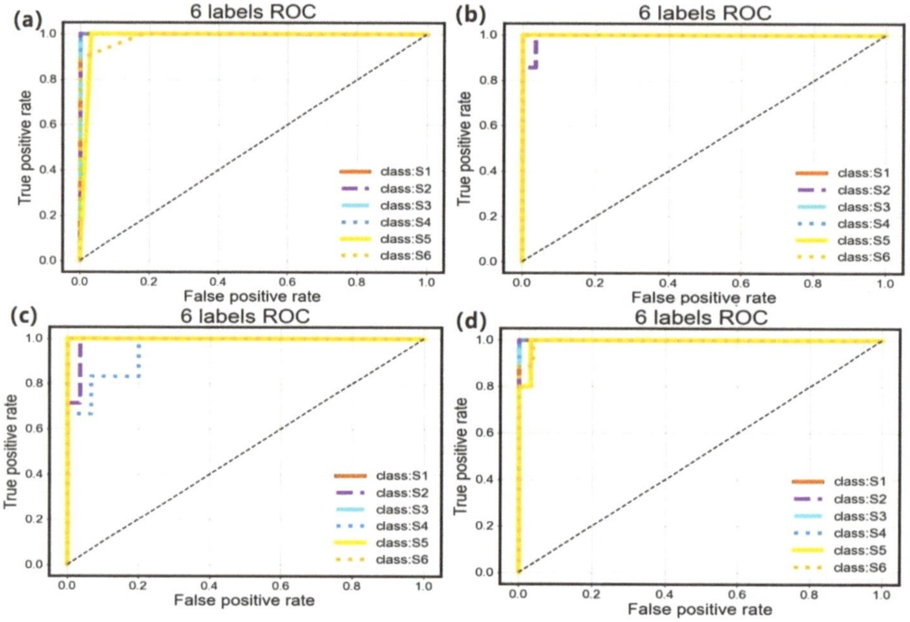

Figure 8. Overall receiver operating characteristics (ROC) curve of XGBT showing the true positive and false-positive: (**a**) based on the original data, (**b**) based on the PCA data, (**c**) based on the ISOMAP data, and (**d**) based on the UMAP data. Class S1 refers to LRW samples, class S2 refers to LE, class S3 refers to ICRE, class S4 refers to AeroE, class S5 refers to ANE, and class S6 refers to MBRE.

For XGBT models, the data training took 20 times to decrease the instability, and the result is shown in Table 4. The accuracy rates of XGBT models were lower than those of the lightGBM-based models. From Table 4, it can be concluded that the UMAP-XGBT model had the best classification performance, with a 99.95% accuracy rate in the training set and a 95.83% accuracy rate in the testing set.

Table 4. The classification results in the training set and testing set based on XGBT.

Model	Accurate Rate in the Training Set (%)	Accurate Rate in the Testing Set (%)
Original-XGBT	100	94.72
PCA-XGBT	100	93.61
ISOMAP-XGBT	100	95.28
UMAP-XGBT	99.95	95.83

From Tables 3 and 4, models of original-XGBT, PCA-XGBT, and ISOMAP-XGBT always had a satisfying performance with a 100% accuracy rate in the training set in the 20 times it was run, while they fell short in the testing set. This might be because the models were overfit in the modeling, so the results in the testing set were not very good.

3.5. Chemical Parameter Prediction Based on EN Signals
3.5.1. Prediction Results Based on LightGBM

As an ensemble learning program, lightGBM aims to build a comprehensive model by parallelizing and serializing weak learners (CART). For lightGBM models, a histogram-based algorithm and tree leaf-wise growth strategy with a maximum depth limit are adopted to increase the training speed. For lightGBM, the max-features was set 4, and the tree leaf-wise was set 3 to simplify the lightGBM model in preliminary work. Then, the number of decision trees was the most important parameter in the later modeling. The number of CARTs was optimized, and the optimization procedure is carried out in Figure 9 (taking COD prediction as an example) according to the R^2s and $EMSE$s in the training set and testing set. The lightGBM was 20 times for each number of CARTs to minimize the contingency. As shown in Figure 9, the results in the training set have stable precision with the increasing decision tree numbers. While the result precision in the testing set was not stable regardless of which data set was applied. All in all, the lightGBM model with the UMAP data set had a slight advantage (not so obvious) when compared to the other three data sets. When the number of CARTs was 25, the lightGBM model showed a satisfactory result. The best number of CARTs for the lightGBM model was decided.

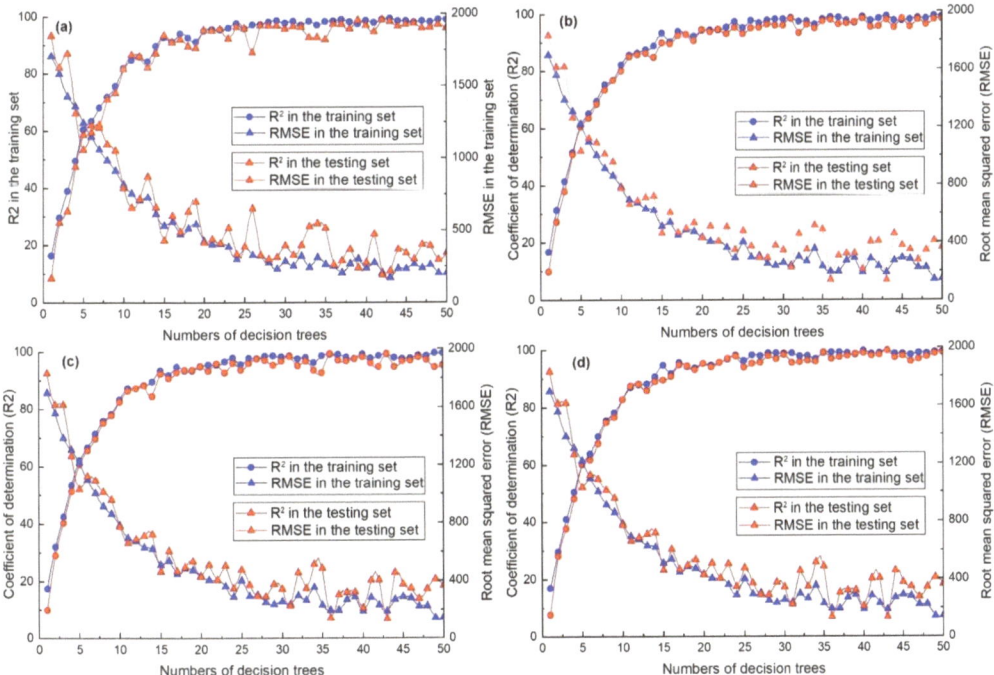

Figure 9. The performance of lightGBM according to the numbers of decision trees: (**a**) based on the original data set, (**b**) based on the PCA data set, (**c**) based on the ISOMAP data set, and (**d**) based on the UMAP data set.

EN signals offered the entirety of the information on leachate headspace gas, which mainly consisted of volatile organic compounds (VOCs), including hydrogen sulphide, methyl mercaptan, acetylene, and so on. The materials in the headspace gas were most closely correlated to the value of COD. The EN combined with the data mining method could predict the contents of COD in leachate samples. Table 5 summarizes the prediction results based on different data reductions for five chemical parameters (pH, COD, BOD_5, AN, TN, and TP). According to the R^2s and $EMSE$s in the training set and testing set, the PCA process had no effect on the lightGBM models compared to models based on original data. LightGBM models based on ISOMAP and UMAP showed satisfactory outcomes. The prediction of five chemical parameters based on the UMAP-lightGBM model showed the best performance with an R^2 higher than 0.98 in the training set and testing set ($RMSE$s are not comparable when it comes to different parameters and units).

Table 5. Comparison of the LightGBM prediction models based on different manifold learning methods.

Data Set		R² (Training)	RMSE (Training)	R² (Testing)	RMSE (Testing)	Data Set		R² (Training)	RMSE (Training)	R² (Testing)	RMSE (Testing)
Original data	pH	0.9721	0.2278	0.7217	0.6870	PCA	pH	0.9258	0.3716	0.8690	0.4521
	COD	0.9987	120.72	0.9779	492.01		COD	0.9968	189.74	0.9916	302.91
	BOD	0.9991	27.82	0.9843	110.01		BOD	0.9968	50.89	0.9893	94.20
	AN	0.9991	40.66	0.9753	208.42		AN	0.9974	68.20	0.9957	87.06
	TN	0.9953	52.35	0.9785	110.42		TN	0.9857	90.85	0.9694	132.52
	TP	0.9978	0.5781	0.9347	3.11		TP	0.9952	0.8652	0.9739	2.02
ISOMAP	pH	0.9211	0.3834	0.8793	0.4286	UMAP	pH	0.9806	0.1339	0.9803	0.1721
	COD	0.9968	189.29	0.9963	202.01		COD	0.9989	113.43	0.9881	356.88
	BOD	0.9967	51.98	0.9921	80.49		BOD	0.9991	27.82	0.9947	66.30
	AN	0.9973	68.55	0.9959	85.75		AN	0.9991	40.52	0.9938	105.89
	TN	0.9837	96.96	0.9701	130.60		TN	0.9952	52.45	0.9933	62.07
	TP	0.9953	0.8512	0.9844	1.51		TP	0.9982	0.5229	0.9895	0.5742

3.5.2. Prediction Results Based on XGBT

As with lightGBM, the parameters of the XGBT models, including the number of trees, maximal depth, and minimum rows, were optimized. Finally, the max-features was set 4, max-depth was set 2, and min-split was set 2 to simplify the XGBT model. The number of decision trees (CART) was the most important parameter in the later modeling. The optimization procedure is carried out in Figure 10 (taking COD prediction as an example) according to the R^2s and EMSEs in the training set and testing set, 20 times for each modeling step to minimize the contingency. As shown in Figure 10, the overall prediction performance of XGBT models was much better than lightGBM in Figure 10. XGBT models have very strong robustness and stability, particularly for the ISOMAP-XGBT models in Figure 10c. When the number of decision trees was 25, the XGBT models had a relatively satisfactory result, meanwhile the model was not so big, the same as the lightGBM models

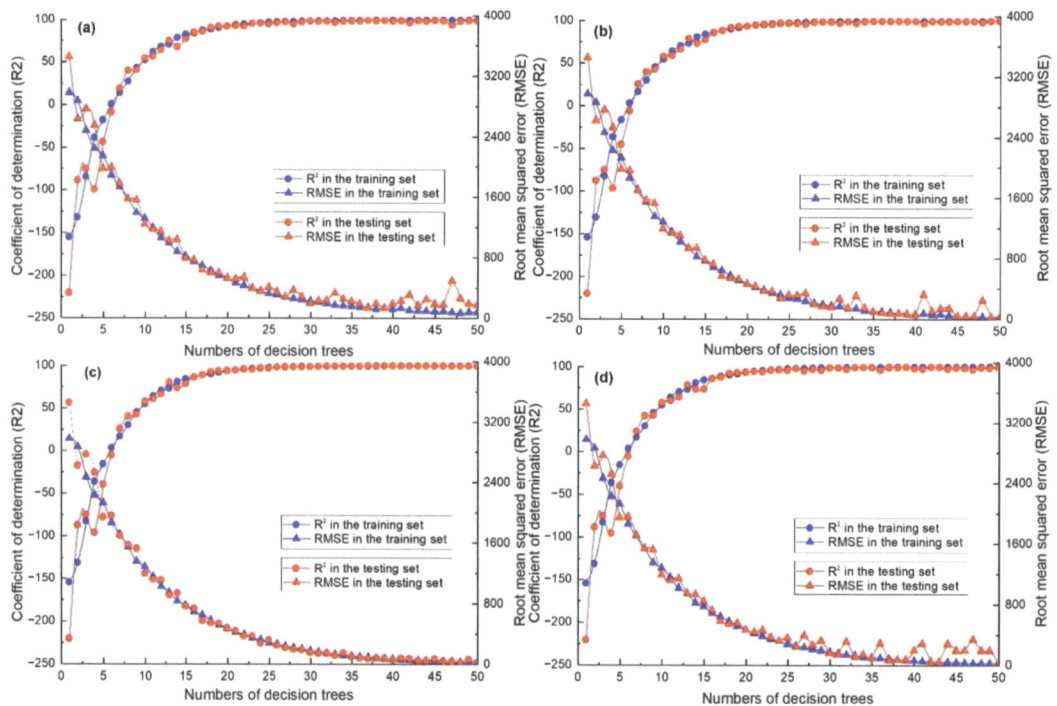

Figure 10. The performance of XGBT according to the numbers of decision trees: (**a**) based on the original data set, (**b**) based on the PCA data set, (**c**) based on the ISOMAP data set, (**d**) based on the UMAP data set.

Table 6 summarizes the prediction results based on different data reductions for five chemical parameters. All in all, prediction results based on XGBT have achieved better results compared to lightGBM models. Different from PCA-lightGBM models, PCA-XGBT models have a slight edge over the (original data)-XGBT models in the testing set. The prediction of five chemical parameters based on the UMAP-XGBT model showed the best performance with an R^2 higher than 0.99 in the training set and testing set.

Table 6. Comparison of the XGBT prediction models based on different manifold learning methods.

Data Set		R^2 (Training)	RMSE (Training)	R^2 (Testing)	RMSE (Testing)	Data Set		R^2 (Training)	RMSE (Training)	R^2 (Testing)	RMSE (Testing)
Original data	pH	0.9721	0.2278	0.7217	0.6870	PCA	pH	0.9258	0.3716	0.8690	0.4521
	COD	0.9987	120.72	0.9779	492.01		COD	0.9968	189.74	0.9916	302.91
	BOD	0.9991	27.82	0.9843	110.01		BOD	0.9968	50.89	0.9893	94.20
	AN	0.9991	40.66	0.9753	208.42		AN	0.9974	68.20	0.9957	87.06
	TN	0.9953	52.35	0.9785	110.42		TN	0.9857	90.85	0.9694	132.52
	TP	0.9978	0.5781	0.9347	3.11		TP	0.9952	0.8652	0.9739	2.02
ISOMAP	pH	0.9861	0.1834	0.9803	0.1986	UMAP	pH	0.9806	0.1939	0.9833	0.1721
	COD	0.9968	189.29	0.9963	202.01		COD	0.9989	113.43	0.9901	156.88
	BOD	0.9967	51.98	0.9921	80.49		BOD	0.9991	27.82	0.9947	66.30
	AN	0.9973	68.55	0.9959	85.75		AN	0.9991	40.52	0.9938	105.89
	TN	0.9837	96.96	0.9701	130.60		TN	0.9952	52.45	0.9933	62.07
	TP	0.9953	0.8512	0.9844	1.51		TP	0.9982	0.5229	0.9975	0.5574

For prediction models, the overall performances of XGBT models are better than those of lightGBM, and the data set based on UMAP reduction has a slight advantage both in the training set and the testing set when compared to other three data sets.

4. Conclusions

MSW incineration is regarded as an ideal method for MSW disposal, with many advantages. This study applied an EN detection method for monitoring MSW incinerator leachate combined with manifold learning and ensemble methods. Some conclusions can be drawn:

(1) COD, BOD5, ammonia, TN, and TP of leachate were significantly changed during the processing procedure, especially for COD;
(2) EN sensors offered unique and abundant characteristics of leachate samples in the headspace gas. The signals at the 80th second varied a lot in the first three process periods (LRW, LE, and ICRE), for ANE, AeroE, and MBRE samples, the signals changed not so remarkably;
(3) Manifold learnings (PCA, ISOMAP, and UMAP) were applied to extract the information hidden in the headspace gas of leachate detected by EN. The first three PCs and ICs have extracted the most information from the original data (>85%), and samples of LPW, LE, and ICRE could be easily classified according to the three-dimensional space, while others were not so satisfied. UMAP outperformed the performance of PCA and ISOMAP;
(4) Ensemble methods (LightGBM and XGBT) were applied to mine the relationship between EN signals of leachate headspace gas and chemical parameter changes combined with PCA, ISOMAP, and UMAP. The UMAP-XGBT model had the best classification performance, with a 99.95% accuracy rate in the training set, and a 95.83% accuracy rate in the testing set. The UMAP-XGBT model showed the best prediction ability for the leachate chemical parameters R^2 higher than 0.99 in the training and testing sets.

Up until now, there have been few in-depth studies that have been conducted to fetch information from the headspace gas of leachate samples. This is the first study with an EN application for leachate monitoring based on manifold learning and ensemble methods, offering an easier and quicker monitoring method than traditional instrumental measurements. Future work will focus on the potential relationship between microorganisms and headspace gas in the leachate based on EN technology to fully understand the MSW incineration leachate chemical parameter changes, which is quite important for leachate disposal.

Author Contributions: Conceptualization, S.Q. and J.Z.; methodology, Z.Z.; software, Z.Z.; validation, J.H.; data curation, Z.Z.; writing—original draft preparation, S.Q.; writing—review and editing, S.Q. and Z.Z.; funding acquisition, S.Q. All authors have read and agreed to the published version of the manuscript.

Funding: The research was funded by the National Key R&D Program of China [2019YFE0124600].

Data Availability Statement: Not Applicable.

Acknowledgments: The authors acknowledge the financial support of the National Key R&D Program of China (2019YFE0124600).

Conflicts of Interest: The authors declare no conflict of interest.

References

1. Kaza:, S.; Yao, L.C.; Bhada-Tata, P.; Van Woerden, F. *What a Waste 2.0: A Global Snapshot of Solid Waste Management to 2050*. Urban Development; World Bank: Washington, DC, USA, 2018. Available online: https://openknowledge.worldbank.org/handle/10986/30317 (accessed on 1 November 2022).
2. Lippi, M.; Ley, M.B.R.G.; Mendez, G.P.; Cardoso Junior, R.A.F. State of Art of Landfill Leachate Treatment: Literature Review and Critical Evaluation. *Ciência Nat.* **2018**, *40*, e78. [CrossRef]
3. Cudjoe, D.; Han, M.S. Economic feasibility and environmental impact analysis of landfill gas to energy technology in African urban areas. *J. Clean. Prod.* **2021**, *284*, 125437. [CrossRef]
4. Shah, A.V.; Srivastava, V.K.; Mohanty, S.S.; Varjani, S. Municipal solid waste as a sustainable resource for energy production: State-of-the-art review. *J. Environ. Chem. Eng.* **2021**, *9*, 105717. [CrossRef]
5. Ren, X.; Liu, D.; Chen, W.; Jiang, G.; Wu, Z.; Song, K. Investigation of the characteristics of concentrated leachate from six municipal solid waste incineration power plants in China. *RSC Adv.* **2018**, *8*, 13159–13166. [CrossRef]
6. Chen, W.; He, C.; Zhuo, X.; Wang, F.; Li, Q. Comprehensive evaluation of dissolved organic matter molecular transformation in municipal solid waste incineration leachate. *Chem. Eng. J.* **2020**, *400*, 126003. [CrossRef]
7. Jiang, F.; Qiu, B.; Sun, D. Degradation of refractory organics from biologically treated incineration leachate by VUV/O3. *Chem. Eng. J.* **2019**, *370*, 346–353. [CrossRef]
8. Hu, W.; Wan, L.; Jian, Y.; Ren, C.; Jin, K.; Su, X.; Bai, X.; Haick, H.; Yao, M.; Wu, W. Electronic Noses: From Advanced Materials to Sensors Aided with Data Processing. *Adv. Mater. Technol.* **2019**, *4*, 1800488. [CrossRef]
9. Eusebio, L.; Derudi, M.; Capelli, L.; Nano, G.; Sironi, S. Assessment of the Indoor Odour Impact in a Naturally Ventilated Room. *Sensors* **2017**, *17*, 778. [CrossRef]
10. Bieganowski, A.; Józefaciuk, G.; Bandura, L.; Guz, Ł.; Łagód, G.; Franus, W. Evaluation of Hydrocarbon Soil Pollution Using E-Nose. *Sensors* **2018**, *18*, 2463. [CrossRef]
11. Tonacci, A.; Sansone, F.; Conte, R.; Domenici, C. Use of Electronic Noses in Seawater Quality Monitoring: A Systematic Review. *Biosensors* **2018**, *8*, 115. [CrossRef]
12. Jońca, J.; Pawnuk, M.; Arsen, A.; Sówka, I. Electronic Noses and Their Applications for Sensory and Analytical Measurements in the Waste Management Plants—A Review. *Sensors* **2022**, *22*, 1510. [CrossRef] [PubMed]
13. Tasaki, H.; Lenz, R.; Chao, J. Dimension Estimation and Topological Manifold Learning. In Proceedings of the 2019 International Joint Conference on Neural Networks (IJCNN), Budapest, Hungary, 14–19 July 2019; pp. 1–7.
14. Zounemat-Kermani, M.; Stephan, D.; Barjenbruch, M.; Hinkelmann, R. Ensemble data mining modeling in corrosion of concrete sewer: A comparative study of network-based (MLPNN & RBFNN) and tree-based (RF, CHAID, & CART) models. *Adv. Eng. Inform.* **2020**, *43*, 101030.
15. HJ 1147-2020; Ministry of Ecology and Environment of the People's Republic of China. Water Qulity—Determination of pH—Electrode Method. Ministry of Ecology and Environment of the People's Republic of China. Available online: https://max.book118.com/html/2020/1129/8117023002003022.shtm (accessed on 1 November 2022).
16. HJ/T 70-2001; High-Chlorine Wastewater—Determination of Chemical Oxygen Demand—Chlorine Emendation Method. Ministry of Ecology and Environment of the People's Republic of China. Available online: https://www.doc88.com/p-9982565679330.html?r=1 (accessed on 1 November 2022).
17. HJ 535-2009; Water Quality—Determination of Ammonia Nitrogen—Nessler's Reagent Spectrophotometry. Ministry of Ecology and Environment of the People's Republic of China. Available online: http://www.doc88.com/p-6836770291709.html (accessed on 1 November 2022).
18. HJ 636-2012; Water Quality—Determination of Total Nitrogen—Alkaline Potassium Persulfate Digestion UV Spectrophotometric Method. Ministry of Ecology and Environment of the People's Republic of China. Available online: http://www.doc88.com/p-7187319550717.html (accessed on 1 November 2022).

19. GB/T 11893-1989; Water Quality—Determination of Total Phosphorus—Ammonium Molybdate Spectrophotometric Method. Ministry of Ecology and Environment of the People's Republic of China. Available online: https://www.doc88.com/p-6764771874050.html?r=1 (accessed on 1 November 2022).
20. Wilson, A.D. Review of Electronic-nose Technologies and Algorithms to Detect Hazardous Chemicals in the Environment. *Procedia Technol.* **2012**, *1*, 453–463. [CrossRef]
21. Dey, A. Semiconductor metal oxide gas sensors: A review. *Mater. Sci. Eng. B* **2018**, *229*, 206–217. [CrossRef]
22. Nair, A.T.; Senthilnathan, J.; ShivaNagendra, S.M. Emerging perspectives on VOC emissions from landfill sites: Impact on tropospheric chemistry and local air quality. *Process Saf. Environ. Prot.* **2019**, *121*, 143–154. [CrossRef]
23. Abdi, H.; Williams, L.J. Principal component analysis. *Wiley Interdiscip. Rev. Comput. Stat.* **2010**, *2*, 433–459. [CrossRef]
24. Gao, S.; Zhang, S.; Zhang, Y.; Gao, Y. Operational reliability evaluation and prediction of rolling bearing based on isometric mapping and NoCuSa-LSSVM. *Reliab. Eng. Syst. Saf.* **2020**, *201*, 106968. [CrossRef]
25. Becht, E.; McInnes, L.; Healy, J.; Dutertre, C.-A.; Kwok, I.W.H.; Ng, L.G.; Ginhoux, F.; Newell, E.W. Dimensionality reduction for visualizing single-cell data using UMAP. *Nat. Biotechnol.* **2019**, *37*, 38–44. [CrossRef]
26. Kumari, P.; Toshniwal, D. Extreme gradient boosting and deep neural network based ensemble learning approach to forecast hourly solar irradiance. *J. Clean. Prod.* **2021**, *279*, 123285. [CrossRef]
27. Taha, A.A.; Malebary, S.J. An Intelligent Approach to Credit Card Fraud Detection Using an Optimized Light Gradient Boosting Machine. *IEEE Access* **2020**, *8*, 25579–25587. [CrossRef]
28. Chang, Y.-C.; Chang, K.-H.; Wu, G.-J. Application of eXtreme gradient boosting trees in the construction of credit risk assessment models for financial institutions. *Appl. Soft Comput.* **2018**, *73*, 914–920. [CrossRef]

Article

Detection of Low-Level Adulteration of Hungarian Honey Using near Infrared Spectroscopy

Zsanett Bodor [1], Mariem Majadi [2], Csilla Benedek [1,*], John-Lewis Zinia Zaukuu [3], Márta Veresné Bálint [1], Éva Csajbókné Csobod [1] and Zoltan Kovacs [2]

[1] Department of Dietetics and Nutrition, Faculty of Health Sciences, Semmelweis University, 17 Vas Street, H-1088 Budapest, Hungary
[2] Department of Measurements and Process Control, Institute of Food Science and Technology, Hungarian University of Agriculture and Life Sciences, 14–16 Somlói Street, H-1118 Budapest, Hungary
[3] Department of Food Science and Technology, Kwame Nkrumah University of Science and Technology (KNUST), Kumasi AK-039-5028, Ghana
* Correspondence: benedek.csilla@se-etk.hu

Abstract: Honey adulteration is a worldwide problem; however, its detection is a challenge for researchers and authorities. There are numerous ways of honey counterfeiting; amongst them, direct adulteration is one of the most common methods. Correlative techniques, such as near-infrared spectroscopy (NIRS), are useful tools in the detection of honey adulteration; however, this method has not been applied to Hungarian honeys. The aim of this research was to investigate the performance of NIRS for the detection of sugar syrup addition to Hungarian honeys at lower concentration levels (<10% w/w). Acacia, rape, forest, sunflower, and linden honeys were mixed with high-fructose-content sugar syrup, rice syrup, or self-made glucose fructose syrup in 3%, 5%, and 10% w/w. NIRS analysis was performed in the spectral range of 950–1650 nm. Principal component analysis was coupled with linear discriminant analysis and partial least square regression models were built for the classification and prediction of adulteration levels, respectively. Our results showed that the performance of NIRS highly depends on both type of syrup and honey. PCA-LDA models provided the 100% correct classification of control in the case of all the models, while PLSR results could predict the added sugar syrup content in the case of rice and F40 syrup models, obtaining >2.2 RPDCV value.

Keywords: honey; fraud; sugar syrup; chemometrics; NIRS

1. Introduction

Honey is a well-known nutritious food, mostly used as a sweetener and nutraceutical product. It is a supersaturated solution of sugars in water matrix. Fructose, glucose, monosaccharides, disaccharides, and other sugars make up more than 80% of the total weight of honey [1–3]. In addition to sugars, water is the other principal compound; according to legislation, its amount should generally not exceed 20% (with some exceptions) [4–6]. Honey has a high value for consumers and on the market, owing to its high nutritional value and health benefits [7].

Recently, the increasing market value and price of some foods have accelerated the problem of food adulteration. Amongst these foods, honey is considered to be one of the most frequently counterfeited foods [8]. Honey fraud has many aspects, such as mislabeling of botanical or geographical origin, direct adulteration with sugar syrups, indirect adulteration by feeding bees with sugar in the collection period, and resin filtration [9,10]. During direct adulteration, different types of sugar syrups are used, such as corn, rice, invert sugar, glucose, beet, and other syrups. The exact methods and protocols used for the detection of this honey fraud are challenging; meanwhile, accuracy highly depends on the type of syrups and method. Lately, different analytical techniques have been tested and found to be useful in honey authentication; however, there is still demand for a reliable and accurate method [9,10]. Recently, the use of correlative techniques, such as near-infrared spectroscopy (NIRS) has increased in the field of food products, including honey adulteration detection [11,12]. NIRS coupled with chemometric analysis methods, such as principal component analysis (PCA), PCA-coupled linear discriminant analysis (PCA-LDA), artificial neural network (ANN) or partial least square regression (PLSR) have been used in studies outside of Hungary for the detection and quantification of honey adulteration. Chinese researchers analyzed mixtures of pure acacia, vitex, and jujube honey with rice and corn syrup at 5%, 10%, 20%, and 40% w/w. In this study the NIRS provided 100% correct classification of all the samples and levels [13]. Bázár et al. [14], studied the effect of high-fructose-content corn syrup addition, and found that the structure of water in honey sample changes as a result of adulteration. In another Chinese study, the NIRS was found to be effective in the detection of adulterants in honey; the average correct classification was 85.71% [15]. Spanish researchers could also predict the level of syrup addition in honey with high accuracy using PLSR ($R^2 > 0.9$) using NIRS in the range of 5–40% w/w adulteration [16]. These results show that the NIRS is a promising technology in the detection of honey adulteration. Moreover, it is non-destructive and does not require any reagents; therefore, it can be considered a sustainable green technology [13,17–21]. In Hungary, the application of NIRS for the detection and quantification of honey adulteration has not been studied deeply. On the other hand, studies around the world focusing on the detection and quantification of honey adulteration have generally used higher syrup concentration levels ($\geq$5% w/w). Based on these findings, it can be concluded that there is a lack of research in adulteration detection of specific Hungarian honey types, especially below the 5% w/w adulteration level.

Therefore, in this study, our aim was to analyze the applicability of NIRS for the detection (PCA-LDA) and quantification (PLSR) of glucose–fructose, high-fructose-content corn syrup, and rice syrup addition to acacia, forest, sunflower, rape, and linden honeys from Hungary. A further aim was to apply concentration levels of 3%, 5%, and 10% that had not been studied in depth in Hungary prior to this research.

2. Materials and Methods

The whole experimental setup was performed in two sets, including a preliminary set (PS), and an extended set (ES).

2.1. Samples and Sample Preparation

In the preliminary set acacia (RP—*Robinia pseudoacacia*) and linden (TI—*Tilia* spp.) honey were mixed with F40 high-fructose-content corn syrup, (HFCS—Kall Ingredients, Tiszapüspöki, Hungary) and rice syrup (RI—dm; drogerie markt GmbH & Co. KG, Karlsruhe, Germany) in 3%, 5%, and 10%. Prior to the current study, the concentration range of 0% ,5%, 50% was studied and—based on those results—the levels below 10% were chosen [22]. The concentrations were chosen based on the results of Bodor et al. (2019), where the levels below 10% were most challenging ones to detect. Moreover, based on the available studies, adulterations below 5% w/w were not typically applied [13–16], as this was mentioned in the introduction.

The extended study set was completed with rape (BN—*Brassica napus*), sunflower (HA—*Helianthus annuus*), and forest (HD) honeys. The authentic samples were mixed with the same rice and HFCS syrups and, additionally, with a self-made glucose–fructose syrup (GS). The GS syrup was prepared as follows: using an analytical balance, 240 g of fructose and 160 g of glucose were weighed and transferred into a volumetric flask of 500 mL. The flask was filled up to volume with distilled water, before putting it in a water bath at 60 °C to ensure and force the solvation of the sugars. The solution was then cooled down to room temperature prior to further steps.

In the case of both sets, a total of 20 g of the sample were prepared to provide three replicates (R1, R2, R3), resulting in a total of 141 samples, including three replicates of the controls and the sugar syrups. The coding of the samples was generated with the concatenation of the type of honey (RP, TI, BN, HA, and HD), the syrup added (RI, F40, GF), and the level of syrup (3%, 5%, and 10%).

2.2. Determination of the Main Physical Properties

The determination of the main physical parameters, such as moisture, pH, and electrical conductivity was performed according to the guidebook of the International Honey Commission [23]. Moisture content was determined using Abbé-type refractometer, pH and electrical conductivity were measured with a Mettler Toledo SevenMulti analyzer with the respective probes (Mettler Toledo, Columbus, OH, USA). Moisture content readings were performed in two replicates, the pH and electrical conductivity in three–three replicates per sample.

2.3. Near-Infrared Spectroscopy Measurements

Near-infrared spectra (NIRS) were recorded using the benchtop MetriNIR spectrophotometer (MetriNIR Research, Development and Service Co., Budapest, Hungary) in the spectral range of 740–1700 with 2 nm spectral resolution. The instrument is equipped with an InGaAs two-beam detector. The spectral recording was performed using a transflectance setup with a temperature-controlled cuvette at 25 °C. The layer thickness of the sample was 0.5 mm in the cuvette. All the samples were measured in randomized order. In the preliminary set five and in the extended set three consecutive scans were recorded per sample. This resulted in 15 and 9 spectra for the preliminary and the extended sample set, respectively, in the case of all adulteration levels.

2.4. Statistical Evaluation

2.4.1. Evaluation of the Physical Parameters

Statistical evaluation of the quality indicators was performed with descriptive statistics: mean and standard deviation of the different adulteration levels were calculated separately for the different honey types. One-way analysis of variance (ANOVA) was performed to check the significant differences among the different adulteration levels and control. The evaluation was performed separately for the different honey types. Prior to the analysis, the assumptions of ANOVA were tested such as the normality (Shapiro–Wilk test) and the homogeneity of the variances (Levene's test). As in most of the cases the homogeneity of the variances did not assume, the pairwise comparison was tested using the Games–Howell test, which is not sensitive to inhomogeneity [24].

2.4.2. Spectral Preprocessing

Evaluation of the NIR spectra was performed in the spectral range of 950–1650 nm. The pretreatment of the spectra was optimized using different pretreatments and their combinations. Single, double, or triple combination of

- Savitzky–Golay smoothing (different window sizes 13, 17, 21, and/or derivation levels—0, 1st, 2nd);
- Multiplicative scatter correction (MSC);
- Standard normal variate (SNV) or;
- Detrending (detr).

The list and abbreviation of the pretreatments can be found in Supplementary Table S1. In total 41 pretreatment combinations were tested during the model optimization of the following methods.

For the spectral presentation, the averaged spectra of the different sample groups were chosen, where the spectra of one sample was averaged (coming from the 9 or 15 consecutives). In the case of the preprocessed spectra, the averaging was performed after the pretreatment.

2.4.3. Chemometric Analysis of the near-Infrared Spectra

During the chemometric analysis—in the case of all the models—consecutive spectra of different samples were used (without averaging).

As a first step, principal component analysis (PCA) was used for pattern recognition and extreme outlier detection of the spectral data. The flow of the chemometric analysis can be found in Figure 1.

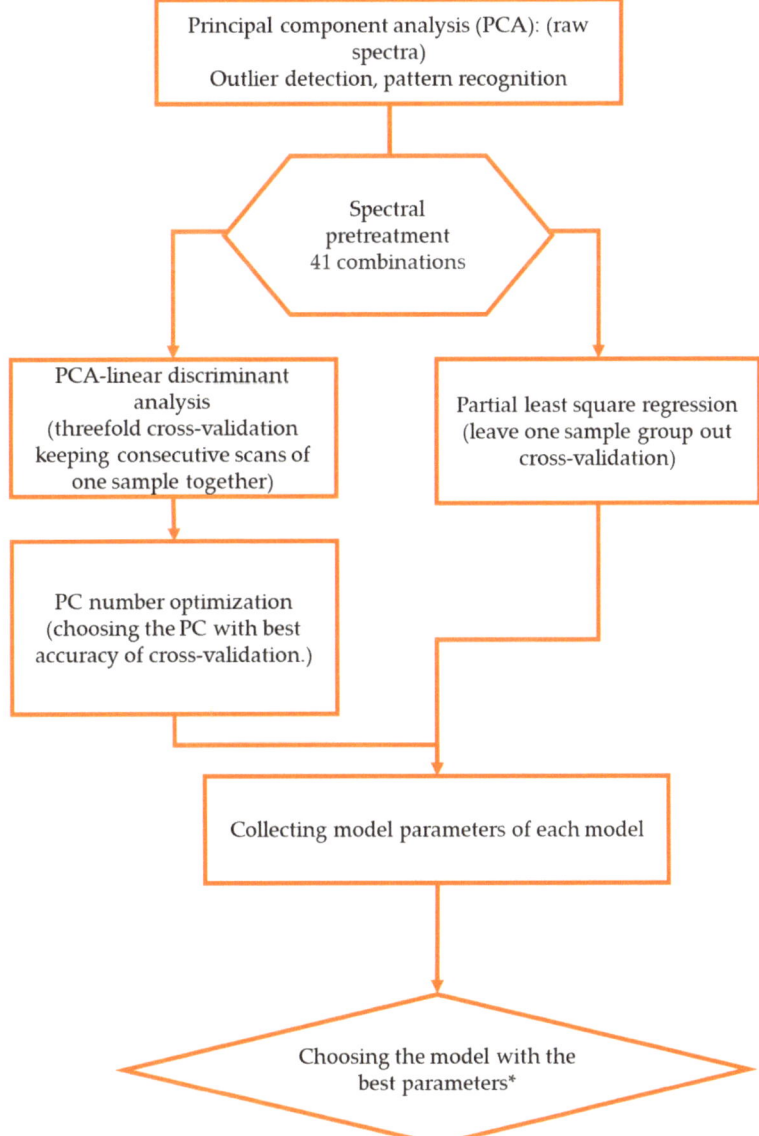

*PCA-LDA: highest accuary of validation
*PLRS: highest R^2CV & RPDCV and at the same time lowest RMSECV

Figure 1. Flowchart of the chemometric analysis of NIRS data.

Classification models were built using PCA-coupled linear discriminant analysis (PCA-LDA). The models were built for the different honey types separately. In the case of each honey type, models were constructed for the classification of different adulteration levels. Two types of models were built:

(1) including the results of all the syrups (all syrup models, one model per honey type)
(2) and separately, including only data of one syrup mixture set and control (resulting in two models in the case of linden and acacia, and three models in the case of rape, sunflower, and forest honeys).

During the analysis, threefold cross-validation was applied. Each model was built using all the pretreatments. After the pretreatment, the PCA-LDA model was built using principal component (PC) number optimization. The model with the highest cross-validation accuracy after choosing the optimal pretreatment was chosen for data interpretation. During the analysis, the correct classification of control after training and validation, and the average correct classification of all the groups for training and validation were computed.

Partial least square regression models (PLRS) were built for the prediction of the added syrup concentration. Models were also computed using the data of different honey types separately. Predictions were performed on the total data set of one honey type (all syrup models, one model per honey type) and separately for the different sugar syrups (resulting in two models in the case of linden and acacia, and three models in the case of rape, sunflower, and forest honeys). During the model construction leave-one-group-out validation was used where one replicate of one sample was left out for validation in each iteration. During the evaluation, the following model parameters were calculated:

- Determination coefficient of training (R^2C) and validation (R^2CV)—the higher the value, the better the model;
- Root mean square error of training and RMSEC validation (RMSECV)—the lower the value, the better the model;
- Residual prediction deviation training (RPDC) and validation (RPDCV).

The benefit of RPD is that it considers the variation of the dataset at the same time with the prediction error. Therefore, RPD provides more detailed information about the reliability, robustness, and validity of the model [25]. The higher the RPD value, the better the model. Based on the work of Muncan et al. [26,27], we considered PLSR satisfactory if the RPD values were above 1.5 [28].

Statistical evaluation was performed using Microsoft Office Excel and R-project software with the package Aquap2 designed for NIRS analysis [29].

3. Results and Discussion

3.1. Results of the Physical Analysis

The moisture contents of the three syrups were the following, in decreasing order: F40 syrup: 21.88 ± 0.44%, GF syrup: 18.4 ± 0.54%, and RI syrup: 17.53 ± 0.66%. In the case of all the honey types, an increasing tendency in moisture content was observed with the increase in the amount of the added syrup. The exception was the forest honey mixed with the RI syrup where a slight, but not significant decrease was observed compared to the control (17.30%) Table 1. However, it can be seen that at these levels the moisture content did not increase above 20%, which is the legislation limit [4,5]. Moreover, the increase of moisture content of acacia, forest, and linden honeys was not significant compared to the authentic honeys. In the case of rape honey, only the honey mixed with the F40 syrup in 10% showed a significant increase, while in the case of sunflower honey, almost all the mixed honeys had significantly higher moisture contents.

The pH of sugar syrups was similar to the pH of honey samples in general. The F40 syrup had a pH of 4.02 ± 0.12 and the RI syrup had pH of 4.48 ± 0.37. Higher pH was observed in the case of the GF syrup (5.33 ± 0.19). Despite this, the pH of the different honey samples did not increase significantly in most of the cases (Table 1).

Table 1. Results of the physical properties of different honey types.

Honey	Sample	Moisture %	pH	Electrical conductivity µS/cm
Acacia	RP Control	16.60 ± 0.14 a	3.59 ± 0.01 a	141.40 ± 0.03 a
	RP F40 3%	17.15 ± 0.35 a	3.59 ± 0.02 a	**137.81 ± 0.40 b**
	RP F40 5%	16.95 ± 0.07 a	3.61 ± 0.01 a	136.12 ± 1.65 abcd
	RP F40 10%	17.40 ± 0.00 a	3.64 ± 0.04 a	**132.11 ± 1.66 bcd**
	RP RI 3%	17.00 ± 0.14 a	3.60 ± 0.02 a	**137.69 ± 0.27 bc**
	RP RI 5%	17.05 ± 0.07 a	3.60 ± 0.02 a	135.77 ± 0.52 c
	RP RI 10%	17.10 ± 0.00 a	3.60 ± 0.02 a	**130.12 ± 0.44 d**
Linden	TI Control	16.60 ± 0.14 a	4.34 ± 0.02 a	627.44 ± 3.50 a
	TI F40 3%	17.15 ± 0.35 a	4.32 ± 0.01 a	**607.44 ± 1.90 b**
	TI F40 5%	16.95 ± 0.07 a	4.32 ± 0.01 a	**593.89 ± 6.26 bcd**
	TI F40 10%	17.40 ± 0.00 a	4.32 ± 0.01 a	568.00 ± 1.00 c
	TI RI 3%	17.00 ± 0.14 a	4.33 ± 0.01 a	607.22 ± 1.50 b
	TI RI 5%	17.05 ± 0.07 a	4.31 ± 0.01 a	597.78 ± 0.96 d
	TI RI 10%	17.10 ± 0.00 a	4.34 ± 0.01 a	570.33 ± 3.61 c
Rape	BN Control	17.83 ± 0.08 a	3.56 ± 0.01 a	180.98 ± 1.16 a
	BN F40 3%	17.83 ± 0.05 a	3.56 ± 0.01 a	**175.46 ± 0.47 bc**
	BN F40 5%	17.97 ± 0.15 ab	3.57 ± 0.01 a	172.72 ± 0.74 d
	BN F40 10%	**18.18 ± 0.04 b**	3.57 ± 0.02 ab	165.76 ± 1.46 e
	BN GF 3%	17.82 ± 0.13 a	3.56 ± 0.01 a	175.06 ± 0.46 b
	BN GF 5%	17.83 ± 0.05 a	**3.61 ± 0.02 b**	172.18 ± 0.66 d
	BN GF 10%	17.83 ± 0.05 a	3.58 ± 0.04 ab	165.17 ± 2.51 ef
	BN RI 3%	17.82 ± 0.04 a	3.55 ± 0.01 ac	176.06 ± 0.60 c
	BN RI 5%	17.92 ± 0.1 a	3.53 ± 0.02 cd	172.68 ± 0.22 d
	BN RI 10%	18.22 ± 0.31 ab	3.52 ± 0.01 d	167.88 ± 0.40 f
Sunflower	HA Control	16.65 ± 0.08 a	3.20 ± 0.01 a	416.78 ± 0.67 ab
	HA F40 3%	**17.02 ± 0.04 bc**	3.19 ± 0.01 ab	421.11 ± 1.27 c
	HA F40 5%	**17.17 ± 0.20 bcd**	3.19 ± 0.01 ab	415.78 ± 1.09 a
	HA F40 10%	**17.25 ± 0.12 b**	3.19 ± 0.01 ab	395.44 ± 0.73 d
	HA GF 3%	**16.85 ± 0.08 d**	3.18 ± 0.01 b	418.56 ± 1.24 b
	HA GF 5%	16.82 ± 0.18 acd	3.18 ± 0.01 ab	407.00 ± 2.60 e
	HA GF 10%	**17.07 ± 0.10 bcd**	3.19 ± 0.01 ab	387.33 ± 1.32 f
	HA RI 3%	**16.90 ± 0.11 cd**	3.28 ± 0.08 abc	425.56 ± 1.33 g
	HA RI 5%	**16.75 ± 0.18 acd**	**3.23 ± 0.02 c**	417.78 ± 0.97 b
	HA RI 10%	**16.98 ± 0.04 cd**	**3.23 ± 0.02 c**	397.78 ± 1.20 h
Forest	HD Control	17.30 ± 0.17 abcd	3.87 ± 0.01 a	508.78 ± 3.07 a
	HD F40 3%	17.25 ± 0.12 abc	3.88 ± 0.00 ab	**493.33 ± 1.58 b**
	HD F40 5%	17.38 ± 0.04 abd	3.88 ± 0.00 a	**483.56 ± 1.51 c**
	HD F40 10%	17.60 ± 0.18 ad	3.88 ± 0.01 ab	461.22 ± 2.17 d
	HD GF 3%	17.27 ± 0.10 abc	**3.89 ± 0.01 bc**	496.89 ± 2.15 e
	HD GF 5%	17.37 ± 0.08 abd	3.88 ± 0.01 ab	485.89 ± 2.52 cf
	HD GF 10%	17.53 ± 0.10 d	3.88 ± 0.01 ab	462.11 ± 2.89 d
	HD RI 3%	17.13 ± 0.10 c	3.89 ± 0.01 abc	496.00 ± 2.55 be
	HD RI 5%	17.17 ± 0.15 bc	3.87 ± 0.01 a	488.22 ± 1.39 f
	HD RI 10%	17.13 ± 0.15 bc	**3.90 ± 0.01 c**	472.11 ± 1.05 g

Letters denote the significant differences among the different samples within one honey type due to the ANOVA Games–Howell post-hoc test. Bold letters denote the samples of values significantly different from control honeys. RI: rice syrup; F40: high fructose-content corn syrup; GF: self-made glucose–fructose syrup.

On the contrary, the electrical conductivity results showed a significant decrease in most of the cases due to the fact that the electrical conductivity of the sugar syrups was very low: F40 18.72 ± 1.42 µS/cm, GF 3.42 ± 0.55 µS/cm, and RI 10.73 ± 1.33 µS/cm. It can be observed that the highest decrease occurred at the highest sugar syrup addition level.

In a Romanian study, a similar increase in moisture and a decrease in electrical conductivity was observed in acacia honey upon the addition of sugar syrups at 5% and 10% w/w level [30]. Moreover, in another Hungarian study, similar results were found, where 30% and 40% w/w of sugar syrup was mixed with acacia honey [31].

3.2. Results of Near-Infrared Spectroscopy

In this section, the results of the chemometric analysis are going to be introduced.

3.2.1. Introduction of the Spectra

For the interpretation of the raw spectra, the entire wavelength range was used (Figure 2a). In the case of the individual honey types (Figure 2b–f), the same preprocessing method was chosen. In this case, the Saviztky–Golay smoothing (2nd order polynomial with 21-window size) coupled with multiplicative scatter correction was chosen, as this preprocessing had the best results in the case of the all syrup models of PCA-LDA for all the honey types.

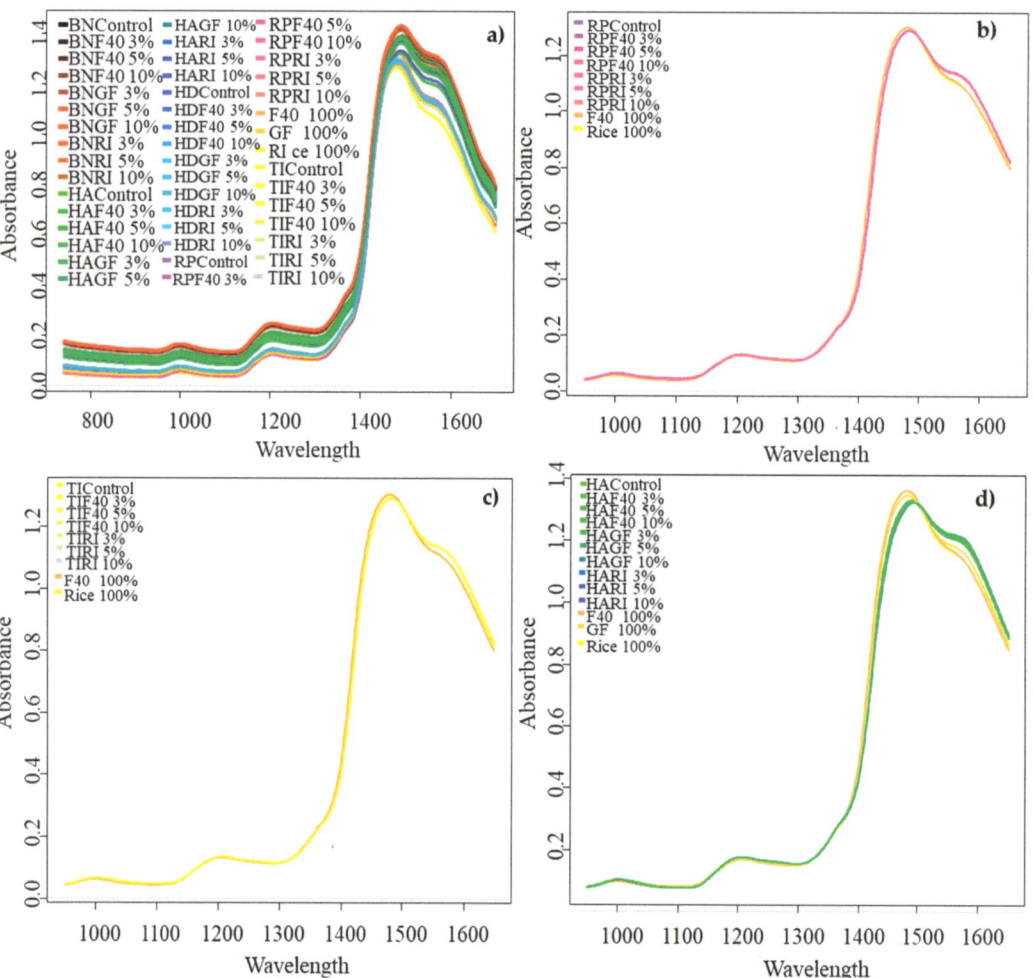

Figure 2. *Cont.*

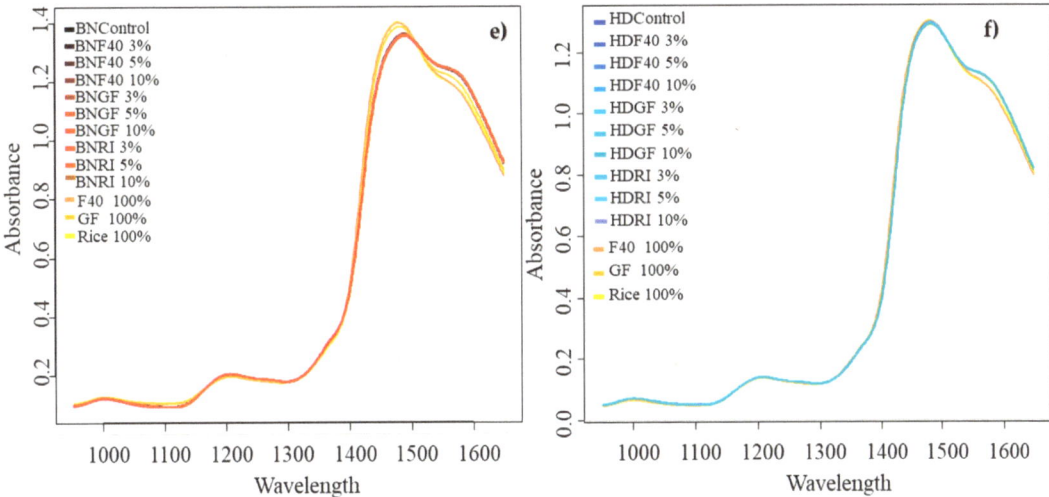

Figure 2. Near-infrared average spectra of the analyzed samples: (**a**) raw spectra (740–1700 nm) and spectra after Savitzky–Golay smoothing (2nd order polynomial, 21 window size, no derivation) and multiplicative scatter correction (950–1650 nm) in the case of (**b**) acacia samples (RP) (**c**) linden samples (TI), (**d**) sunflower samples (HA), (**e**) rape samples (BN), and (**f**) forest honey samples and their mixtures with rice, F40, and self-made glucose–fructose syrup (GF).

The recorded spectra of the different samples showed similar pattern in the case of all the honey types and their mixtures. Peaks were found around wavelength range of 950–1050 nm, 1200 nm, and 1450 nm (Figure 2). Another underlying peak was found around 1550–1600 nm. The spectral assignation is discussed in Section 3.2.4. Based on the raw and the pretreated spectra, the large water peak can be found in the range of 1400–1500 nm [14]. In this region, the highest absorbance values were found for the sugar syrups, followed by the sunflower (Figure 2d), rape (Figure 2e), forest (Figure 2f), acacia (Figure 2b), and linden (Figure 2c) honeys. This can be attributed to the higher water content of the sugar syrups. Moreover, in the case of all the honeys, it can be observed that the sugar syrups had a lower absorbance value in the range of the 1550–1600 nm region, which was the region assigned to the sugar content of the samples [14].

3.2.2. Results of the PCA-LDA Analysis

The results of the PCA-LDA analysis performed on the different honeys with different sugar syrup additions provided promising results. Different honeys and sugar syrup models needed different spectral pretreatment in most of the cases. For the interpretation of the results, we chose models containing all the syrups and models where 100% classification was not achieved. The model of acacia honey provided 100% correct classification of all the groups in the case of all the models. This shows that not only the adulterant, but also the type of adulterant could be discriminated based on the model including both syrups (Figure 3a).

Similar results were obtained for the linden honey in the case of the models of F40 and rice syrup mixtures. However, when the two syrups were included in the model, the average correct classification was 97.62% in both training and validation (Figure 3b). The confusion tables showed correct classification of the control, 3%, and 10% mixtures. The honey mixed with rice syrup in 5% (TI Rice 5%) was misclassified as belonging to the 5% F40 mixture in 16.67% (TI F40 5%).

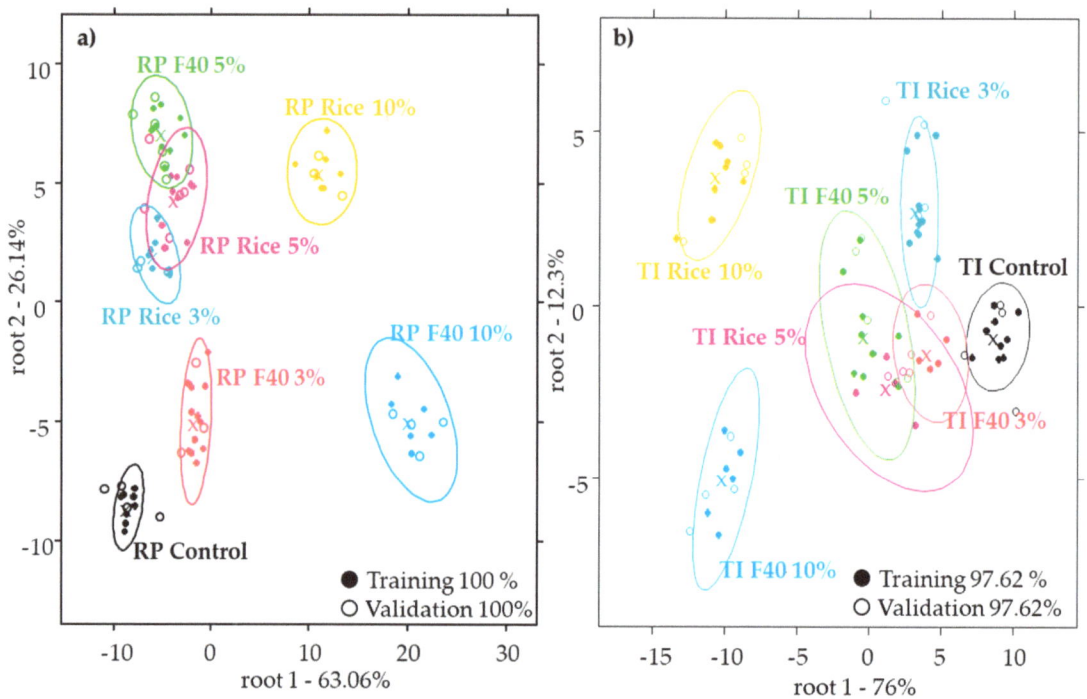

Figure 3. PCA-LDA model score plot of the (**a**) acacia (RP) and (**b**) linden (TI) honey samples for the classification of rice and high-fructose corn (F40) syrup addition levels including the results of all the samples.

The model of the forest honey was like the model of acacia honey, where all the models showed 100% correct classification. Figure 4a shows the model built for the classification of forest honeys including all the sugar syrups. In the figure, the control group separates from the rest of the models, also providing 100% classification of the control. The result of the rape honey model was similar to the model of linden where the models built separately for the different syrups provided the 100% classification. The model including results of all the syrups altogether provided the average correct classification of the training model and 98.89% correct classification of validation model, as can be seen in Figure 4b. A misclassification was found in the case of the sample mixed with F40 syrup in 3% (BNF40 3%), which was misclassified as belonging to the 5% F40 mixture (BNF40 5%) in 11.11%.

The "weakest" model was obtained for the sunflower honey where the models of honeys mixed with the rice and F40 syrups provided 100% correct classification for all the sample groups. The model of the GF syrup mixtures and model including data of all the samples (Figure 4c) provided a correct classification of the control, but misclassifications were found for the mixtures. The model of the GF syrup mixtures showed average correct classification of 94.45%; its score plot can be seen in Figure 4d. The honey mixtures of 3% w/w and 5% w/w were classified correctly, while the honey mixed with 10% w/w using the GF syrup (HA GF 10% w/w) was classified as honey mixed with 5% w/w GF syrup (HA GF 5% w/w) in 22.22%. Similar results were obtained by Chinese researchers, where honeys mixed with 5% w/w and 10% w/w sugar syrups were misclassified as belonging to each other [13]. The PCA-LDA of sunflower honey built for all the samples classified all the samples correctly during the training; however, in the case of validation, the average correct classification was 96.67%. In this model, the sample containing F40 syrup at 3% w/w lead to misclassification as belonging to the honey mixed with 5% w/w of F40 syrup at 11.11%. In our study, compared with the Chinese and other studies, the detection of adulterants could be achieved not only on the levels presented in those studies (5–40% w/w), but also at a lower level of 3% w/w [13–16]. Moreover, for the results of PCA-LDA, in all cases, the control could be discriminated from the adulterated samples, which was not assumed in the case of the physical parameters (Table 1).

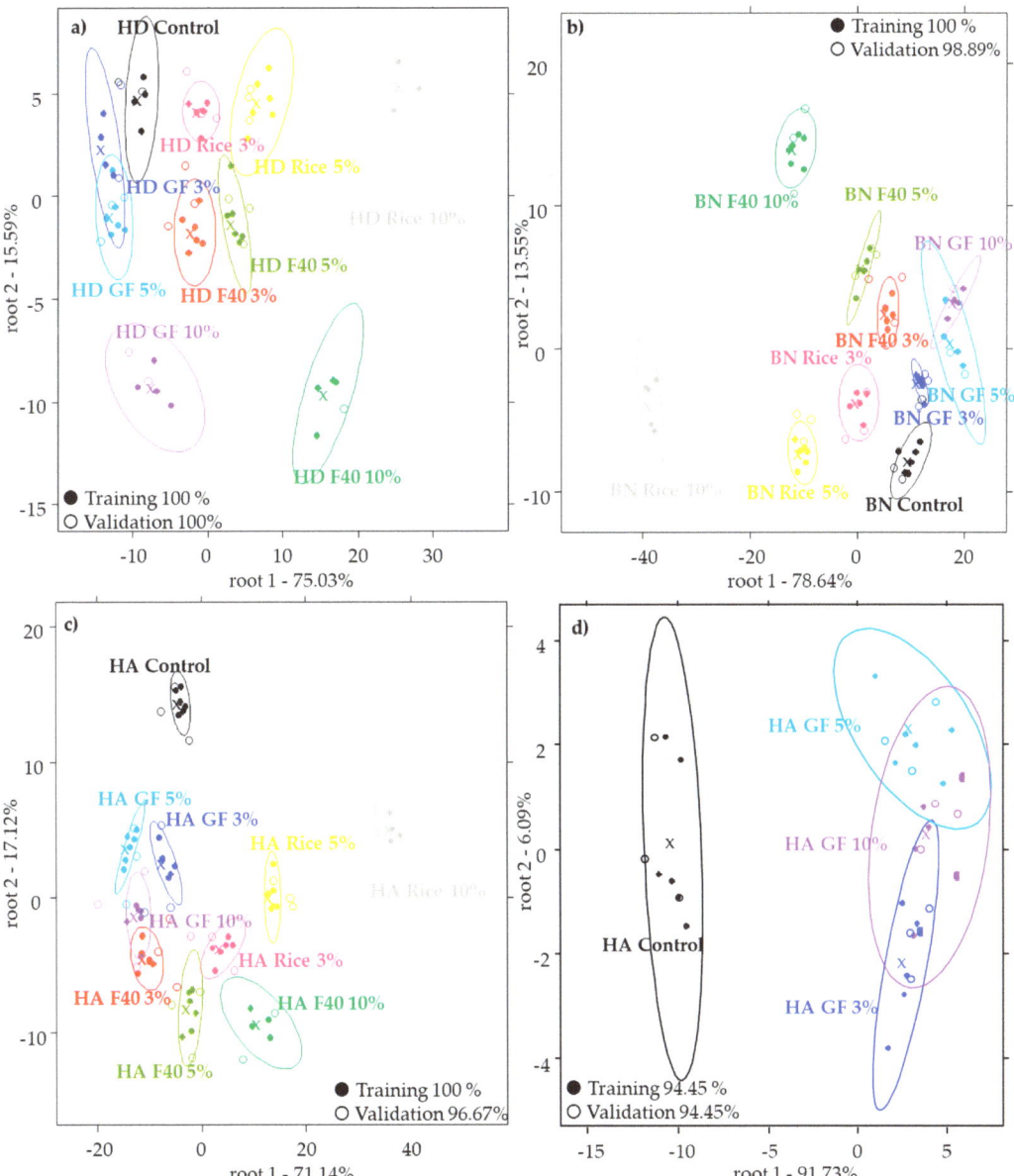

Figure 4. PCA-LDA model score of the classification model of rice, high-fructose corn (F40), and self-made glucose–fructose (GF) syrup addition levels in the case of models containing results of all the syrups of (**a**) forest (HD), (**b**) rape (BN), and (**c**) sunflower (HA) honey, and (**d**) classification of sunflower (HA) honeys blended with self-made glucose–fructose syrup.

3.2.3. Results of the Partial Least Square Regression Models

Similar to the PCA-LDA results, the model parameters were different in the case of the different honeys and syrup models within honeys (Table 2). Moreover, the accuracy highly depended on the applied pretreatment, which was also different in the case of the models.

Table 2. Result of the partial least square regression models built to regress on the added syrup concentration of sugar syrups adapted from Bodor Z. (2022) [32].

Honey	Syrup	Pretreatment	Number of Latent Variables	R^2C	RMSEC %	RPDC	R^2CV	RMSECV %	RPDCV
Acacia	All syrups	sgol@2-13-0+sgol@2-17-1	4	0.99	0.29	11.42	0.98	0.49	6.70
	Rice	sgol@2-13-0+deTr	2	0.98	0.52	6.79	0.94	0.84	4.22
	F40	sgol@2-17-0+sgol@2-21-1	4	0.99	0.36	9.69	0.98	0.50	6.83
Linden	All syrups	deTr+snv	4	0.97	0.68	5.49	0.92	1.07	3.50
	Rice	sgol@2-21-0	3	0.96	0.71	5.25	0.80	1.64	2.27
	F40	deTr	3	0.94	0.93	4.03	0.86	1.37	2.73
Forest	All syrups	sgol@2-17-0+snv	4	0.92	0.89	3.53	0.88	1.07	2.95
	Rice	sgol@2-17-0+sgol@2-17-2	4	1.00	0.22	14.70	0.95	0.72	4.52
	F40	sgol@2-21-0+sgol@2-21-1	4	0.99	0.36	9.32	0.97	0.54	6.20
	GF	sgol@2-21-0+sgol@2-13-2	4	0.98	0.54	6.46	0.71	1.85	1.89
Rape	All syrups	msc	3	0.77	1.53	2.09	0.68	1.80	1.78
	Rice	sgol@2-17-0+sgol@2-21-2	3	0.96	0.69	4.80	0.88	1.13	2.93
	F40	deTr	2	0.98	0.49	7.14	0.96	0.72	4.91
	GF	sgol@2-13-0+sgol@2-21-2	4	0.92	0.99	3.70	0.46	2.62	1.39
Sunflower	All syrups	msc	4	0.60	2.11	1.60	0.36	2.69	1.26
	Rice	sgol@2-17-0+sgol@2-17-1	4	0.99	0.40	9.29	0.92	1.01	3.67
	F40	sgol@2-13-0	4	0.99	0.41	8.36	0.94	0.80	4.27
	GF	msc	4	0.83	1.50	2.49	0.39	2.89	1.29

R^2C and R^2CV determination coefficient of training and validation; RMSEC and RMSECV root mean square error of training and validation; RPDC and RPDCV residual prediction deviation training and validation.

Results of acacia, linden, and forest honeys were better than models obtained for the rape and sunflower honeys.

The prediction model of acacia provided >0.94 R^2CV with RMSECV values lower than 0.9% for the syrup addition. Moreover, the RPDCV values were also good, with >4.22 values. The best model was obtained in this case for the prediction of F40 and all syrups together, while the model built for the prediction of rice syrup was slightly weaker, however still very good.

The results obtained in the case of the linden honey were similarly good; during the cross validation, the R^2CV values were >0.80, with an RMSECV error lower than 1.65%. Similar to acacia, the models built for the prediction of syrup concentration including results of all the samples and prediction of F40 was better than in the case of the rice syrup model.

The models of the forest honey provided satisfactory results based on the RPDCV values. In this case, the best model was obtained for the prediction of the rice and the F40 model with $R^2CV \geq 0.95$ and RPDCV higher than 4.5. The model built for all the syrups was also quite satisfactory; in this case, the R^2CV value was higher than 0.8, with an RMSECV of 1.07%. The worst results were obtained in the case of the GF syrup prediction model, this resulting in R^2CV of 0.71 and RMSECV of 1.85%.

In the case of the PLSR models of rape honey, the trend was similar to the results of the forest honey. The best results were obtained for the prediction of the F40 syrup, where the R^2CV was above 0.96 and the RPDCV value was 4.91, and lower than 1% RMSECV. Slightly worse values were obtained for the prediction of rice syrup, while the worst model was obtained for the GF syrup prediction. In this case, based on the RPDCV values, the models could not be considered satisfactory.

Similar results were provided by the model of the sunflower honey. The prediction of rice and F40 syrup could be considered good, with $R^2CV > 0.92$ and RPDCV > 3.5. The error of prediction based on the RMSECV values was lower than 1.01%. The models built for the prediction of all the syrups and for the prediction of GF syrup reached lower performance. In these cases, the RPDCV values were below 1.5; therefore, these could not be considered satisfactory.

Results showed that the accuracy of prediction highly depends on the type of honey and type of syrup. PLRS model parameters of acacia, linden, and forest honeys were better than models obtained for the rape and sunflower honeys. Moreover, the results obtained for the prediction of rice and F40 were also better than those obtained for the GF syrup. Chinese researchers also found different accuracies of prediction between HFCS and maltose syrups [31].

3.2.4. Regression Vectors and Spectral Assignations

Based on the regression vectors of the different PLSR models and spectra (Figure 2), it can be concluded that in the case of the models, similar wavelength ranges contributed to the PLSR models (Table 3), which are in line with the peaks found during the observation of the raw and pretreated spectra (Figure 2). Mostly all of the models showed the contributing range of 950–1000 nm, which can be assigned to the O–H stretches of the second overtone. The second overtone O–H (1000–1130) and C–H stretches (1150–1220) were also affected in most of the models. Moreover, all the models provided the changes in the range of the 1300–1600 nm region, which is known as the region of the O–H stretches of the first overtone. Based on the latest aquaphotomics-related studies, the water spectral pattern changes can be well-discovered in the 1300–1600 nm range. A wavelength range of 1320–1420 nm can be assigned to the water with less or no hydrogen bonds. Within this range, the water solvation shells OH–$(H_2O)_{1,2,4}$ (1360–1366 nm), free water (1412 nm), and free OH– bond were affected (1398–1418 nm) the most [33–35].

The spectral range of 1432–1444 nm can be assigned to water molecules with one hydrogen bond, and the range of 1458–1495 nm is connected to water molecules with 2–4 hydrogen bonds [35–37]. The range of around 1490–1520 nm is assigned to $\nu 1$, $\nu 2$, symmetrical stretching fundamental vibrations and strongly bonded water (1516 nm). Moreover, as honey is mainly composed of sugars and water, the changes in the sugar composition were also reflected in the spectra. Studies showed that the 1439 nm wavelength can be assigned to the sucrose molecules, and higher wavelengths, such as 1520 nm and 1590 nm, to the glucose [37]. The changes in the spectra can be explained by the fact that the sugar profile of the sugar syrups differs highly from that of the honeys, and also the water structure of these sugars is different. Therefore, the addition of these sugars syrups is well-reflected in the spectra [14,33,35–43]. Previous Hungarian research has shown that the addition of sugar syrup to honey results an in the increase in the absorbance in the range of 1342–1480 nm, while the decrease in absorbance can be observed at 1512 nm. This shows that the addition of sugar syrups increased the free water and decreased the amount of highly bonded water molecules, which was also proven in the current study [22].

Table 3. Spectral assignation based on the regression vectors contributing to the PLSR models of all the models and syrups.

Spectral Region nm	Models	Spectral Assignation Based on [14,33,37–43]
950–1000	Acacia—all syrups, F40 Linden—Rice syrup Forest—All models Rape—all syrups, GF Sunflower—Rice	N–H stretches of second overtone
1000–1130	All the models except sunflower GF model	O–H stretches of second overtone
1150–1220	Acacia—all syrups, F40 Linden, forest, rape, all models Sunflower—all syrups, rice, and GF	C–H (CH_2, CH_3) stretches of second overtone, C–H combination stretch of first overtone (CH_2, CH_3)
1300–1600	All the models	1st overtone O–H stretches

4. Conclusions

Our study focused on the adulteration detection of different honey types using three types of sugar syrup. The changes resulting from the syrup addition were evaluated with physical methods and near-infrared spectroscopy.

The results showed that at these levels (3%, 5%, and 10% w/w) the moisture and the pH did not provide sufficient support for detection of adulteration while, conductivity seemed to be a good indicator. The electrical conductivities of sugar-blended and control honeys were significantly different from each other in most of the cases (Table 1). Still, these differences are not relevant enough, as in the Hungarian and international legislation there are no specific limits for all the honey types in terms of electrical conductivity.

The results of the NIRS analysis showed that pretreatment and PC number optimization have a high importance in the model parameters and robustness. Following correct model optimization, promising results could be achieved for both the detection and quantification of adulteration even at the lowest level (3% w/w), which is lower than the most commonly studied sugar syrup concentration in such studies. The results of the PCA-LDA models showed a high accuracy of classification in the case of the optimized models. Control honeys were classified correctly in the case of all the models. The worst average classification accuracy was ~94%, obtained for the model of the sunflower honeys containing GF sugar syrup. However, in this case misclassifications were found among the 5% and 10% w/w mixtures. With the model optimization 100% correct classification could be achieved for the control for all the studied syrup concentrations (3, 5, 10% w/w). PLSR results of the acacia, linden, and forest honeys could also be considered good based on the obtained model parameters: the R^2CV values were above 0.80 for the prediction of F40 and rice syrup, and the added syrup concentration was predicted with lower than 1% of error based on the RMSECV values. Similarly good models were obtained for the prediction of the F40 and rice syrup addition (R^2CV >0.8 and RPDCV > 2.97) of the rape, sunflower, and forest honeys. Based on the results, we can conclude that the model accuracy highly depends on both the honey and syrup type. These observations were more obvious in the case of the PLSR models. In most of the cases the worst model parameters were obtained for the self-made glucose–fructose syrups. Based on our research, it can be concluded that it is very important to build specific models using the NIRS method on different honey types and with different sugar syrup adulterants. The main limitation of the study was that the applied NIR instruments worked only in the spectral range of 740–1700 nm, therefore lacking the region of higher wavelengths where chemical bounds can be found for the sugars (1800 nm) [14]. Another limitation and possibility of further work could be the analysis of more physicochemical parameters and sugar properties of the honeys. Moreover, the applicability of this in an authority level could be achieved only with a large database. This should be built containing numerous spectra of authentic honeys; furthermore, spectra of honeys that were syrup added with different type of syrups and at different concentrations. Nonetheless, based on the results, NIRS applied together with well-optimized models can be an effective tool in the detection of adulteration with sugar syrups even at low levels. In the future, these models could be further improved by the extension of the database with more honey and syrup types.

Supplementary Materials: The following supporting information can be downloaded at: https://www.mdpi.com/article/10.3390/chemosensors11020089/s1, Table S1. Near-infrared spectroscopy pretreatment combinations used for the spectral preprocessing.

Author Contributions: Conceptualization, Z.B., Z.K., J.-L.Z.Z. and C.B.; methodology, Z.B., Z.K., M.M. and C.B; resources, Z.K., Z.B., C.B., M.V.B. and É.C.C.; writing—original draft preparation, Z.B., Z.K. and C.B.; writing—review and editing, Z.B., Z.K., C.B., M.V.B., É.C.C., J.-L.Z.Z. and M.M.; visualization, Z.B.; supervision, Z.K. and C.B.; project administration, Z.K.; funding acquisition, Z.B., Z.K. and C.B. All authors have read and agreed to the published version of the manuscript.

Funding: This paper was supported by the TKP2021-NVA-22 project of the Hungarian University of Agriculture and Life Sciences. This project was supported by the Doctoral School of Food Science of MATE.

Institutional Review Board Statement: Not applicable.

Informed Consent Statement: Not applicable.

Data Availability Statement: Not applicable.

Acknowledgments: Authors are grateful to Vanessza Nóra Tamás for her contribution to the analysis.

Conflicts of Interest: The authors declare no conflict of interest.

References

1. Ciulu, M.; Solinas, S.; Floris, I.; Panzanelli, A.; Pilo, M.I.; Piu, P.C.; Spano, N.; Sanna, G. RP-HPLC Determination of Water-Soluble Vitamins in Honey. *Talanta* **2011**, *83*, 924–929. [CrossRef]
2. Zamora, M.C.; Chirife, J. Determination of Water Activity Change Due to Crystallization in Honeys from Argentina. *Food Control* **2006**, *17*, 59–64. [CrossRef]
3. Da Silva, P.M.; Gauche, C.; Gonzaga, L.V.; Costa, A.C.O.; Fett, R. Honey: Chemical Composition, Stability and Authenticity. *Food Chem.* **2016**, *196*, 309–323. [CrossRef] [PubMed]
4. Codex Alimentarius Hungaricus 1-3-2001/110 számú Előírás Méz (1-3-2001/110 Regulation Honey). In *Codex Alimentarius Hungaricus*; Codex Alimentarius Hungaricus (Ed.) Magyar Élelmiszerkönyv Bizottság: Budapest, Hungary, 2002; pp. 1–7.
5. The European Council. Council Directive 2001/110/EC of 20 December 2001 Relating to Honey. *Off. J. Eur. Union* **2001**, *L 10*, 47–52.
6. Codex Alimentarius Commission. *Codex Standard for Honey, CODEX STAN 12-1981*; FAO: Rome, Italy, 2001; Volume 11, p. 7.
7. Benedek, C.; Zaukuu, Z.J.-L.; Bodor, Z.; Kovacs, Z. Honey-Based Polyphenols: Extraction, Quantification, Bioavailability, and Biological Activities. In *Plant-Based Functional Foods and Phytochemicals*; Goyal, R.M., Nath, A., Suleria, R.H.A., Eds.; Apple Academic Press: New York, NY, USA, 2021; pp. 35–63.
8. National Food Crime Unit; Scottish Food Crime Unit. *Food Crime Strategic Assesment 2020*; Food Standards Agency: London, UK, 2020.
9. Zábrodská, B.; Vorlová, L. Adulteration of Honey and Available Methods for Detection—A Review. *Acta Vet. Brno* **2014**, *83*, S85–S102. [CrossRef]
10. European Commission. *Technical Round Table on Honey Authentication*; European Commission: Brussels, Belgium, 2018.
11. Lohumi, S.; Lee, S.; Lee, H.; Cho, B.-K. A Review of Vibrational Spectroscopic Techniques for the Detection of Food Authenticity and Adulteration. *Trends Food Sci. Technol.* **2015**, *46*, 85–98. [CrossRef]
12. Aouadi, B.; Zaukuu, J.L.Z.; Vitális, F.; Bodor, Z.; Fehér, O.; Gillay, Z.; Bazar, G.; Kovacs, Z. Historical Evolution and Food Control Achievements of Near Infrared Spectroscopy, Electronic Nose, and Electronic Tongue—Critical Overview. *Sensors* **2020**, *20*, 5479. [CrossRef] [PubMed]
13. Gan, Z.; Yang, Y.; Li, J.; Wen, X.; Zhu, M.; Jiang, Y.; Ni, Y. Using Sensor and Spectral Analysis to Classify Botanical Origin and Determine Adulteration of Raw Honey. *J. Food Eng.* **2016**, *178*, 151–158. [CrossRef]
14. Bázár, G.; Romvári, R.; Szabó, A.; Somogyi, T.; Éles, V.; Tsenkova, R. NIR Detection of Honey Adulteration Reveals Differences in Water Spectral Pattern. *Food Chem.* **2016**, *194*, 873–880. [CrossRef]
15. Huang, F.; Song, H.; Guo, L.; Guang, P.; Yang, X.; Li, L.; Zhao, H.; Yang, M. Detection of Adulteration in Chinese Honey Using NIR and ATR-FTIR Spectral Data Fusion. *Spectrochim. Acta Part A Mol. Biomol. Spectrosc.* **2020**, *235*, 118297. [CrossRef]
16. Aliaño-González, M.J.; Ferreiro-González, M.; Espada-Bellido, E.; Palma, M.; Barbero, G.F. A Screening Method Based on Visible-NIR Spectroscopy for the Identification and Quantification of Different Adulterants in High-Quality Honey. *Talanta* **2019**, *203*, 235–241. [CrossRef] [PubMed]
17. Grassi, S.; Jolayemi, O.S.; Giovenzana, V.; Tugnolo, A.; Squeo, G.; Conte, P.; De Bruno, A.; Flamminii, F.; Casiraghi, E.; Alamprese, C. Near Infrared Spectroscopy as a Green Technology for the Quality Prediction of Intact Olives. *Foods* **2021**, *10*, 1042. [CrossRef] [PubMed]
18. Lang, J.; McNitt, L.; Inc, P.; CT, S. *Detection of Honey Adulteration Using FT-NIR Spectroscopy*; PerkinElmer, Inc.: Waltham, MA, USA, 2015.
19. Shafiee, S.; Polder, G.; Minaei, S.; Moghadam-Charkari, N.; van Ruth, S.; Kuś, P.M. Detection of Honey Adulteration Using Hyperspectral Imaging. *IFAC-PapersOnLine* **2016**, *49*, 311–314. [CrossRef]
20. Longin, L.; Jurinjak Tusek, A.; Valinger, D.; Benkovic, M.; Jurina, T.; Gajdos Kljusuric, J. Application of Artificial Neural Networks (ANN) Coupled with Near-InfraRed(NIR) Spectroscopy for Detection of Adulteration in Honey. *Biodivers. Inf. Sci. Stand.* **2019**, *3*, e38048. [CrossRef]
21. Yang, X.; Chen, J.; Jia, L.; Yu, W.; Wang, D.; Wei, W.; Li, S.; Tian, S.; Wu, D. Rapid and Non-Destructive Detection of Compression Damage of Yellow Peach Using an Electronic Nose and Chemometrics. *Sensors* **2020**, *20*, 1866. [CrossRef]
22. Bodor, Z.; Zaukuu, J.Z.; Aouadi, B.; Benedek, C.; Kovacs, Z. Application of NIRS and Aquaphotomics for the Detection of Adulteration of Honey, Paprika and Tomato Paste. In Proceedings of the SZIEntific Meeting for Young Researchers—Ifjú Tehetségek Találkozója, Szent István University, Budapest, Hungary, 9 December 2019; pp. 76–91. Available online: http://itt.budaicampus.szie.hu/sites/default/files/files/ITT_2019_konferencia.pdf (accessed on 20 January 2020).
23. Bogdanov, S. Harmonised Methods of the International Honey Commission. Swiss Bee Research Centre, FAM: Liebefeld, Switzerland, 2009; ISBN 9780874216561.
24. Tabachnick, B.G.; Fidell, L.S. *Using Multivariate Statistics*, 6th ed.; Pearson Education: London, UK, 2013; ISBN 9780205849574.

25. Luedeling, E. RPD: Residual Prediction Deviation (RPD) in ChillR: Statistical Methods for Phenology Analysis in Temperate Fruit Trees. Available online: https://rdrr.io/cran/chillR/man/RPD.html (accessed on 28 July 2021).
26. Muncan, J.; Tsenkova, R. Aquaphotomics—From Innovative Knowledge to Integrative Platform in Science and Technology. *Molecules* **2019**, *24*, 2742. [CrossRef]
27. Muncan, J.; Kovacs, Z.; Pollner, B.; Ikuta, K.; Ohtani, Y.; Terada, F.; Tsenkova, R. Near Infrared Aquaphotomics Study on Common Dietary Fatty Acids in Cow's Liquid, Thawed Milk. *Food Control* **2021**, *122*, 107805. [CrossRef]
28. Munawar, A.A.; Zulfahrizal; Meilina, H.; Pawelzik, E. Near Infrared Spectroscopy as a Fast and Non-Destructive Technique for Total Acidity Prediction of Intact Mango: Comparison among Regression Approaches. *Comput. Electron. Agric.* **2022**, *193*, 106657. [CrossRef]
29. Pollner, B.; Kovacs, Z. R-Package Aquap2—Multivariate Data Analysis Tools for R Including Aquaphotomics Methods. Available online: https://www.aquaphotomics.com/aquap2/ (accessed on 28 July 2021).
30. Ciursa, P.; Oroian, M. Influence of Corn and Inverted Sugar Adulteration on Physicochemical Properties of Romanian Acacia Honeys. *Sci. Bulletin. Ser. F. Biotechnol.* **2020**, *XXIV*, 85–93.
31. Czipa, N.; Phillips, C.J.C.; Kovács, B. Composition of Acacia Honeys Following Processing, Storage and Adulteration. *J. Food Sci. Technol.* **2019**, *56*, 1245–1255. [CrossRef]
32. Bodor, Z. Application of Classical and Correlative Analytical Methods for Authentication of Honey. Ph.D. Thesis, Hungarian University of Agriculture and Life Sciences, Budapest, Hungary, 2022. Available online: https://uni-mate.hu/documents/20123/336900/Bodor_Zsanett-ertekezes.pdf/1908f699-17cd-9985-a058-2bc19dbdb5c7?t=1659596636531 (accessed on 22 December 2022).
33. Li, S.; Zhang, X.; Shan, Y.; Su, D.; Ma, Q.; Wen, R.; Li, J. Qualitative and Quantitative Detection of Honey Adulterated with High-Fructose Corn Syrup and Maltose Syrup by Using near-Infrared Spectroscopy. *Food Chem.* **2017**, *218*, 231–236. [CrossRef]
34. Xantheas, S.S. Ab Initio Studies of Cyclic Water Clusters (H_2O)n, N = 1–6. III. Comparison of Density Functional with MP2 Results. *J. Chem. Phys.* **1995**, *102*, 4505. [CrossRef]
35. Tsenkova, R.; Muncan, J.; Pollner, B.; Kovacs, Z. Essentials of Aquaphotomics and Its Chemometrics Approaches. *Front. Chem.* **2018**, *6*, 363. [CrossRef] [PubMed]
36. Bodor, Z.; Benedek, C.; Aouadi, B.; Zsom-Muha, V.; Kovacs, Z. Revealing the Effect of Heat Treatment on the Spectral Pattern of Unifloral Honeys Using Aquaphotomics. *Molecules* **2022**, *27*, 780. [CrossRef] [PubMed]
37. Siesler, H.W.; Ozaki, Y.; Kawata, S.; Heise, H.M. (Eds.) *Near-Infrared Spectroscopy: Principles, Instruments, Applications*; Wiley-VCH Verlag GmbH: Weinheim, Germany, 2001; ISBN 9783527612666.
38. López, M.G.; García-González, A.S.; Franco-Robles, E. Carbohydrate Analysis by NIRS-Chemometrics. In *Developments in Near-Infrared Spectroscopy*; IntechOpen: London, UK, 2017; ISBN 978-953-51-3018-5.
39. Ozaki, Y.; Genkawa, T.; Futami, Y. Near-Infrared Spectroscopy. In *Encyclopedia of Spectroscopy and Spectrometry*; Academic Press: Cambridge, MA, USA, 2016; pp. 40–49.
40. Zhang, C.; Liu, F.; He, Y. Identification of Coffee Bean Varieties Using Hyperspectral Imaging: Influence of Preprocessing Methods and Pixel-Wise Spectra Analysis. *Sci. Rep.* **2018**, *8*, 2166. [CrossRef] [PubMed]
41. Yang, X.; Guang, P.; Xu, G.; Zhu, S.; Chen, Z.; Huang, F. Manuka Honey Adulteration Detection Based on Near-Infrared Spectroscopy Combined with Aquaphotomics. *LWT* **2020**, *132*, 109837. [CrossRef]
42. Farkas, J.; Dalmadi, I. Near Infrared and Fluorescence Spectroscopic Methods and Electronic Nose Technology for Monitoring Foods. *Prog. Agric. Eng. Sci.* **2009**, *5*, 1–29. [CrossRef]
43. Muncan, J.; Kuroki, S.; Moyankova, D.; Morita, H.; Atanassova, S.; Djilianov, D.; Tsenkova, R. Protocol for Aquaphotomics Monitoring of Water Molecular Structure in Leaves of Resurrection Plants during Desiccation and Recovery. *Protoc. Exch.* **2019**. [CrossRef]

Disclaimer/Publisher's Note: The statements, opinions and data contained in all publications are solely those of the individual author(s) and contributor(s) and not of MDPI and/or the editor(s). MDPI and/or the editor(s) disclaim responsibility for any injury to people or property resulting from any ideas, methods, instructions or products referred to in the content.

Article

A Paper-Chip-Based Phage Biosensor Combined with a Smartphone Platform for the Quick and On-Site Analysis of *E. coli* O157:H7 in Foods

Chaiyong Wu [1], Dengfeng Li [2], Qianli Jiang [3] and Ning Gan [2,*]

[1] School of Material Science and Chemical Engineering, Ningbo University, Ningbo 315211, China
[2] School of Oceanography, Ningbo University, Ningbo 315211, China
[3] Southern Hospital of Nanfang Medical University, Guangzhou 510515, China
* Correspondence: ganning@nbu.edu.cn

Abstract: The rapid and specific point-of-care (POC) analysis of virulent pathogenic strains plays a key role in ensuring food quality and safety. In this work, a paper-based fluorescent phage biosensor was developed for the detection of the virulent *E. coli* O157:H7 strain (as the mode of virulent pathogens) in food samples. Firstly, phages that can specifically combine with *E. coli* O157:H7 (*E. coli*) were stained with SYTO-13 dye to prepare a novel fluorescent probe (phage@SYTO). Simultaneously, a micro-porous membrane filter with a pore size of 0.45 μm was employed as a paper chip so as to retain the *E. coli*-phage@SYTO complex (>1.2 μm) on its surface. The phage@SYTO (200 nm in size) was able to pass through the pores of the chip, and the complex could be retained on the paper chip using the free phage@SYTO probes. The *E. coli*-phage@SYTO could emit a visual fluorescent signal (excited at 365 nm; emitted at 520 nm) onto the chip, which could be detected by a smartphone to reflect the concentration of *E. coli*. Under optimized conditions, the detection limit was as low as 50 CFU/mL (S/N = 3) and exhibited a wide linear range from 10^2 to 10^6 CFU/mL. The sensor has potential application value for the quick and specific POCT detection of virulent *E. coli* in foods.

Keywords: POCT assay; paper-chip-based phage biosensor; smartphone; *E. coli* O157:H7; foods

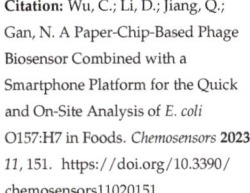

Citation: Wu, C.; Li, D.; Jiang, Q.; Gan, N. A Paper-Chip-Based Phage Biosensor Combined with a Smartphone Platform for the Quick and On-Site Analysis of *E. coli* O157:H7 in Foods. *Chemosensors* **2023**, *11*, 151. https://doi.org/10.3390/chemosensors11020151

Academic Editors: Vahid Hamedpour and Andreas Richter

Received: 25 January 2023
Revised: 15 February 2023
Accepted: 16 February 2023
Published: 20 February 2023

Copyright: © 2023 by the authors. Licensee MDPI, Basel, Switzerland. This article is an open access article distributed under the terms and conditions of the Creative Commons Attribution (CC BY) license (https://creativecommons.org/licenses/by/4.0/).

1. Introduction

Threats to food safety caused by pathogenic bacteria infections have given rise to serious damage worldwide and resulted in serious losses in livestock husbandry, agriculture, and aquaculture [1–4]. To preserve food security, early diagnostic methods for foodborne pathogens should be developed so as to prevent pathogenic infections in advance [5–7]. Moreover, there is a great need to develop rapid on-site assays for the identification of pathogenic bacterial strains because the virulence of different strains of the same bacteria varies greatly; for example, *E. coli* O157:H7 is significantly more virulent than other *E. coli* strains.

The conventional culture-based bacterial detection assays are cumbersome and tedious [8,9]. Due to their high equipment and laboratory environment requirements, nucleic-acid-based polymerase chain reaction (PCR) detection methods can discriminate different strains of the same bacteria, but are not conducive to large-scale promotion, especially in resource-limited areas [10,11]. The fabrication of a commercial point-of-care (POC) diagnostic platform has become a cutting-edge and challenging task in the field of rapid on-site analysis [1]. It is hoped that such a platform will be able to perform quick, user-friendly analysis through a cheap, automatic, and easily operated setup. The paper-chip-based biosensor is a promising alternative for the detection of samples because of its visual and sensitivity properties [12,13]. As the base of the analytical platform, paper has the advantages of a low cost, liquid flow features, biodegradability, and good chemical properties, and it has been widely used in various analytical systems. Compared with traditional advantages of a low cost, liquid flow features, biodegradability, and good chemical properties, and it has been widely used in various analytical systems. Compared with traditional analytical chips, the paper-chip-based analytical platform does not require external fluid pumping equipment or a complex chip design, and can be used for sample sampling and delivery through the capillaries of paper fibers, and additional paper chips with different pore sizes can also be employed as a filter membrane. In one paper chip, paper filters with different pore sizes can also be used to filter components present in the sample, which has attracted much attention. As this method is suitable for combination with a smartphone, it can fulfill the requirements for fabricating field-adaptive sensors with good sensitivity and specificity [14–17]. Being easy to operate and carry, and the phones have a cloud function [18,19]. With the excellent analytical platform of the digital imaging capabilities of these smartphones, the ideal platform for the development of potential sensors. Compared to other methods for the detection of the pathogenic microorganism in the sample, the proposed detection is proportional to the concentration of the target microorganism, which is not interfered with by various heterogeneous matrices and has a function that avoids the occurrence of false positive results, thereby improving the detection accuracy and extending the dynamic detection range for biological food samples [20]. Aiming at preparing a paper-chip-based bacterial sensing biological probes that are essential for fabricating a paper chip is composed of nucleic acids or their nanoproducts, the phage can specifically combine with the receptor on the surface of the host bacteria and its genetic material, DNA or RNA, can be injected into the host bacteria. Compared with antibodies, phages exhibit much higher specificity towards bacterial strains [22–24]. Moreover, they have a higher affinity for live strains. Therefore, phage probes can be employed to establish paper chips for the POC detection of E. coli O157:H7 [15,16]. Then signal changes marking the emergence of pathogenic bacteria can be monitored by the biosensors and detected by the smartphone [17,25–27].

Hence, herein, we propose a novel paper-based phage sensor using a phage probe integrated with a smartphone for the visual detection of *E. coli* O157:H7 based on RGB values. The fabrication of the phage probe and the sensor is shown in Figure 1. Firstly, the phage, which can specifically combine with *E. coli* O157:H7 (*E. coli*), was stained with SYTO 13 florescence dye, a universal DNA embedding dye [17,18], to prepare the phage@SYTO probe. The probes were mixed with samples containing the bacteria that would specifically combine with the target *E. coli* of the *E. coli*-phage@SYTO complex (>1.2 nm) and be retained on the paper chip (the filter membrane with a pore size of 0.45 µm). The free phage@SYTO probes, with a size of approximately 200 nm, can pass through the pores of the paper chip. Thus, the conjugate complex (*E. coli* phage@SYTO) can easily be separated from the free probes and enriched on the paper chips. In the next step, the complex was excited by 365 nm light using a green fluorescence light with an emission wavelength of 520 nm, which can be visually observed and recorded by a smartphone. The G value from RGB was employed for the quantification of the target bacteria. The sensor was also successfully employed to detect *E. coli* in food samples.

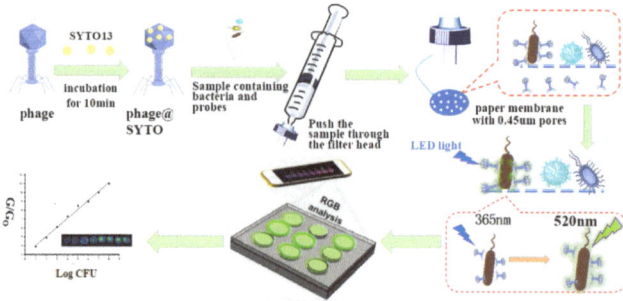

Figure 1. The preparation of the phage@SYTO probe and detection of *E. coli* O157:H7. based on a paper chip biosensor.

2. Materials and Methods

2.1. Chemicals

SYTO-13 (99.5% wt) was obtained from Macklin Co., Ltd. (Shanghai, China) (250 µL, 5 mmol). Phages of E. coli O157:H7 were screened at Dr Dengfeng Li's lab. The strains used in this study included *Vibrio parahaemolyticus* (*V. P*, ATCC 17802), *Salmonella typhimurium* (*S. T*, ATCC 14028), and *Staphylococcus aureus* (*S. A*, ATCC 43300), which were purchased from Shanghai Luwei Technology Co., Ltd. (Shanghai, China); *Listeria monocytogenes* (*L. M*, CICC 21662), which was purchased from the China Industrial Culture Collection (CICC); and *Escherichia coli* O157:H7 (ATCC 25922), obtained from Dengfeng Li's group at Ningbo University. Cellulose membrane filters with pore sizes of 0.10, 0.22, 0.45, 0.65, and 1.00 µm were purchased from Shanghai Luwei Technology Co., Ltd. (Shanghai, China). The LB broth and agar powder (bacterial grade) were derived from Hangzhou Microbial Reagent Co., Ltd. (Hangzhou, China).

2.2. Apparatus

The fluorescence, probe excitation, and emission spectra were measured using a fluorescence spectrophotometer which was purchased from Shimadzu Co., Ltd. (Kyoto, Japan). The centrifuge was purchased from Sigma Laborzentrifugen GmbH (Berlin, Germany). The medical constant temperature incubator was purchased from Panasonic Appliances Cold Chain Co., Ltd. (Tokyo, Japan). The desktop constant temperature oscillator was purchased from Jiemei Electronics Co., Ltd. (Suzhou, China). We used a transmission electron microscope (TEM, JEM 2100F, Tokyo, Japan). The FL images were captured via an Android smartphone (model number: Xiaomi 11X hypercharge, with camera specifications of 64 MP/26 mm wide). A 365 nm LED light was placed above the sample, positioned over one side. The distance from the excited LED light source to the sample remained constant (5 cm) during the measurement. Color analysis software was used to divide the color fluorescence image of the sample into red (R), green (G), and blue (B) pixels. In the required analysis, we added different concentrations of bacteria ($10 \sim 10^8$ CFU/mL) to the solution, each on a different piece of paper. The software used was Colour Reader, a free and open-source software downloaded from the Android mobile app market.

2.3. Culture and Preparation of Phages

The phages were purified using the bilayer medium method. First, the already purified bacterial stock solution was diluted to 10^7 CFU/mL, and 100 µL of 10^7 CFU/mL bacteria was added to 1 mL of LB broth until the broth was slightly cloudy. The already purified phage solution was diluted 10-fold to obtain a concentration of 10^2–10^9 PFU/mL of phage solution for further use. The semi-solid (0.7% agar by volume) was heated in the medium using a muffle until it was completely melted, and then it was dispensed into 3 mL centrifuge tubes. In total, 100 µL of diluted bacteria and 100 µL of phage solution of different concentrations were placed in 3 mL centrifuge tubes. The solution was well shaken and poured onto the plate.

The densities of the phage spots in the Petri dishes were dependent on the concentration of phage. The cultured phage Petri dishes were placed in a biosafety cabinet, 3 mL LB bouillon was added, and the medium containing phage, as well as the LB bouillon, was scraped into a 50 mL centrifuge tube using a sterilized glass spatula. The scraped mixture was centrifuged at 8000 rpm for 30 min, and the liquid was then passed through a disposable paper membrane filter with a pore size of 0.22 um. Two purifications were required to obtain the final phage for the experiment. The concentration of phage was determined by the plate counting method.

2.4. Culture and Preparation of Bacteria

In the biological safety cabinet, a small amount of *E. coli*, *S.a*, and *S.T* was scraped from the bacteriophage plate using an alcohol-sterilized catch ring and placed into 3 mL

LB broth, which was sterilized in an autoclave (120 °C, 2 h) and incubated overnight in a shaker (37 °C, 180 rmp) until it gradually became turbid. The concentration of bacteria was calculated using the plate counting method. First, the bacteria were serially diluted 108 times using PBS (the dilution was performed 10 times), and 100 µL of the diluted solution was taken and inoculated in a semi-solid Petri dish and incubated in a 37 °C thermostat for 24 h. The bacterial concentration of the original solution was calculated by counting the plaque numbers on the dish.

2.5. Phage Specificity and Probe Activity

A total of 100 µL 10^{11} PFU/mL of phage stock solution was added to PBS (pH = 7.4, 0.01 M) and diluted stepwise to 10^8 CFU/mL. Then, 100 µL 10^8 PFU/mL of phage was taken; 10 µL SYTO13 (5 mM, 250 µL) was added to the experimental group; 10 µL water was added to the control group; and 100 µL 10^7 CFU/mL *E. coli*, *S.a*, and *S.T* was treated as the phage culture and incubated overnight in Petri dishes in a 37 °C incubator.

A total of 100 µL 10^{11} PFU/mL of phage stock solution was added to PBS (pH = 7.4, 0.01 M) and diluted stepwise to 10^8 CFU/mL. Then, 100 µL 10^8 PFU/mL of phage was taken, and 10 µL SYTO13 (5 mM, 250 µL) was added to the experimental group. Simultaneously, 10 µL water and 100 µL 10^7 CFU/mL *E. coli*, *S.a*, and *S.T* were spiked into the control group and proliferated overnight in a 37 °C incubator. Then, the phage specificity and probe activity were discussed.

2.6. Synthesis of Phage @ DNA Probes

A total of 450 µL 10^{10} PFU/mL phage was mixed with 50 µL 0.1 mg/mL dye in a 1 mL centrifuge tube. Afterwards, it was shaken for 30 s and incubated for 10 min. The tube was then placed into a 50 kd dialysis tube and centrifuged for 5 min at 8000 rpm. The phage's molecular weight (MW) was much larger than 50 kd, and it could not pass through the dialysis membrane. Though the free SYTO dye, with a MW of approximately 320, could pass through the membrane pores, the membrane retained the dye-modified phage, as the free dye had no fluoresce, and the dye bound to the phage emitted green fluorescence under 360 nm excitation light. Therefore, after ultrafiltration, the upper solution of the filter membrane was taken out, and the emitted light at 520 nm was measured using a fluorometer under 365 nm excitation. Then, 50 µM of another phage@SYTO (*E. coli*) was taken and incubated with *S.A*, *S.T*, *V.P*, *E. coli*, and the mixture for 20 min, respectively, and excited by blue light from a fluorescence microscope.

2.7. Detection of Escherichia coli O157:H7 (E. coli) in Fishes

The food samples (shrimp, fish, and others) were mixed with 100 mL 100 µM PBS (pH 7.5) for 10 min. Then, the supernatant was collected for further detection. Firstly, 10 µL of 1 µM pahge@SYTO was mixed with the 100 mL supernatant solution and then agitated (200 rpm on the vortex machine) for 10 min at room temperature. After that, the mixture solution was passed through a micropore membrane filter with a pore size of 450 nm using a syringe. The *E. coli*-phage@SYTO complex (whose size was larger than 1.2 µm) could be separated and retained on the membrane, whereas the free probes, whose size was approximately 200 nm, could pass through the pores. The paper chip was first dried, and then the fluorescent (FL) detection was performed on the paper chip after excitation at 360 nm. All the test samples were placed in a black box, being unaffected by ambient light, and the FL intensity was recorded using a smartphone. Different color intensities can be employed for different concentrations of analytes to determine the presence of some bacterial strains without using any analytical apparatus. We also designed an improved version of the portable fluorimeter by separating the different concentrations into different sections in order to introduce the respective phage@SYTO probes. Images of the spiked samples, as well as the real samples, were taken with an Android smartphone. The distance between the excitation light source and the sample was kept constant (30 cm) during the measurements. For each sample, the data for five points were detected from the center to the edge of the same sample according to the five-point sampling method. The photographed photos were also imported into the analytical software, which enabled the color fluorescence image of the sample to be divided into red (R), green (G), and blue (B) channels by reading the G value of the paper chip's surface and comparing it to the G_0 read from the paper chip with the uncaptured bacteria. The G/G_0 was employed for the qualification of the targets.

3. Results and Discussion

3.1. Characterization of the Phage@SYTO Probes

SYTO 13 is a specific phage dye which cannot emit fluorescent light by itself, but can emit fluorescence light after the staining of the phage. From Figure 2, it can be seen that 10 µg/mL of phage@SYTO could emit strong fluorescence light after being excited at 365 nm, whereas the dye itself could only emit negligible light and the phage did not emit any light. All these findings illustrated that the phage@SYTO probes had been successfully prepared. From the FL intensity of the phage@SYTO, it can be seen that the SYTO could penetrate the phage's DNA [18].

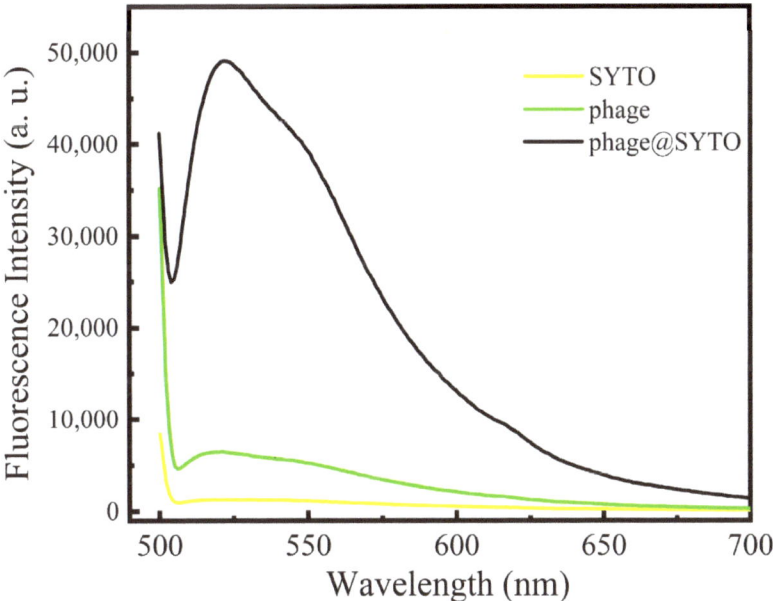

Figure 2. The fluorescent light of 10 µg/mL phage@SYTO probes, SYTO, or phage only.

3.2. The Phage Specificity and Probe Activity

The activity of the probe had a significant effect on the experimental results. Therefore, a comparison of the activity of the phage@SYTO and pure phage was conducted. The comparison shown in Figure 3A,D revealed no significant change in the number of phage spots after the SYTO 13 modification, indicating that the dye had no effect on the activity of the phage. Moreover, the probe was still able to identify the specific bacteria. In addition, as shown in Figure 3A,D, it was found that the plate with spiked *E. coli* produced phage spots. By comparison, Figure 3B,E and Figure 3C,F, which represent the *S.A* and *V.P* plate with the addition of the phage@SYTO probes, depicted no significant plaques. All of these findings indicated that the phage@SYTO was specific to *E. coli*, but not other bacteria.

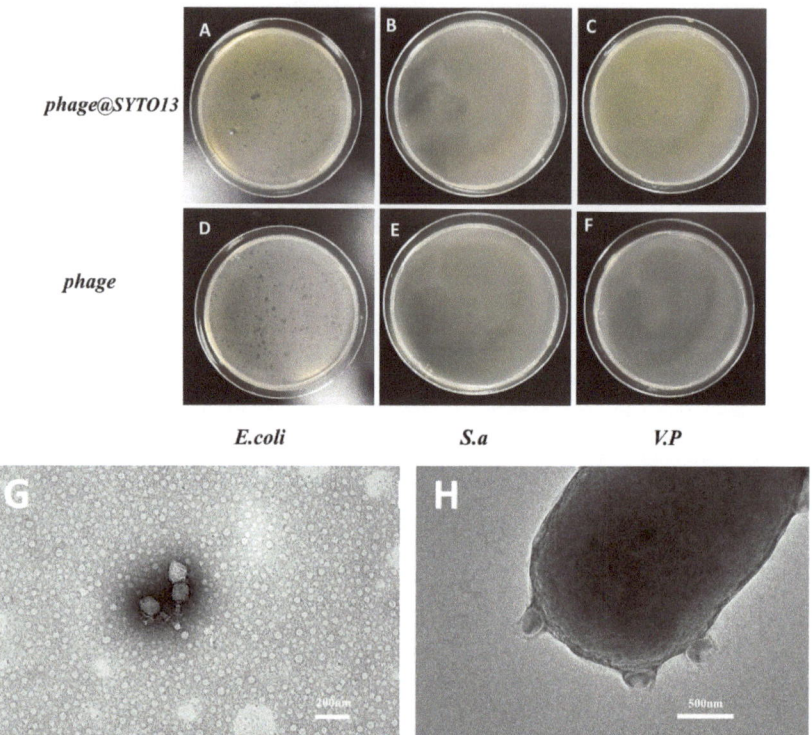

Figure 3. The plaques produced after mixing the 10^5 PFU/mL phage and phage@SYTO on plates containing (**A,D**) *E. coli* O157:H7, (**B,E**) *S.a*, or (**C,F**) *V.P*. The TEM of the (**G**) phage and (**H**) phages surrounding *E. coli* O157:H7.

Furthermore, transmission electron microscopy (TEM) was employed to reveal the formation process of the plaques. As shown in Figure 3G, phages with heads and tails of approximately 200 nm in length could be found. In Figure 3H, there are a large number of phages adhered to the outer cells of *E. coli* O157. The results indicated that the phage has a strong affinity for *E. coli* O157:H7.

3.3. Principle of the Assay

From Figure 4, it can be seen that the phage@SYTO could only bind with *E. coli* to emit fluorescence light. Moreover, the mixture of *E. coli* with other bacteria could also emit fluorescence light. These findings demonstrated that the phage@SYTO has enough specificity and discrimination capacity to identify the target bacteria [19,20]. The labeling of the SYTO amount on the probes was very important for the detection sensitivity. An amount of 100 µL SYTO-13 at different concentrations was mixed with 100 µL 10^9 PFU/mL phage. The changes in fluorescence light intensity (I) were continuously recorded with the change in the dye concentration. Because SYTO-13 only emitted fluorescence light when combined with the phage to form the phage@SYTO, the labeled amount of SYTO-13 on the phage@SYTO could reach saturation after the fluorescence intensity reached its highest level. According to the amount of SYTO-13 material at this time point, the amount of SYTO-13 in each phage could be calculated. According to the highest peak intensity of SYTO-13 at 0.1 µM, we calculated approximately 60,200 SYTO on each phage (equation: 10^{-7} M × 10^{-4} L × $6.02 × 10^{23}/10^8$ PFU = 60,200 SYTO/phage).

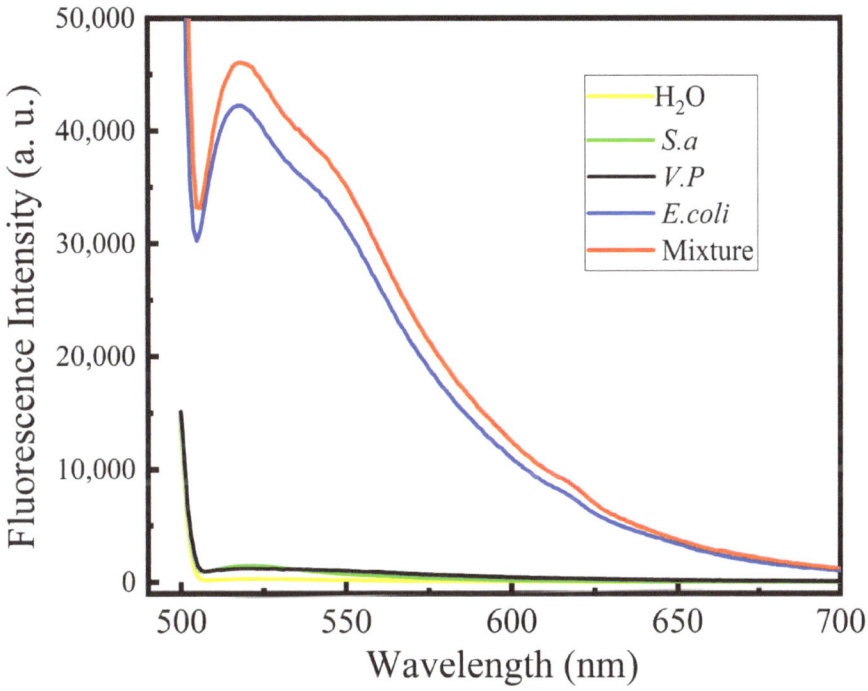

Figure 4. The fluorescence light of the phage@SYTO probe for different targets.

3.4. The Optimization of the Assay

In order to improve the detection performance of this strategy, several reaction conditions were optimized (the concentration of *E. coli* in the optimization experiment was 10^5 CFU/mL). We first optimized the binding time for the phage@SYTO to bind to the pathogenic bacteria *E. coli*. It can be seen from Figure 5A that the optimal binding time was 20 min. The bacteria's surface combined more easily with the phage@SYTO, which meant that the phage did not lose its bioactivity during labeling with SYTO. In the detection process, the concentration of phage@SYTO was also crucial, as a probe concentration that is too low will prevent the phage from fully combining with the target bacteria when the concentration of bacteria is high. However, if the probe concentration is too high, they will be stained on the membrane easily; thus, the background signal will be too high. As can be seen from Figure 5B, the optimum concentration of phage@SYTO was 10^8–10^9 PFU/mL, and the signal to noise ratio (I/I_0) reached its highest value. If the concentration is too high, there will be a strong background signal, even if the signal from *E. coli*-phage@SYTO is high. Taken together, the amount of 10^8 PFU/mL phage@SYTO was selected for the detection. The membrane pore size was also very important for the filtering of the unbound phage@SYTO. From Figure 5C, we can see that the optimum membrane pore size was 0.45 μm, whereas 0.25 μm retained too much phage@SYTO in the solution. Moreover, with a 1 μm pore size, the *E. coli*-phage@SYTO could also pass through the pores, leading to a lower signal. The reaction pH was also optimized, and we found that pH 5.0 to 8.0 was the suitable reaction pH. From Figure 6D, we can see that a pH that was too low or high resulted in a lower FL signal. This may be due to the fact that the phage@SYTO could not combine well with the bacteria. The usual pH of the sample solution was between 5 and 8; therefore, the assay was suitable for many types of real samples.

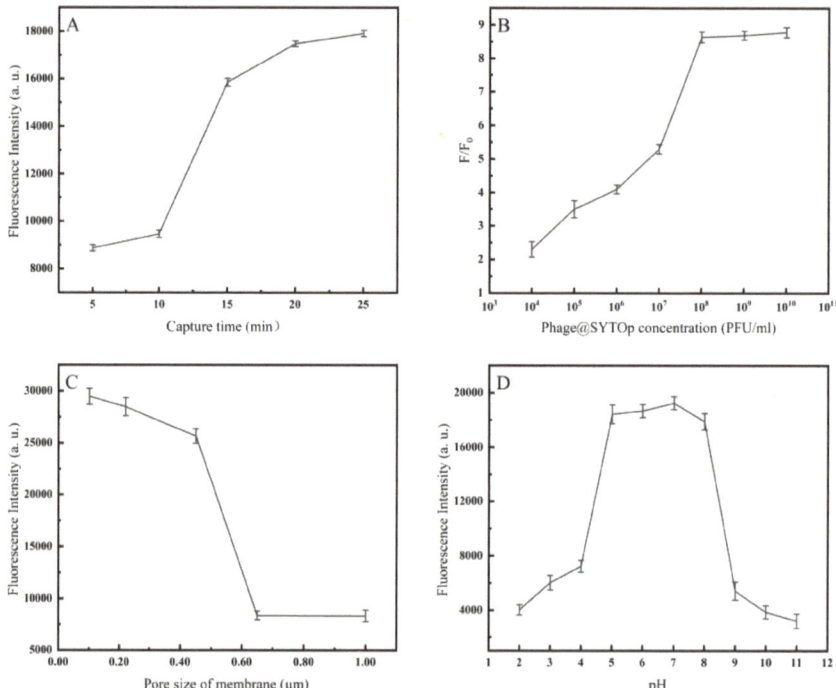

Figure 5. Optimization of experimental conditions: (**A**) phage@SYTO capture time; (**B**) phage@SYTO13 concentration; (**C**) pore size of membrane; (**D**) reaction pH.

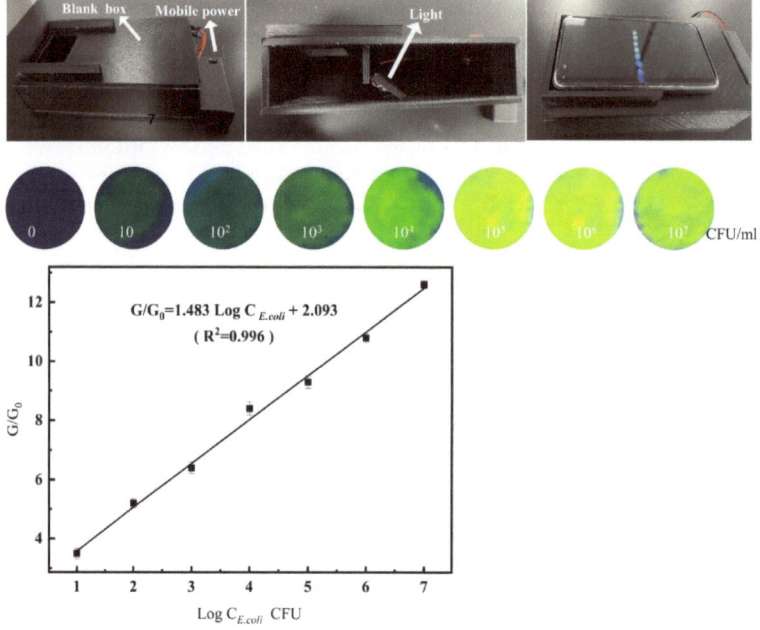

Figure 6. The calibration line of *E. coli* at different concentrations and its FL signals detected by a smartphone.

3.5. Determination of E. coli in the Real Samples

Under the optimal experimental conditions, *E. coli* was diluted with PBS to different concentrations (10^6, 10^5, 10^4, 10^3, 10^2, 10 CFU/mL) for visual fluorescence detection and subsequent microfluidic chip analysis. For visual fluorescence detection, we used different concentration gradients of *E. coli* Figure 6). It can be found that as the concentration of *E. coli* increased, the visualized fluorescence gradually increased. It was easy to observe 10^3 CFU/mL *E. coli* with the naked eye, and no fluorescence could be observed with the naked eye at 100 and 10 CFU/mL. Therefore, the lower limit of detection for visualizing fluorescence with the naked eye was 10^3 CFU/mL.

In addition, this method was compared with some reported *E. coli* detection methods (Table 1). Despite the shortcomings of the drop method, the prepared phage DNA probes still had a better detection range and lower limit of detection compared with the other methods. This is because phages are highly specific to the target bacteria. At the same time, rolling circle amplification (RCA) of the DNA on the phage can amplify the length of the DNA chain by tens of thousands of times, greatly reducing the detection limit.

Table 1. Comparison with *E. coli* detection methods in the literature.

Method	Linear Range (CFU/mL)	LOD (CFU/mL)	Time	Ref.
Nanogap network electrochemical method	10^3–10^8	10^2	>3 h	[28]
SERS	1.0×10^3–5.0×10^7	10^2	~30 min	[29]
The fiberoptic surface plasmon resonance method	1.5×10^3–1.5×10^5	5.0×10^2	45 min	[30]
Field environment monitoring	10^3–10^8	200	40 min	[31]
Fluorescence method	10–10^6	14	2 h	[32]
This work (fluorimeter)	10^3–10^6	300	45 min	[33]
This work	10^2–10^6	50	10 min	

3.6. Specificity, Stability, and Precision of the Sensor

To assess the method's anti-interfering capacity for *E. coli*, we included *Salmonella typhimurium* (*S.T*), *Staphylococcus aureus* (*S.A*), *Listeria monocytogenes* (*L.M*), *Vibrio vulnificus* (*V.V*), and *Vibrio parahaemolyticus* (*V.P*) in the comparison, and the concentration of the other bacteria was 10^7 CFU/mL, except for *E. coli*, which was 10^5 CFU/mL. As shown in Figure 7A, the difference in the signals indicated that other bacteria did not interfere with the detection of *E. coli*. In addition, some common ions in the sample solution were selected to perform anti-interference experiments. As shown in Figure 7B, the effects of the above substances on the detection of *E. coli* were negligible. The above results show that the phage sensor has a good specificity and anti-interference ability for the detection of the target bacteria. The assay was used to detect 10^6 CFU/mL of *E. coli* five times using the same batch of phage@SYTO probes. The RSD was lower than 7%, which proved that the sensor showed high stability. We also tested the same sample over 5 days of storage using the same batch of phage@SYTO probes, and the RSD was 5.8%. This indicated that the probe showed approximately 5 days of stability. Moreover, the inter- and intra-precision were detected using 10^6 CFU/mL of *E. coli*. The RSD values were 5.3% and 6.2%, respectively. All of these findings demonstrated that the sensor exhibited high precision.

To evaluate the potential application of this strategy, some samples mixed with different concentrations of *E. coli* were tested. The results are summarized in Table 2. The recovery rates of *E. coli* in the pork samples ranged from 95.2% to 102%. Compared with the ELISA method for the rapid detection of *E. coli*, the detection limit of this method is lower, which further illustrates the superiority of this method. The above conclusion demonstrates that the proposed sensing platform can be used for *E. coli* in food samples.

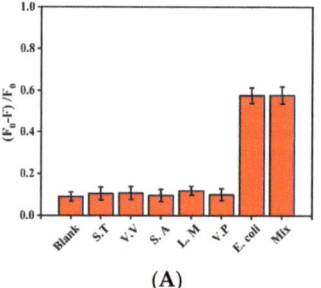 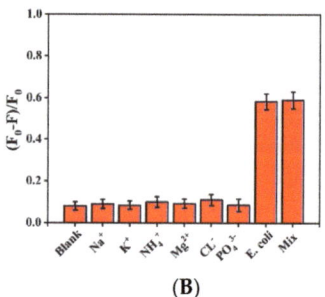

Figure 7. (**A**) The bacterial specificity based on paper chip analysis. (**B**) The resistance ability formicrofluidic chip bacteria analysis based on matrix interference.

Table 2. Detection of *E. coli* in real samples using the proposed method ($n = 5$).

Samples	Spiked (CFU/mL)	Measured (CFU/mL)	Recovery (%)	RSD (%)	ELISA (CFU/mL)
Fish 1	0	ND			ND
	10^3	$(0.95 \pm 0.06) \times 10^3$	95.2	6.5	ND
	10^5	$(0.98 \pm 0.05) \times 10^5$	98.1	5.4	$(1.18 \pm 0.10) \times 10^5$
Fish 2	0	ND			ND
	10^3	$(0.96 \pm 0.05) \times 10^3$	96.3	5.9	ND
	10^5	$(1.02 \pm 0.07) \times 10^5$	102	6.8	$(0.94 \pm 0.19) \times 10^5$
Shrimp 1	0	ND			ND
	10^3	$(1.01 \pm 0.05) \times 10^3$	101	4.7	ND
	10^5	$(0.99 \pm 0.06) \times 10^5$	99.4	6.1	$(1.10 \pm 0.14) \times 10^5$

ND: No detection.

4. Conclusions

In conclusion, the fabricated FL phage sensor can be used to POC quantify *E. coli* O157 bacteria in foods. Its particular merits illustrate the simplicity of its manipulation. It avoids the need for the employment of any precise instrumentation or trained persons. It also has sufficient sensitivity, a wide detection range, and high selectivity. The LOD for the bacteria is below 100 CFU/mL, which meets the standards established by governmental agencies for *E. coli* O157. Although it is comparable to ELISA, the new paper-chip-based platform has the potential to be developed as a simple, portable, sensitive, and low-cost analytical tool for immediate bacterial diagnosis and POC tests of food safety.

Author Contributions: Conceptualization, writing—original draft, investigation, validation, writing—review and editing, C.W.; methodology, conceptualization, supervision, writing, funding acquisition, N.G.; conceptualization, writing—review and editing, D.L.; writing—review and editing, Q.J. All authors have read and agreed to the published version of the manuscript.

Funding: This research was funded by the China Natural Science Foundation, grant number 21974074.

Institutional Review Board Statement: Not applicable.

Informed Consent Statement: Not applicable.

Data Availability Statement: Not applicable.

Acknowledgments: This work was supported by the National Natural Science Foundation of China (grant no. 21974074, Ningbo), Ningbo Public Welfare Technology Plan Project of China (grant no. 2021Z056, 2022S011), Zhejiang Provincial Natural Science Foundation (Y23B050013, Y23C200022), Major Scientific and Technological Tasks of Ningbo (grant no. 2022Z170), and the K. C. Wong Magna Fund of Ningbo University.

Conflicts of Interest: The authors declare no conflict of interest.

References

1. Shi, C.; Qing, X.; Yue, G.; Ling, J.; He, H. Luciferase-Zinc-Finger System for the Rapid Detection of Pathogenic Bacteria. *J. Agric. Food Chem.* **2017**, *65*, 6674–6681. [CrossRef] [PubMed]
2. Oyedeji, A.B.; Ezekiel, G.; Janet, A.A.; Opeolu, M.O.; Sefater, G.; Martins, A.A.; Samson, A.O.; Oluwafemi, A.A. Metabolomic Approaches for the Determination of Metabolites from Pathogenic Microorganisms: A Review. *Food Res. Int.* **2020**, *140*, 110042. [CrossRef] [PubMed]
3. Martinović, T.; Uroš, A.; Martina, Š.G.; Dina, R.; Djuro, J. Foodborne Pathogens and Their Toxins. *J. Proteom.* **2016**, *147*, 226–235. [CrossRef] [PubMed]
4. Rajapaksha, P.; Elbourne, A.; Gangadoo, S.; Brown, R.; Cozzolino, D.; Chapman, J. A Review of Methods for the Detection of Pathogenic Microorganisms. *Analyst* **2018**, *144*, 396–411. [CrossRef] [PubMed]
5. Carrillo, J.D.; Mayorquin, J.S.; Stajich, J.E.; Eskalen, A. Probe-Based Multiplex Real-Time PCR as a Diagnostic Tool to Distinguish Distinct Fungal Symbionts Associated with *Euwallacea kuroshio* and *Euwallacea whitfordiodendrus* in California. *Plant Dis.* **2019**, *104*, 227–238. [CrossRef] [PubMed]
6. Lv, X.; Huang, Y.; Liu, D.; Liu, C.; Shan, S.; Li, G.; Duan, M.; Lai, W. Multicolor and Ultrasensitive Enzyme-Linked Immunosorbent Assay Based on the Fluorescence Hybrid Chain Reaction for Simultaneous Detection of Pathogens. *J. Agric. Food Chem.* **2019**, *67*, 9390–9398. [CrossRef]
7. Richter, M.F.; Drown, B.S.; Riley, A.P.; Garcia, A.; Shirai, T.; Svec, R.L.; Hergenrother, P.J. Predictive compound accumulation rules yield a broad-spectrum antibiotic. *Nature* **2017**, *545*, 299–304. [CrossRef]
8. Zhang, Y.; Wu, Y.; Wu, Y.; Chang, Y.; Liu, M. CRISPR-Cas systems: From gene scissors to programmable biosensors. *Trends Anal. Chem.* **2021**, *137*, 116210. [CrossRef]
9. Shi, R.; Zou, W.; Zhao, Z.; Wang, G.; Guo, M.; Ai, S.; Zhou, Q.; Zhao, F.; Yang, Z. Development of a sensitive phage-mimotope and horseradish peroxidase based electrochemical immunosensor for detection of O,O-dimethyl organophosphorus pesticides. *Biosens. Bioelectron.* **2022**, *218*, 114748. [CrossRef]
10. Jiang, H.; Yang, J.; Wan, K.; Jiang, D.; Jin, C. Miniaturized Paper-Supported 3D Cell-Based Electrochemical Sensor for Bacterial Lipopolysaccharide Detection. *ACS Sens.* **2020**, *5*, 1325–1335. [CrossRef]
11. Baoqing, Z.; Qinghua, Y.; Fan, L.; Xinran, X.; Yuting, S.; Chufang, W.; Yanna, S.; Liang, X.; Jumei, Z.; Juan, W.; et al. CRISPR/Cas12a based fluorescence-enhanced lateral flow biosensor for detection of *Staphylococcus aureus*. *Sens. Actuators B Chem.* **2021**, *351*, 130906.
12. Bu, S.; Wang, K.; Wang, C.; Li, Z.; Hao, Z.; Liu, W.; Wan, J. Immunoassay for foodborne pathogenic bacteria using magnetic composites Ab@Fe$_3$O$_4$, signal composites Ap@PtNp, and thermometer readings. *Microchim. Acta* **2020**, *187*, 679. [CrossRef]
13. Dieckhaus, L.; Park, T.S.; Yoon, J.-Y. Smartphone-Based Paper Microfluidic Immunoassay of *Salmonella* and *E. coli*. *Methods Mol. Biol.* **2020**, *2182*, 83–101.
14. Liu, S.; Lu, S.; Sun, S.; Hai, J.; Meng, G.; Wang, B. NIR II Light-Response Au Nanoframes: Amplification of a Pressure- and Temperature-Sensing Strategy for Portable Detection and Photothermal Therapy of Cancer Cells. *Anal. Chem.* **2021**, *93*, 14307–14316. [CrossRef]
15. Lee, W.-I.; Park, Y.; Shrivastava, S.; Jung, T.; Meeseepong, M.; Lee, J.; Jeon, B.; Yang, S.; Lee, N.-E. A fully integrated bacterial pathogen detection system based on count-on-a-cartridge platform for rapid, ultrasensitive, highly accurate and culture-free assay. *Biosens. Bioelectron.* **2020**, *152*, 112007. [CrossRef] [PubMed]
16. Laliwala, A.; Svechkarev, D.; Sadykov, M.R.; Endres, J.; Bayles, K.W.; Mohs, A.M. Simpler Procedure and Improved Performance for Pathogenic Bacteria Analysis with a Paper-Based Ratiometric Fluorescent Sensor Array. *Anal. Chem.* **2022**, *94*, 2615–2624. [CrossRef] [PubMed]
17. Shin, J.H.; Hong, J.; Go, H.; Park, J.; Kong, M.; Ryu, S.; Kim, K.-P.; Roh, E.; Park, J.-K. Multiplexed Detection of Foodborne Pathogens from Contaminated Lettuces Using a Handheld Multistep Lateral Flow Assay Device. *J. Agric. Food Chem.* **2017**, *66*, 290–297. [CrossRef] [PubMed]
18. Luo, K.; Kim, H.-Y.; Oh, M.-H.; Kim, Y.-R. Paper-based lateral flow strip assay for the detection of foodborne pathogens: Principles, applications, technological challenges and opportunities. *Crit. Rev. Food Sci. Nutr.* **2018**, *60*, 157–170. [CrossRef] [PubMed]
19. Li, W.; Coulon, F.; Singer, A.; Zhu, Y.-G.; Yang, Z. Paper-Based Devices as a New Tool for Rapid and on-Site Monitoring of "Superbugs". *Environ. Sci. Technol.* **2021**, *55*, 12133–12135. [CrossRef]
20. Yizhong, S.; Yunlong, W.; Chunlei, Z.; Jinxuan, C.; De-Man, H. Ratiometric fluorescent signals-driven smartphone-based portable sensors for onsite visual detection of food contaminants. *Coord. Chem. Rev.* **2022**, *458*, 214442.
21. Lin, B.; Guan, Z.; Song, Y.; Song, E.; Lu, Z.; Liu, D.; An, Y.; Zhu, Z.; Zhou, L.; Yang, C. Lateral flow assay with pressure meter readout for rapid point-of-care detection of disease-associated protein. *Lab A Chip* **2018**, *18*, 965–970. [CrossRef] [PubMed]

22. Ji, T.; Liu, D.; Liu, F.; Li, J.; Ruan, Q.; Song, Y.; Tian, T.; Zhu, Z.; Zhou, L.; Lin, H.; et al. A pressure-based bioassay for the rapid, portable and quantitative detection of C-reactive protein. *Chem. Commun.* **2016**, *52*, 8452–8454. [CrossRef]
23. Wei, X.; Tian, T.; Jia, S.; Zhu, Z.; Ma, Y.; Sun, J.; Lin, Z.; Yang, C.J. Microfluidic Distance Readout Sweet Hydrogel Integrated Paper-Based Analytical Device (µDiSH-PAD) for Visual Quantitative Point-of-Care Testing. *Anal. Chem.* **2016**, *88*, 2345–2352. [CrossRef] [PubMed]
24. Torres-Barceló, C. The disparate effects of bacteriophages on antibiotic-resistant bacteria. *Emerg. Microbes Infect.* **2018**, *7*, 168. [CrossRef]
25. Sha, L.; Xuliang, H.; Tao, Z.; Kaixuan, Z.; Changhu, X.; Zengrui, T.; Lian, J.; Nongyue, H.; Yan, D.; Song, L.; et al. Highly sensitive smartphone-based detection of Listeria monocytogenes using SYTO9. *Chin. Chem. Lett.* **2021**, *33*, 1933–1935.
26. You, S.-M.; Jeong, K.-B.; Luo, K.; Park, J.-S.; Park, J.-W.; Kim, Y.-R. Paper-based colorimetric detection of pathogenic bacteria in food through magnetic separation and enzyme-mediated signal amplification on paper disc. *Anal. Chim. Acta* **2021**, *1151*, 338252. [CrossRef] [PubMed]
27. Yuwei, R.; Lulu, C.; Xiyan, Z.; Rui, J.; Dexin, O.; Yang, W.; Danfeng, Z.; Yizhong, S.; Na, L.; Yingwang, Y. A novel fluorescence resonance energy transfer (FRET)-based paper sensor with smartphone for quantitative detection of Vibrio parahaemolyticus. *Food Control* **2022**, *145*, 109412.
28. Lei, J.; Shi, L.; Li, B.; Yang, C.J.; Jin, Y. Ultrasensitive and portable assay of mercury (II) ions via gas pressure as readout. *Biosens. Bioelectron.* **2018**, *122*, 32–36. [CrossRef]
29. Niazi, M.; Alizadeh, E.; Zarebkohan, A.; Seidi, K.; Ayoubi-Joshaghani, M.H.; Azizi, M.; Dadashi, H.; Mahmudi, H.; Javaheri, T.; Jaymand, M.; et al. Advanced Bioresponsive Multitasking Hydrogels in the New Era of Biomedicine. *Adv. Funct. Mater.* **2021**, *31*, 2104123. [CrossRef]
30. Jung, I.Y.; Kim, J.S.; Choi, B.R.; Lee, K.; Lee, H. Hydrogel Based Biosensors for In Vitro Diagnostics of Biochemicals, Proteins, and Genes. *Adv. Healthc. Mater.* **2017**, *6*, 1601475. [CrossRef]
31. Pinelli, F.; Magagnin, L.; Rossi, F. Progress in hydrogels for sensing applications: A review. *Mater. Today Chem.* **2020**, *17*, 100317. [CrossRef]
32. Liang, Y.; He, J.; Guo, B. Functional Hydrogels as Wound Dressing to Enhance Wound Healing. *ACS Nano* **2021**, *15*, 12687–12722. [CrossRef] [PubMed]
33. He, X.; Yang, Y.; Guo, Y.; Lu, S.; Du, Y.; Li, J.-J.; Zhang, X.; Leung, N.L.C.; Zhao, Z.; Niu, G.; et al. Phage-Guided Targeting, Discriminated Imaging and Synergistic Killing of Bacteria by AIE Bioconjugates. *J. Am. Chem. Soc.* **2020**, *142*, 3959–3969. [CrossRef] [PubMed]

Disclaimer/Publisher's Note: The statements, opinions and data contained in all publications are solely those of the individual author(s) and contributor(s) and not of MDPI and/or the editor(s). MDPI and/or the editor(s) disclaim responsibility for any injury to people or property resulting from any ideas, methods, instructions or products referred to in the content.

Article

Application of an Electronic Nose Technology for the Prediction of Chemical Process Contaminants in Roasted Almonds

Marta Mesías [1], Juan Diego Barea-Ramos [2], Jesús Lozano [3,*], Francisco J. Morales [1] and Daniel Martín-Vertedor [2]

[1] Institute of Food Science, Technology and Nutrition (ICTAN-CSIC), Jose Antonio Novais 6, 28040 Madrid, Spain; mmesias@ictan.csic.es (M.M.); fjmorales@ictan.csic.es (F.J.M.)
[2] Technological Institute of Food and Agriculture (CICYTEX-INTAEX), Junta de Extremadura, Avda. Adolfo Suárez s/n, 06007 Badajoz, Spain; juandiego.barea@juntaex.es (J.D.B.-R.); daniel.martin@juntaex.es (D.M.-V.)
[3] Industrial Engineering School, University of Extremadura, 06006 Badajoz, Spain
* Correspondence: jesuslozano@unex.es

Abstract: The purpose of this study was to investigate the use of an experimental electronic nose (E-nose) as a predictive tool for detecting the formation of chemical process contaminants in roasted almonds. Whole and ground almonds were subjected to different thermal treatments, and the levels of acrylamide, hydroxymethylfurfural (HMF) and furfural were analysed. Subsequently, the aromas were detected by using the electronic device. Roasted almonds were classified as positive or negative sensory attributes by a tasting panel. Positive aromas were related to the intensity of the almond odour and the roasted aroma, whereas negative ones were linked to a burnt smell resulting from high-intensity thermal treatments. The electronic signals obtained by the E-nose were correlated with the content of acrylamide, HMF, and furfural ($R^2_{CV} > 0.83$; $R^2_P > 0.76$ in whole roasted almonds; $R^2_{CV} > 0.88$; $R^2_P > 0.95$ in ground roasted almonds). This suggest that the E-nose can predict the presence of these contaminants in roasted almonds. In conclusion, the E-nose may be a useful device to evaluate the quality of roasted foods based on their sensory characteristics but also their safety in terms of the content of harmful compounds, making it a useful predictive chemometric tool for assessing the formation of contaminants during almond processing.

Keywords: roasted almond; sensory analysis; electronic nose; acrylamide; hydroxymethylfurfural; furfural

1. Introduction

The temperatures applied during the production and cooking of foods can have a significant impact on the quality and nutritional aspects of these products. Such changes are related to the development of the Maillard reaction, which leads to the formation of compounds responsible for the taste, aroma, and appearance of food products [1], as well as desirable compounds with potential health benefits [2]. However, concurrently, other harmful and undesirable compounds may also be generated, such as acrylamide and some emerging process contaminants, including hydroxymethylfurfural (HMF) and furfural [3,4]. Acrylamide is considered a potential carcinogen for humans by the International Agency for Research on Cancer [5]. It is formed when free asparagine and reducing sugars react at temperatures above 120 °C. The European Food Safety Agency (EFSA) has identified the presence of acrylamide in food as a public health concern [6]. HMF and furfural are intermediate products of the Maillard reaction, and HMF can also be produced by the caramelization of sugars at high temperatures [7,8]. These compounds have been evaluated as emerging processing contaminants due to the potential genotoxic and mutagenic effects of HMF [9–11] and the association of furfural with hepatotoxicity outcomes [12].

Normalized analytical protocols are applied to monitor the occurrence of regulated chemical contaminants in processed foods. These methods involve robust, highly specialized instrumentation, time-consuming and expensive procedures, and predominating chromatographic techniques, such as GC–MS, GC–MS-MS, HPLC–MS, and LC–MS–MS [13]. To facilitate the determination, there is an increasing demand for rapid and precise alternative methods that allow a fast screening of these harmful compounds in foods. In this regard, the European regulation for acrylamide allows its analysis to be replaced by measuring product attributes or process parameters as long as a statistical correlation can be demonstrated between them and the acrylamide level [14]. Most of these methodologies rely on linking the colour of processed foods to the acrylamide levels [15,16]. However, few studies have established a connection between the formation of chemical process contaminants and the development of aromatic compounds [17].

The electronic nose (E-nose) is a cost-effective and powerful electronic device able to conduct fast measurements to discriminate flavours. It is widely used in many disciplines as a commodity, such as environmental, medicinal and food sciences, mainly with the aim of evaluating the quality of the products and detecting off-flavours that can be attributed to food deterioration [18,19]. Since molecules responsible for the aromas released during food processing are recognizable by the human nose, it may be possible to use the E-nose to detect changes that occur in processed food products. Moreover, due to the relationship between the formation of aromatic compounds and the generation of chemical contaminants during processing, the E-nose could also be used as a chemometric tool to predict the formation of the toxic compounds. In a previous study, Martín-Tornero et al. [17] used an E-nose to analyse the aroma of table olives, demonstrating the effectivity of this device to indirectly quantify the acrylamide levels in these foods. Based on these results, it would be worthwhile to verify the effectiveness of the E-nose in other food matrices.

Almonds are among the most consumed tree nuts in the world [20], and they are recommended as a healthy food and a source of beneficial constituents for human health [21]. Nevertheless, they contain amino acids and reducing sugars in appreciable amounts, and therefore, the roasting temperature and physical form (whole, sliced, and cut) of the almond kernel strongly promote the development of the Maillard reaction, leading then to the formation of aromatic compounds but also to the generation of chemical process contaminants [22]. Previous studies have examined the development of acrylamide and furanic compounds during the roasting of almonds at various temperature and time settings [22–25]. However, none of them have explored the possible correlation between these compounds and the resulting aroma, which could be measured with chemometric tools such as the electronic nose. In this context, the objective of this work was to evaluate the feasibility of using an E-nose as a predictive tool for the formation of acrylamide and furanic compounds in almonds that were processed under different roasting conditions and in different structural formats (whole and ground). The findings would provide a straightforward and efficient approach to manage the development of contaminants during the processing of this particular food matrix.

2. Materials and Methods
2.1. Reagents and Chemicals

Acrylamide standard (99%) [CAS. 79-06-01], potassium hexacyanoferrate (II) trihydrate (98%, Carrez-I) [CAS. 14459-95-1], and zinc acetate dihydrate (>99%, Carrez-II) [CAS. 5970-45-6] were purchased from Sigma (St. Louis, MO, USA). Formic acid (98%) [CAS. 64-18-6], methanol (99.5%) [CAS. 67-56-1] and hexane [CAS. 110-54-3] were obtained from Panreac (Barcelona, Spain). [$^{13}C_3$]-acrylamide (isotopic purity 99%) [CAS.287399-26-2] was obtained from Cambridge Isotope Labs (Andover, MA, USA). The Milli-Q water used was produced using an Elix 3 Millipore water purification system coupled to a Milli-Q module (model Advantage A10) (Millipore, Molsheim, France). All other chemicals, solvents, and reagents were of analytical grade.

2.2. Samples and Roasting Treatments

Two formats (whole and ground) of raw almonds (Prunus dulcis) from the Marcona cultivar were purchased from Spanish markets between March/2022 and April/2022. Fifteen bags of the same batch were acquired for each format to ensure that there was a sufficient quantity of the same sample to conduct the complete kinetic study of the roasting experiments.

Roasting experiments were carried out in a convective and air-forced oven (Memmert UFE400, Scwabach, Germany). Two batches of whole almonds (60 g) and ground almonds (30 g) were roasted simultaneously in aluminium trays (21.9 × 15.8 × 3.8 cm; length × wide × height). Samples were placed on the tray laying in a uniform layer. A calibrated K-type thermocouple data logger (Delta Ohm, Caselle di Selvazzano, Italy) was used to monitor and register the temperature of hot air within and above the almonds (°C/s). Samples were roasted at 120, 135, 150, 165, 180, and 200 °C for 20 min. Temperature/time combinations would resemble two levels at moderate and three at more intense heat load above a reference. The temperature/time combination at 150 °C/20 min was selected as reference. After roasting, samples were cooled until the temperature within the tray stayed below 30 °C. A portion was placed in a sealed bottle and kept under refrigeration (4–6 °C) until analysis.

2.3. Analysis

2.3.1. Sensory Analysis

Sensory analysis was carried out by a tasting committee made up of eight tasters in the tasting room of the Research Centre of Extremadura (CICYTEX). The panel members were trained to evaluate agri-food products. The samples, consisting of 3 g of whole and ground roasted almonds, were placed in a standard glass on a heating plate at 25 °C. The tasting panel identified almond and roasted aromas as positive attributes, while burnt attributes were classified as negative in both whole and ground almond samples. The roasted aroma was evaluated on a scale of 0 to 10 points based on its intensity, with scores categorized as either positive or negative. The test was considered valid only if the coefficient of variation was less than 20% [19].

2.3.2. LC-ESI-MS/MS Determination of Acrylamide

Liquid chromatography-electrospray ionization-tandem mass spectrometry (LC-ESI-MS/MS) was used to determine the levels of acrylamide, following the method described by González-Mulero et al. [26]. Acrylamide standard was added to the samples to determine the recovery rate, obtaining values between 95% and 107%. Relative standard deviations (RSD) for precision (2.8%), repeatability (1.2%), and reproducibility (2.5%) were calculated to confirm the precision of the analysis. Analytical determination was checked through the analysis of different food matrices provided as proficiency tests by the Food Analysis Performance Assessment Scheme (FAPAS) program. The z-scores obtained from these tests demonstrated the accuracy of the results: −0.1 for crispbread (test 30,118, January 2022) and 0.4 for potato crisps (test 30,120, April 2022). The limit of quantitation (LOQ) was set at 15 µg/kg, and the results were expressed as µg/kg sample.

2.3.3. HPLC Determination of Hydroxymethylfurfural and Furfural

HMF and furfural were determined by HPLC according to Mesias et al. [27]. The LOQ for HMF and furfural were estimated to be 0.6 mg/kg and 0.3 mg/kg, respectively, and the results were expressed as mg/kg sample.

2.3.4. E-Nose Analysis

Researchers from the School of Industrial Engineering at the University of Extremadura were the designers of the handmade electronic device (E-nose). This device contains four electronic chips with a total of eleven sensors that emit different signals related to the aroma of the sample headspace [28]. Samples (~3 g) were placed in the tasting glasses in the same way as the tasting panel and E-nose was placed on the top of the cup. In a first phase, known as the adsorption phase, the E-nose was allowed to adsorb the headspace of the sample for 60 s. Once this time elapsed, the electronic device was moved to an empty cup to carry out the desorption phase for 30 s to obtain the baseline. The electronic nose received data in one-second intervals, and eight measurements were carried out for each roasted almond sample. Data analysed were transferred from the E-nose to a smart device via Bluetooth, and a multivariate analysis was performed.

2.4. Chemometric Analysis

Different methods were performed using MATLAB software and PLS_Toolbox: (i) Principal component analysis (PCA); (ii) Partial least squares (PLS). First, a PCA was performed to show the discrimination between the different samples based on the profile of volatile components [29]. The original variables were autoscaled because the study variables were measured in different units. Finally, for the detection of acrylamide, furfural, and hydroxymethylfurfural with the E-nose data, a PLS model was carried out [30].

2.5. Statistical Analysis

Statistical analyses were performed using SPSS version 18.0 (SPSS Inc., Chicago, IL, USA). Data were expressed as mean ± standard deviation (SD). One-way ANOVA followed by Tukey's method, or Student's *t*-test, was used to identify the overall significance of differences. The homogeneity of variances was determined using Levene's test. All statistical parameters were evaluated at $p < 0.05$ significance level.

3. Results and Discussion

3.1. Sensory Analysis

A tasting committee made up of eight tasters conducted a sensory analysis for the discrimination of the aromas of the roasted almond samples (Table 1). The samples were classified as having positive or negative attributes. Positive aroma was associated with the intensity of the almond odour and the roasted aroma caused by the thermal treatment. Conversely, the negative aroma was related to the burnt smell resulting from the over-roasting of the samples. It should be noted that the aroma of almond in both whole and ground samples was more intense when roasted at 150 °C, with a score of 3.8 for whole almonds and 3.2 for ground almonds. From this treatment onwards, the almond aroma started to decrease until it was undetectable in samples roasted at 200 °C. Regarding the roasted odour, it was not detected until Treatment 3 (heating at 150 °C), being of lower intensity in ground almonds. In this case, both type of samples showed the highest intensity at temperatures of 165 °C, with 3.5 points for whole almonds and 2.0 for ground almonds, decreasing with more intense treatments. Again, samples roasted at 200 °C did not exhibit a roasted aroma. In contrast to the positive attributes, burnt aroma was not present in samples heated at lower temperatures, but it was detected when the temperature that was applied was 180 °C, and it reached the maximum value at 200 °C (4.5 points in whole almonds and 4.0 points in ground samples). In conclusion, it can be inferred that most of the whole and ground almond samples roasted up to 165 °C exhibited the most desirable aromas, without any burnt notes.

In a similar study, Schlörmann et al. [31] conducted a sensory evaluation of roasted nuts assigning points ranging from five (no deviation of expected quality) to one (clear defects) for attributes such as appearance, texture, odour, and taste. Nuts roasted at low (120–140 °C) or middle temperatures (140–160 °C) obtained the highest scores (4.8 and 4.7, respectively), followed by those heated at temperatures between 160–180 °C (4.4). In contrast, a hedonic analysis established points between one (excellent) and five (highly disliked). In this case, nuts roasted at low or middle temperatures received the best scores (1.8 and 2.1, respectively), whereas higher temperatures resulted in lower scores (2.8). Based on these results, temperatures of up to 160 °C were recommended for obtaining almond products suitable for consumption, as higher roasting conditions may result in bitter and over-roasted flavours. Amrein et al. [22] also noted that 165 °C is the upper limit for almond roasting, due to the formation of a bitter and over-roasted flavour, in agreement with findings of the present study and with those reported by other authors [23].

Table 1. Descriptive sensory olfactory evaluation of roasted whole and ground almond aroma evaluated by the tasting panel. All samples were roasted at different temperatures for 20 min.

Treatments	T (°C)	Whole Almond Aroma			Ground Almond Aroma		
		Almond	Roasted	Burnt	Almond	Roasted	Burnt
1	120	2.0 ± 0.1 b	<LOD	<LOD	1.0 ± 0.1 a	<LOD	<LOD
2	135	3.5 ± 0.2 c	<LOD	<LOD	2.0 ± 0.2 c	<LOD	<LOD
3	150	3.8 ± 0.2 c	2.5 ± 0.2 b	<LOD	3.2 ± 0.2 e	1.5 ± 0.2 b	<LOD
4	165	3.0 ± 0.3 c	3.5 ± 0.2 c	<LOD	2.5 ± 0.3 d	2.0 ± 0.2 c	<LOD
5	180	1.5 ± 0.2 a	1.5 ± 0.3 a	3.0 ± 0.3 a	1.5 ± 0.2 b	1.0 ± 0.3 a	2.5 ± 0.3 a
6	200	<LOD	<LOD	4.5 ± 0.2 b	<LOD	<LOD	4.0 ± 0.2 b

Results are expressed as mean ± standard deviation (SD) of three replicates. Different lowercase letters mean statistically significant differences between samples ($p < 0.05$). LOD: limit of detection.

3.2. Effect of Thermal Treatments on Chemical Process Contaminant Content in Roasted Almonds

The temperature-dependent formation of acrylamide, HMF, and furfural in roasted almond is shown in Table 2. As expected, an increase in temperature promoted the formation of the chemical process contaminants during roasting. Several studies have suggested that the formation of acrylamide in almonds is primarily influenced by the roasting temperature rather than the duration of the thermal treatment [22–24]. In that sense, an increase of 15 °C resulted in an almost 3-fold increase in acrylamide content in whole almonds and up to 2.5-fold in ground almonds. Whole and ground almonds exhibited different profiles for the acrylamide levels, ranging from 25 to 466 µg/kg in whole samples and from 19 to 397 µg/kg in ground nuts. Acrylamide concentration in whole almonds reached the highest level when roasted at 165 °C and drastically ($p < 0.05$) decreased to 285 and 170 µg/kg when the temperature was raised to 180 and 200 °C, respectively. In the case of ground almonds, the maximum formation was observed at 180 °C, decreasing to 301 µg/kg with roasting at 200 °C. This type of kinetic pattern was similar to that observed in roasted coffee, sesame seeds or chia seeds, as reported in previous studies [32–34]. As is known, during the thermal treatment of foods, both the formation and the elimination of acrylamide occur simultaneously. Acrylamide formation will predominate in the presence of its precursors (asparagine and carbonyl compounds), while its elimination becomes dominant in the absence of any precursor. Almonds contain acrylamide precursors in high quantity in the range of 98–641 mg/100 g for free asparagine [35], 3100–4680 mg/100 g for sucrose, and 70–134 mg/100 g for the total content of glucose and fructose [36]. Among them, strong correlations between asparagine levels and acrylamide formation in almonds have been reported, suggesting that asparagine may be the limiting factor in this food [22,24,25]. Based on the results, it can be inferred that at temperatures of 165 °C (for whole almonds) and 180 °C (for ground almonds), the precursor content (particularly asparagine) would be significantly reduced, leading to a higher prevalence of elimination over the formation of acrylamide, resulting in decreased levels of the contaminant in the roasted samples.

The acrylamide content reported by other authors has shown high variability, probably due to the different content of the precursor associated with the almond variety or even with the growing region and harvest year in addition to the processing conditions. Similar vales have been described in almonds experimentally roasted for 22.5 min at 130 °C (94 µg/kg), but higher after roasting for 20 min at 145 °C (around 500 µg/kg) and 165 °C (more than 1500 µg/kg). Industrial samples heated for 22 min at 150 °C also exhibited greater values (885 µg/kg) [22]. In contrast, Lasekan and Abbas [37] presented results between 42 and 65 µg/kg in almonds roasted at 150 °C for 15 and 25 min, respectively, increasing only up to 60 and 86 µg/kg at 200 °C. Regardless of the levels reached, all the studies found in the literature indicate that acrylamide formation begins when the temperature exceeds 130 °C [25], with low levels at 145 °C, and reaching a maximum in the range of 165–180 °C, decreasing afterward [24], and leading to unacceptable bitter off-flavours [25], in agreement with observations of this research.

Table 2. Acrylamide (μg/kg), HMF (mg/kg) and furfural (mg/kg) content of whole and ground almonds roasted under different temperatures for 20 min. Descriptive sensory olfactory evaluation of roasted whole and ground almonds aromas evaluated by the tasting panel. All samples were roasted at different temperatures (range 120–200 °C) for 20 min.

Treatments	T (°C)	Whole Almond			Ground Almond		
		Acrylamide	HMF	Furfural	Acrylamide	HMF	Furfural
1	120	25 ± 0.71 a	0.99 ± 0.02 a	0.35 ± 0.02 a	19 ± 0.01 a	0.75 ± 0.07 a	0.42 ± 0.02 a
2	135	82 ± 4.24 b	4.21 ± 0.10 ab	0.72 ± 0.01 a	51 ± 2.12 a	4.96 ± 0.04 b	1.04 ± 0.01 a
3	150	166 ± 2.12 c	11.09 ± 0.51 b	1.53 ± 0.02 a	128 ± 0.71 b	20.28 ± 0.05 c	2.22 ± 0.09 a
4	165	466 ± 20.51 e	22.58 ± 0.83 c	3.65 ± 0.22 b	265 ± 8.49 c	53.32 ± 0.37 d	6.16 ± 0.03 b
5	180	285 ± 0.01 d	100.14 ± 3.39 d	13.21 ± 0.71 c	397 ± 24.04 d	145.58 ± 1.15 e	11.94 ± 0.97 c
6	200	170 ± 0.71 c	98.92 ± 1.43 d	23.85 ± 0.36 d	301 ± 14.14 c	149.77 ± 0.98 f	13.32 ± 0.11 c

Results are expressed as mean ± standard deviation (SD) of three replicates. Different lowercase letters mean statistically significant differences between samples ($p < 0.05$).

The formation of HMF and furfural increased during roasting at all temperatures, except for whole almonds roasted at 200 °C, where it slightly decreased. At the lowest temperatures (120 °C), the levels of these compounds were relatively low, but they significantly increased as the temperature rose to 200 °C. In whole almonds, HMF concentrations ranged between 0.99 and 100.14 mg/kg, with maximum formation at 180 °C, and decreased to 98.92 mg/kg at 200 °C. For ground almonds, HMF levels increased from 0.75 to 149.77 mg/kg, with maximum values at the maximum temperature. Furfural concentrations in whole samples increased from 0.35 to 23.85 mg/kg, and in ground samples from 0.42 to 13.32 mg/kg. The results for HMF were within the range reported for roasted commercial samples (10.52–49.37 mg/kg) [38].

Comparing the two sample formats, ground almonds promoted a greater formation of HMF, reaching maximum levels 200 times higher than those at 120 °C. In contrast, the largest increase for furfural was observed in the whole almonds, with concentrations almost 70 times higher than those at 120 °C. Previous studies have reported that the grinding increases the accessibility of constituents of the inner part of chia seeds, and consequently, their reactivity to promote the formation of acrylamide [34]. This may explain why roasted ground almonds showed the highest levels of furanic compounds compared to whole almonds. However, this was not observed in the acrylamide results. In this regard, it is necessary to point out that the raw materials were not the same, and therefore, they could present a different composition, with differences in the levels of precursors, leading to varying levels of acrylamide. To analyse the content of precursors in the raw material was not the objective of this work; however, it is known that the levels of free asparagine in almonds can vary considerably and to a larger extent than those of the sugars [22], which could explain the variability in the acrylamide levels and the highest concentrations for furanic compounds in ground samples.

3.3. Discrimination of Roasted Almonds by Using E-Nose

Roasted almonds submitted to different thermal treatments were analysed by an electronic devise. The following algorithm feature was used to obtain the raw data for the E-nose: maximum signal value minus minimum, plus 100, and minus one. Then, Principal Component Analysis (PCA) was carried out to observe the clusters formed in the different thermal treatments on both roasted whole and ground almonds (Figure 1). In the two groups of samples, more than 90% of the total variance was explained by PC1. The clustering of the results in both PCA showed that the E-nose is able to discriminate the different thermal treatments. In ground almonds the clusters are better differentiated. However, in the PCA of whole almonds, samples roasted at 200 °C (Treatment 6) are far behind the rest of the treatments, in the negative region. Almonds heated at 165 and 180 °C (Treatments 4 and 5) are found in the quadrant of PC1 positive and PC2 positive, whereas those roasted at the lowest temperatures (120 and 135 °C, Treatments 1 and 2) are in PC1 positive and PC2 negative. It should be noted that these analyses support the tasting panel as neither the almond odour nor the roasted one was detected in Treatment 6 (200 °C), in which the burnt aroma appears with a high intensity. These results demonstrate the effectiveness of the electronic nose to differentiate the quality of the almonds according to the degree of roasting, allowing an adequate classification of these products. Previous studies have also reported the successful use of the electronic olfaction in food quality control, such as in the analysis of edible oils to detect adulteration and deterioration caused by external factors [39–41] or in roasted coffee samples to differentiate the degree of roasting [28,42].

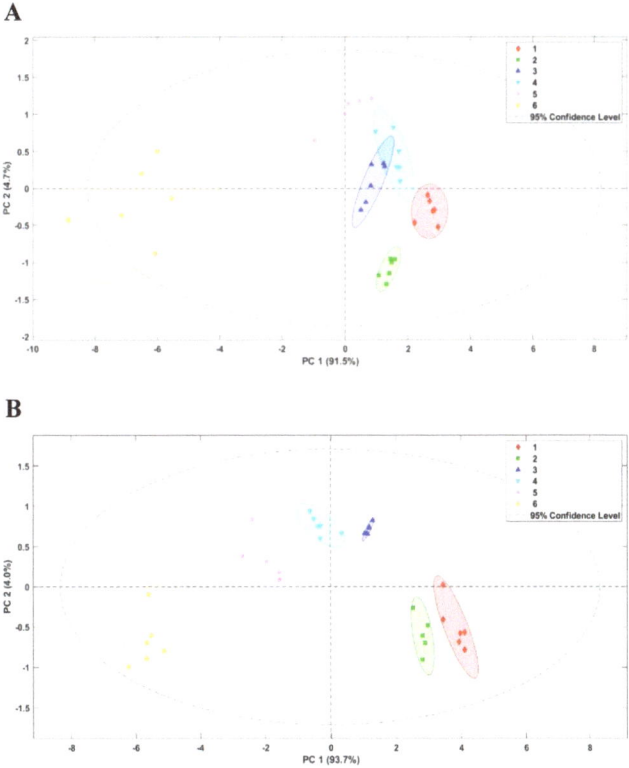

Figure 1. Score plot of the Principal Component Analysis (PCA) for roasted whole (**A**) and ground (**B**) almond aroma submitted to different thermal treatments (1–6), at different temperatures for 20 min. Treatment 1, 120 °C; Treatment 2, 135 °C; Treatment 3, 150 °C; Treatment 4; 165 °C; Treatment 5, 180 °C; Treatment 6, 200 °C.

3.4. Prediction of the Presence of Chemical Process Contaminants in Roasted Almonds by E-Nose

A PLS model was stablished to correlate the data obtained by E-nose and the presence of chemical process contaminants in the roasted almonds (Figure 2). A total of 70% of the sample data were used for calibration and cross-validation of the models, whereas the remaining 30% were used for testing the strength and precision of the generated models. Three distinct models were built for both whole and ground roasted almonds, considering the three compounds evaluated in the present study (acrylamide, HMF, and furfural). In the case of whole roasted almonds, the R^2_{CV} value was 0.89 for the model developed for acrylamide and furfural and 0.83 for HMF. Higher values were found for ground roasted almonds, with results of 0.99 for acrylamide, 0.98 for HMF, and 0.88 for furfural. The RMSECV values were also estimated, being 61.1, 18.4, and 3.0, for acrylamide, HMF, and furfural, respectively, in whole roasted almonds, and 14.0, 9.8, and 1.8 for ground roasted samples. R_P^2 and RMSEP are parameters related to the validation methods. Regarding the R_P^2 values, they were 0.76 for acrylamide, 0.85 for HMF and 0.94 for furfural in whole roasted almonds, increasing up to 0.99, 0.97, and 0.95, respectively, in ground roasted almonds. Finally, the RMSEP values for whole almonds were 78.8, 18.6, and 2.6 for ground almonds and 17.7, 11.3, and 1.2 for the three compounds, respectively. The results show a good correlation between the experimental and the predicted values, suggesting that this model is suitable for prediction and allows the quantification of the chemical process contaminants in roasted almonds.

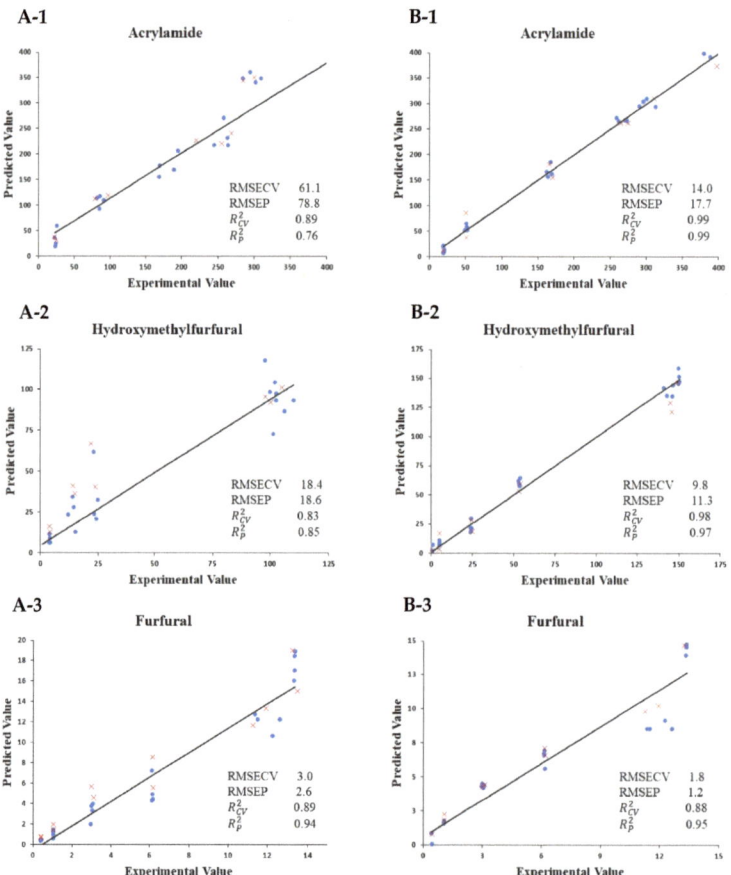

Figure 2. Experimental values for acrylamide (**1**), HMF (**2**) and furfural (**3**) against PLS cross-validation predictions (•) and validation set predictions (×) for roasted whole (**A**) and ground (**B**) almond aromas submitted to different thermal treatments (1–6), at different temperatures for 20 min. Treatment 1, 120 °C; Treatment 2, 135 °C; Treatment 3, 150 °C; Treatment 4; 165 °C; Treatment 5, 180 °C; Treatment 6, 200 °C.

Until now, the use of the electronic nose as a predictive device for the formation of chemical contaminants during processing has been very rare. In a previous study, Martín-Tornero et al. [17] characterized the polyphenol and volatile fractions of Californian-style black olives and demonstrated the innovative application of the E-nose in determining the acrylamide content in these foods. These positive results together with those found in the present study encourage the continuation of the usage of the E-nose not only to control the quality of food but also as a quick and effective chemometric tool to predict the formation of contaminants during processing.

4. Conclusions

The processing of foods promotes the development of the Maillard reaction, leading to the formation of compounds that contribute to the sensory properties of foods, including taste, aroma, and the appearance of foods. However, this reaction can also produce undesirable compounds, known as chemical process contaminants. The determination of compounds such as acrylamide, HMF and furfural requires robust, highly specialized instrumentation, and time-consuming and expensive procedures. Therefore, there is a need to develop other rapid and precise alternative methods that allow a fast screening of these harmful compounds in foods. In the present study, an experimental electronic nose was tested to detect and discriminate between the content of acrylamide, HMF, and furfural in whole and ground almonds roasted under different thermal treatments. The results showed that the E-nose was able to discriminate the positive and negative aromas of different samples and to predict the levels of chemical process contaminants in the roasted almonds. In conclusion, the use of this electronic device, together with a suitable statistical model, allows for the assessment of the quality of roasted foods according to their sensory characteristics, as well as their safety regarding harmful compounds. Therefore, it can be used as a predictive chemometric tool for the formation of contaminants during almond processing.

Author Contributions: Formal analysis, M.M., F.J.M., J.D.B.-R., J.L. and D.M.-V.; investigation, M.M., F.J.M. and D.M.-V.; supervision, M.M., F.J.M. and D.M.-V., writing—original draft, M.M., J.D.B.-R. and D.M.-V.; writing—review and editing, F.J.M., J.L. and D.M.-V.; funding acquisition, F.J.M. and D.M.-V. All authors have read and agreed to the published version of the manuscript.

Funding: This research work was supported by the Community of Madrid and European funding from FSE and FEDER programs (project S2018/BAA-4393, AVANSECAL-II-CM) and by GR21121 and GR21045 Projects, co-funded by FEDER and Junta de Extremadura.

Institutional Review Board Statement: Not applicable.

Informed Consent Statement: Not applicable.

Data Availability Statement: The authors confirm that the data supporting the findings of this study are available within the article and the raw data that support the findings are available from the corresponding author upon reasonable request.

Conflicts of Interest: The authors declare no conflict of interest.

References

1. Starowicz, M.; Zielińsk, H. How Maillard Reaction Influences Sensorial Properties (Color, Flavor and Texture) of Food Products? *Food Rev. Int.* **2019**, *35*, 707–725. [CrossRef]
2. Nooshkama, M.; Varidia, M.; Vermac, D.K. Functional and biological properties of Maillard conjugates and their potential application in medical and food: A review. *Food Res. Int.* **2020**, *131*, 109003. [CrossRef]
3. Stadler, R.H.; Blank, I.; Varga, N.; Robert, F.; Hau, J.; Guy, P.A.; Robert, M.C.; Riediker, S. Acrylamide from Maillard reaction products. *Nature* **2002**, *419*, 449–450. [CrossRef] [PubMed]
4. Morales, F.J. Hydroxymethylfurfural (HMF) and related compounds. In *Process-Induced Food Toxicants: Occurrence, Formation, Mitigation and Health Risks*; Stadler, R.H., Lineback, D.R., Eds.; John Wiley & Sons, Inc., Publications: Hoboken, NJ, USA, 2009; pp. 135–174.
5. IARC (International Agency for Research on Cancer). Some industrial chemicals. In *IARC Monographs on the Evaluation for Carcinogenic Risk of Chemicals to Humans*; IARC: Lyon, France, 1994; Volume 60, pp. 435–453.
6. EFSA (European Food Safety Agency). Scientific Opinion on acrylamide in food. *EFSA J.* **2015**, *13*, 4104.
7. Ames, J.M. The Maillard reaction. In *Biochemistry of Food Proteins*; Hudson, B.J.F., Ed.; Elsevier Science Publishers: London, UK, 1992; pp. 99–153.
8. Kroh, L.W. Caramelisation in food and beverages. *Food Chem.* **1994**, *51*, 373–379. [CrossRef]
9. Glatt, H.; Schneider, H.; Liu, Y. V79-hCYP2E1-hSULT1A1, a cell line for the sensitive detection of genotoxic effects induced by carbohydrate pyrolysis products and other food-borne chemicals. *Mut. Res. Genet. Toxicol. Environ. Mutagen.* **2005**, *580*, 41–52. [CrossRef]

10. Høie, A.H.; Svendsen, C.; Brunborg, G.; Glatt, H.; Alexander, J.; Meinl, W.; Husøy, T. Genotoxicity of three food processing contaminants in transgenic mice expressing human sulfotransferases 1A1 and 1A2 as assessed by the in vivo alkaline single cell gel electrophoresis assay. *Environ. Mol. Mutagen.* **2015**, *56*, 709–714. [CrossRef]
11. Capuano, E.; Fogliano, V. Acrylamide and 5-hydroxymethylfurfural (HMF): A review on metabolism, toxicity, occurrence in food and mitigation strategies. *LWT Food Sci. Technol.* **2011**, *44*, 793–810. [CrossRef]
12. EFSA (European Food Safety Agency). Opinion of the Scientific Panel on food additives, flavourings, processing aids and materials in contact with food (AFC) related to Flavouring Group Evaluation 13 (FGE.13); Furfuryl and furan derivatives with and without additional side-chain substituents and heteroatoms from chemical group 14. *EFSA J.* **2005**, *215*, 1–73.
13. Batra, B.; Pundir, C.S. Detection of Acrylamide by Biosensors. In *Acrylamide in Food. Analysis, Content and Potential Health Effects*; Gökmen, V., Ed.; Elsevier: Amsterdam, The Netherlands, 2016; pp. 497–505.
14. EC (European Commission). Commission regulation (EU) 2017/2158 of 20 November 2017 establishing mitigation measures and benchmark levels for the reduction of the presence of acrylamide in food. *OJ* **2017**, *L304*, 24–44.
15. Aykas, D.P.; Urtubia, A.; Wong, K.; Ren, L.; López-Lira, C.; Rodriguez-Saona, L.E. Screening of Acrylamide of Par-Fried Frozen French Fries Using Portable FT-IR Spectroscopy. *Molecules* **2022**, *27*, 1161. [CrossRef] [PubMed]
16. Sáez-Hernández, R.; Ruiz, P.; Mauri-Aucejo, A.R.; Yusa, V.; Cerver, M.L. Determination of acrylamide in toasts using digital image colorimetry by smartphone. *Food Control* **2022**, *141*, 109163. [CrossRef]
17. Martín-Tornero, E.; Sánchez, R.; Lozano, J.; Martínez, M.; Arroyo, P.; Martín-Vertedor, D. Characterization of Polyphenol and Volatile Fractions of Californian-Style Black Olives and Innovative Application of E-nose for Acrylamide Determination. *Foods* **2021**, *10*, 2973. [CrossRef] [PubMed]
18. Vinaixa, M.; Vergara, A.; Duran, C.; Llobet, E.; Badia, C.; Brezmes, J.; Vilanova, X.; Correig, X. Fast detection of rancidity in potato crisps using e-noses based on mass spectrometry or gas sensors. *Sens. Actuators B* **2005**, *106*, 67–75. [CrossRef]
19. Sánchez, R.; Martín-Tornero, E.; Lozano, J.; Boselli, E.; Arroyo, P.; Meléndez, F.; Martín-Vertedor, D. E-Nose discrimination of abnormal fermentations in Spanish-Style Green Olives. *Molecules* **2021**, *26*, 5353. [CrossRef]
20. Karimi, Z.; Firouzi, M.; Dadmehr, M.; Javad-Mousavi, S.A.; Bagheriani, N.; Sadeghpour, O. Almond as a nutraceutical and therapeutic agent in Persian medicine and modern phytotherapy: A narrative review. *Phytother. Res.* **2021**, *35*, 2997–3012. [CrossRef]
21. Barreca, D.; Nabavi, S.M.; Sureda, A.; Rasekhian, M.; Raciti, R.; Sanches Silva, A.; Annunziata, G.; Arnone, A.; Tenore, G.C.; Süntar, I.; et al. Almonds (Prunus dulcis Mill. D. A. Webb): A Source of Nutrients and Health-Promoting Compounds. *Nutrients* **2020**, *12*, 672. [CrossRef]
22. Amrein, T.M.; Lukac, H.; Andres, L.; Perren, R.; Escher, F.; Amadò, R. Acrylamide in roasted almonds and hazelnuts. *J. Agric. Food Chem.* **2005**, *53*, 7819–7825. [CrossRef]
23. Zhang, G.; Huang, G.; Xiao, L.; Seiber, J.; Mitchell, A.E. Acrylamide Formation in Almonds (Prunus dulcis): Influences of Roasting Time and Temperature, Precursors, Varietal Selection, and Storage. *J. Agric. Food Chem.* **2011**, *59*, 8225–8232. [CrossRef]
24. Amrein, T.M.; Andres, L.; Schönbächler, B.; Conde-Petit, B.; Escher, F.; Amadò, R. Acrylamide in almond products. *Eur. Food Res. Technol.* **2005**, *221*, 14–18. [CrossRef]
25. Lukac, H.; Amrein, T.M.; Perren, R.; Conde-Petit, B.; Amadò, R.; Escher, F. Influence of roasting conditions on the acrylamide content and the color of roasted almonds. *J. Food Sci.* **2007**, *72*, C33–C38. [CrossRef] [PubMed]
26. González-Mulero, L.; Mesías, M.; Morales, F.J.; Delgado-Andrade, C. Acrylamide Exposure from Common Culinary Preparations in Spain, in Household, Catering and Industrial Settings. *Foods* **2021**, *10*, 2008. [CrossRef] [PubMed]
27. Mesías, M.; Holgado, F.; Márquez-Ruiz, G.; Morales, F.J. Effect of sodium replacement in cookies on the formation of process contaminants and lipid oxidation. *LWT Food Sci. Technol.* **2015**, *62*, 633–639. [CrossRef]
28. Barea-Ramos, J.D.; Cascos, G.; Mesías, M.; Lozano, J.; Martín-Vertedor, D. Evaluation of the Olfactory Quality of Roasted Coffee Beans Using a Digital Nose. *Sensors* **2022**, *22*, 8654. [CrossRef] [PubMed]
29. Abdi, H.; Williams, L.J. Principal component analysis. *Wiley Interdiscip. Rev. Comput. Stat.* **2010**, *2*, 433–459. [CrossRef]
30. Geladi, P.; Kowalski, B.R. Partial least-squares regression: A tutorial. *Anal. Chim. Acta* **1986**, *185*, 1–17. [CrossRef]
31. Schlörmann, W.; Birringer, M.; Böhmc, V.; Löber, K.; Jahreis, G.; Lorkowski, S.; Müller, A.K.; Schöne, F.; Glei, M. Influence of roasting conditions on health-related compounds in different nuts. *Food Chem.* **2015**, *180*, 77–85. [CrossRef]
32. Kocadagli, T.; Göncüoglu, N.; Hamzalioglu, A.; Gökmen, V. In Depth Study of Acrylamide Formation in Coffee during Roasting: Role of Sucrose Decomposition and Lipid Oxidation. *Food Funct.* **2012**, *3*, 970–975. [CrossRef]
33. Berk, E.; Hamzalıoğlu, A.; Gökmen, V. Investigations on the Maillard Reaction in Sesame (Sesamum indicum L.) Seeds Induced by Roasting. *J. Agric. Food Chem.* **2019**, *67*, 4923–4930. [CrossRef]
34. Mesias, M.; Gómez, P.; Olombrada, E.; Morales, F.J. Formation of acrylamide during the roasting of chia seeds (Salvia hispanica L.). *Food Chem.* **2023**, *401*, 134169. [CrossRef]
35. Seron, L.H.; Poveda, E.G.; Moya, M.S.P.; Carratalá, M.L.M.; Berenguer-Navarro, V.; Grané-Teruel, N. Characterisation of 19 almond cultivars on the basis of their free amino acids composition. *Food Chem.* **1998**, *61*, 455–459. [CrossRef]
36. Fourie, P.C.; Basson, D.S. Sugar content of almond, pecan and macadamia nuts. *J. Agric. Food Chem.* **1990**, *38*, 101–104. [CrossRef]
37. Lasekan, O.; Abbas, K. Analysis of volatile flavour compounds and acrylamide in roasted Malaysian tropical almond (Terminalia catappa) nuts using supercritical fluid extraction. *Food Chem. Toxicol.* **2010**, *48*, 2212–2216. [CrossRef] [PubMed]

38. Liu, W.; Wang, Y.; Xu, D.; Hu, H.; Huang, Y.; Liu, Y.; Nie, S.; Li, C.; Xie, M. Investigation on the contents of heat-induced hazards in commercial nuts. *Food Res. Int.* **2023**, *163*, 112041. [CrossRef]
39. Loutfi, A.; Coradeschi, S.; Mani, G.K.; Shankar, P.; Rayappan, J.B.B. Electronic noses for food quality: A review. *J. Food Eng.* **2015**, *144*, 103–111. [CrossRef]
40. Majchrzak, T.; Wojnowski, W.; Dymerski, T.; Gębicki, J.; Namieśnik, J. Electronic noses in classification and quality control of edible oils: A review. *Food Chem.* **2018**, *246*, 192–201. [CrossRef]
41. Martín-Torres, S.; Ruiz-Castro, L.; Jiménez-Carvelo, A.M.; Cuadros-Rodríguez, L. Applications of multivariate data analysis in shelf life studies of edible vegetal oils–A review of the few past years. *Food Packag. Shelf Life* **2022**, *31*, 100790. [CrossRef]
42. Rodríguez, J.; Durán, C.; Reyes, A. Electronic nose for quality control of Colombian coffee through the detection of defects in "Cup Tests". *Sensors* **2009**, *10*, 36–46. [CrossRef]

Disclaimer/Publisher's Note: The statements, opinions and data contained in all publications are solely those of the individual author(s) and contributor(s) and not of MDPI and/or the editor(s). MDPI and/or the editor(s) disclaim responsibility for any injury to people or property resulting from any ideas, methods, instructions or products referred to in the content.

Review

Optical Immunosensors for Bacteria Detection in Food Matrices

Dimitra Kourti, Michailia Angelopoulou *, Panagiota Petrou and Sotirios Kakabakos

Immunoassays/Immunosensors Lab, Institute of Nuclear & Radiological Sciences & Technology, Energy & Safety, NCSR "Demokritos", 15341 Agia Paraskevi, Greece; dimkourti96@gmail.com (D.K.); ypetrou@rrp.demokritos.gr (P.P.); skakab@rrp.demokritos.gr (S.K.)
* Correspondence: mikangel@ipta.demokritos.gr; Tel.: +30-210-650-3861

Abstract: Optical immunosensors are one of the most popular categories of immunosensors with applications in many fields including diagnostics and environmental and food analysis. The latter field is of particular interest not only for scientists but also for regulatory authorities and the public since food is essential for life but can also be the source of many health problems. In this context, the current review aims to provide an overview of the different types of optical immunosensors focusing on their application for the determination of pathogenic bacteria in food samples. The optical immunosensors discussed include sensors based on evanescent wave transduction principles including surface plasmon resonance (SPR), fiber-optic-, interferometric-, grating-coupler-, and ring-resonator-based sensors, as well as reflectometric, photoluminescence, and immunosensors based on surface-enhanced Raman scattering (SERS). Thus, after a short description of each transduction technique, its implementation for the immunochemical determination of bacteria is discussed. Finally, a short commentary about the future trends in optical immunosensors for food safety applications is provided.

Keywords: immunosensor; optical detection; bacteria

1. Introduction

Foodborne diseases affect, according to the World Health Organization (WHO), 1 in 10 people every year worldwide [1], with symptoms ranging from mild diarrhea to severe complications and even death [2]. According to the WHO, it has been estimated that cases of food poisoning have reached up to 600 million, with 420,000 deaths worldwide, amongst which 125,000 were children under the age of 5 [1]. In particular, pathogenic bacteria play a crucial role in food poisoning, with the majority of incidents being caused by 15 pathogenic bacteria including *Listeria monocytogenes*, *Escherichia coli O157:H7*, *Clostridium botulinum*, *Legionella pneumophila*, *Campylobacter jejuni*, *Salmonella* spp., *Staphylococcus aureus*, *Shigella*, *Vibrio vulnificus*, and *Bacillus cereus* [3]. Most of these bacteria are detected in dairy products, fresh vegetables, raw products, and undercooked meat and seafood [4,5].

The efficiency and reliability of the techniques employed to detect these pathogens in food matrices are of paramount importance for the prevention of foodborne diseases. [6,7]. The conventional methods for bacteria detection and identification are based on culturing and colony counting. Those methods are reliable, sensitive, and are considered as the "gold standard" for detecting the presence of bacteria; however, prior to detection, they require several steps such as pre-enrichment, selective enrichment, isolation, and confirmation through biochemical and serological tests [8,9]. Thus, there was an urgent need for rapid bacteria detection techniques, which led to the development of new methods such as molecular and immunological ones [10].

Molecular methods rely on the polymerase chain reaction (PCR) [11,12] or isothermal nucleic acid amplification methods such as nucleic acid sequence-based amplification (NASBA), loop-mediated isothermal amplification (LAMP), strand displacement amplification (SDA), recombinase polymerase amplification (RPA), rolling circle amplification

(RCA), helicase-dependent amplification, and others [13] to rapidly and effectively identify various bacteria through their genetic fingerprinting. In particular, isothermal nucleic acid amplification methods have the potential for on-site determinations since they do not require thermal cycling with strict temperature control like the standard PCR. Despite this advantage, isothermal nucleic acid amplification methods can be inhibited by several compounds present in foods, like polysaccharides and polyphenolic compounds, leading to false negative results [13]. Moreover, molecular methods cannot discriminate between dead and live bacteria, and this can lead to false positive results.

On the other hand, immunoassays rely on antibodies that recognize specific proteins or liposaccharides of the bacterial external membrane. Chemiluminescent enzyme immunoassay (CL-EIA) [14] and enzyme-linked immunosorbent assay (ELISA) [15–17] are the immunochemical methods most widely employed for rapid bacterial detection and quantification. Immunoassays are characterized by high sensitivity and accuracy, simple sample preparation, and low costs of instrumentation compared to other instrumental methods such as chromatographic ones. Compared to molecular methods, the immunochemical ones are less prone to false negative results, but they can also produce false positive results due to the cross-reactivity of antibodies used with other bacteria species or sub-species of the targeted one. In addition, they are laboratory bound since they involve multiple processing steps and desktop instrumentation [18,19].

The quest for portable analytical devices that can provide reliable results in a short amount of time and that are suitable for on-site applications has been the driving force behind the development of biosensors. Nowadays, the realization of biosensor systems that combine outstanding analytical performance with portability has moved from the sphere of fiction to reality due to significant progress in the field of nanotechnology [20]. Thus, modern biosensor technologies can provide high detection sensitivity and specificity, high-speed analysis, and quantitative results in real time [21]. Biosensors are defined by the IUPAC as analytical devices that provide quantitative or semi-quantitative information by combining a biological recognition element with a physicochemical transducer, which transforms the biorecognition event into a physically detectable signal [22]. The recognition element can be any bioreceptor [23], such as a nucleic acid probe, aptamer, phage, antibody, antigen, whole cell, enzyme, etc., that can bind the target molecule specifically and with high affinity [24–31] (Figure 1a).

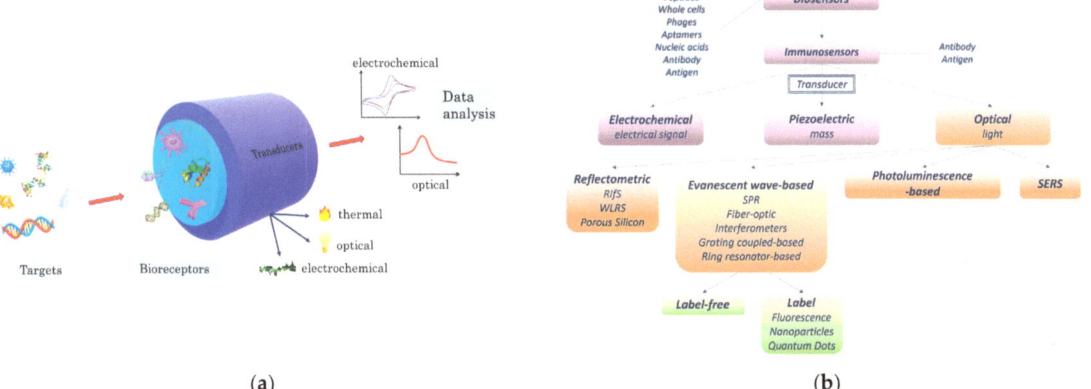

Figure 1. (**a**) Schematic of a biosensor. (**b**) The different categories of biosensors.

Immunosensors, i.e., biosensors that rely on antibody–antigen reactions for analyte detection, are the most abundant category of biosensors due to the indispensable ability of antibodies for the highly selective detection of the targeted analytes in complex media and the versatility of available assay formats that could be employed for the determination of different kinds of analytes, including bacteria [31]. They can be divided according to the transduction principle into electrochemical [32–35], piezoelectric [36,37], or optical ones [32,38], which can detect biological interactions by recording the variations in the electrical signal, mass, or light, respectively (Figure 1b). Electrochemical immunosensors are cost-efficient devices and have the potential for miniaturization. Although their detection sensitivity is high and the detection limits achieved are suitable for bacteria detection at very low concentrations, especially in cases where labels are involved to enhance the signal, they are prone to interferences from the matrix, which can reduce the reproductivity of the readings and lead to either false positive or false negative results [39]. Immunosensors based on piezoelectric phenomena, even though suitable for label-free detection, are characterized in general by relatively low detection sensitivity [40]. On the other hand, optical immunosensors exhibit high sensitivity and offer simple, fast, and accurate detection of a great variety of analytes, including bacteria [41]. Optical detection methods (Figure 1b) are based on different transduction principles such as light absorbance, total internal reflectance, photoluminescence, fluorescence, light polarization, interferometry, Raman scattering, and surface plasmon resonance [42,43].

The scope of this review is to summarize the achievements of optical immunosensors for foodborne bacteria detection. First, short descriptions of the different detection principles reported so far for bacteria detection will be provided. Then, the application of optical immunosensors for bacteria detection in different food matrices will be extensively presented. Finally, a short commentary on the future trends regarding the prospective applications and challenges of optical immunosensors in food analysis is included.

2. Principles of Optical Immunosensors

2.1. Assay Formats

In principle, most immunosensors offer the potential to monitor antigen–antibody binding in real time. Such direct detection although simpler and faster, since there is no need for additional reaction steps, is usually limited to high-molecular-weight analytes, for which the antigen–antibody reaction results in a measurable change in sensor response [32]. Therefore, in most immunosensors, the assay formats usually applied in microtiter plate solid-phase immunoassays are followed, i.e., the competitive or non-competitive assay format. As schematically depicted in Figure 2, competitive immunoassays are based either on an immobilized antibody (Figure 2A), an immobilized analyte, or an analyte–protein conjugate (Figure 2B). In the first approach, the antibody is immobilized onto the transducer surface, and the concentration of the analyte is defined through its competition with a labeled analyte or analyte–protein conjugate for binding to the antibody. In the second approach, the analyte is immobilized onto the transducer surface (either directly or as an analyte–protein conjugate) and competes with the analyte in the sample for the binding sites of the antibody. Although both approaches are applicable for a given analyte–antibody pair, the second might be advantageous regarding the stability of immobilized biomolecule, i.e., the analyte or analyte–protein conjugate, since antibodies are known to lose a great part of their functionality upon immobilization. The non-competitive or sandwich immunoassay format is suitable for analytes that have at least two antigenic determinants or epitopes in their molecule. This means that at least two antibodies recognizing two different parts of the analyte are available. This is essential for the realization of a non-competitive assay since, as depicted in Figure 2C, an antibody immobilized onto the transducer surface (usually referred to as the capture antibody) binds the analyte through one epitope, and a second antibody (referred to as the detection or reporter antibody) is attached to a different epitope forming a "sandwich" with the analyte. The detection antibody can be labeled or not depending on the transduction principle involved.

Figure 2. Schematic presentation of different immunoassay formats applied to optical immunosensors. (**A**) competitive immunoassay based on immobilized antibody; (**B**) competitive immunoassay based on immobilized analyte or analyte–protein conjugate; (**C**) non-competitive sandwich immunoassay.

2.2. Main Optical Transduction Principles

The optical transduction principles can be divided into two main categories: those involving labels and the label-free ones.

2.2.1. Detection Using Labels

A significant number of optical immunosensors are based on the use of labels such as enzymes, fluorescent organic dyes, gold nanoparticles, or quantum dots for the quantification of antibody–analyte reactions and the determination of analyte concentrations in a sample [44]. Using enzymes as labels in immunosensors is endorsed by their successful implementation in standard microtiter plate immunoassays. Enzymes offer significant signal enhancement in a short time; however, their use in combination with optical transducers is rather limited to insoluble substrates that can precipitate on the transducer, thus generating a measurable signal change. Biosensing using fluorescently labeled molecules and optical fibers was amongst the first optical transduction principles to be explored [45]. Apart from the typical cylindrical optical fibers, planar waveguides and capillaries that enable light propagation through total internal reflection have also been employed as transducers for the development of optical fluorescence immunosensors [46,47]. Over the years, metal nanoparticles have been employed as labels and incorporated into immunochromatographic strips or used as liquid phase reagents to obtain semi-quantitative information, relying on visual evaluation, or quantitative results through the implementation of an instrument that could quantitate the color, fluorescence, or chemiluminescence intensity. The latter can be performed using smartphones as detection, signal processing, and transmission apparatus, moving the realization of portable devices a step forward [48].

2.2.2. Label-Free Detection

Label-free optical biosensing, on the other hand, is based almost exclusively on monitoring refractive index changes due to immunochemical reactions taking place on top of the optical transducer [37]. There are two main categories of label-free optical transduction principles: the refractometric and the reflectometric ones [49]. In refractometric immunosensors, the light that is transmitted through a waveguide creates an electromagnetic field, known as an evanescent field (Figure 3), which extends in the medium above the biosensor surface. This field is influenced by refractive index changes over the transducer surface resulting from the biomolecular layer thickness increase due to immunoreactions. Thus, when the evanescent wave field is coupled back to the transducer, a change in the intensity, polarization, or phase of the waveguided light is observed, which is proportional to the concentration of analyte in the sample, enabling its quantitative determination [50–52]. Transducers based on surface plasmon resonance (SPR), fiber optics, grating couplers, interferometers, and ring resonators fall into this category of optical sensors (Figure 4).

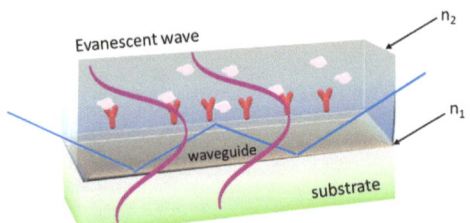

Figure 3. Schematic of the evanescent-field-sensing principle, where n_1 is the refractive index of the waveguide material and n_2 is the refractive index of the covering medium. For effective waveguiding $n_1 > n_2$.

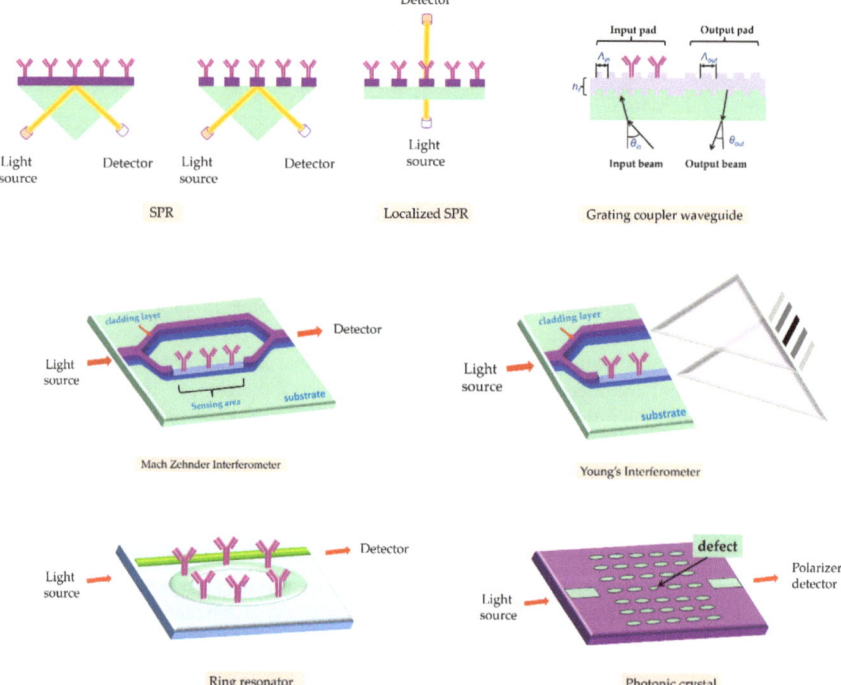

Figure 4. Schematic presentation of basic conformations of refractometric label-free optical transducers including SPR and localized SPR, grating coupler waveguides, interferometers, ring resonators, and photonic crystals.

Reflectometric sensors are based on light reflection by a stack of materials with different refractive indices leading to the creation of an interference spectrum. The most widely explored reflectometric sensing method is the one introduced in 1991 by Gauglitz et al. [53], which is known as reflectometric interference spectroscopy (RIfS). RIfS transducers are made of a glass substrate on top of which a thin layer of transparent dielectric material has been deposited. Immunochemical reactions that take place on top of the dielectric layer are evidenced as shifts in the reflected interference spectrum. This spectrum is created due to the reflection of the light beam at each interface of different refractive indices with a slightly different angle leading to either constructive or destructive interference. The spectrum shift is directly correlated with the increase in the biomolecular layer thickness due to immunoreactions, and therefore, it can be correlated with the analyte concentration in the sample through a calibration curve. Since the first report on RIfS sensors, a lot of progress has been made in the direction of reducing the cost and size of the instrumentation and increasing the multiplexing capabilities, e.g., by monitoring, instead of the whole spectrum, specific wavelengths [54] or even a single wavelength [55]. In addition, materials other than glass have been employed as transducers including porous silicon with or without thermally grown oxide [56], porous silicon–carbon composites [57], other porous materials such as TiO_2 [58], or non-transparent substrates such as silicon with a transparent dielectric composed of silicon dioxide [59–61] or silicon nitride [62].

3. Application of Optical Sensors for Bacteria Detection
3.1. Evanescent-Wave-Based Biosensors
3.1.1. SPR Immunosensors

Surface plasmon resonance (SPR)-based immunosensors are the optical biosensors most frequently used for the single or multiplex, label-free detection of foodborne pathogenic bacteria due to their high detection sensitivity and ability to monitor binding reactions in real-time. The SPR phenomenon relies on the excitation of metal-free electrons (surface plasmons) when polarized light strikes a metal layer (usually gold) deposited on the surface of an optically transparent material (prism, grating coupler, or dielectric waveguide) at a certain angle. The excited plasmons create an evanescent wave field at the solution/gold interface. This wave is very sensitive to refractive index changes at the gold layer surface occurring due to a biomolecular reaction, and as a result, the angle of incident light has to change during the course of the reaction to preserve the surface plasmon wave, providing the means to monitor these reactions in real time. Thus, it is possible to monitor both the immobilization of specific biomolecules (e.g., antibodies) as well as the binding of analytes in real time (Figure 5) [41,44].

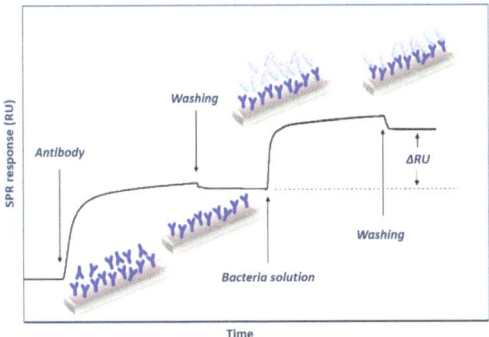

Figure 5. SPR response during the binding of a bacteria-specific antibody onto the biosensor surface followed by capture of the bacteria by the immobilized antibody.

The first report regarding the detection of bacteria with an SPR sensor dates back to 1998 [63], and it was based on a sandwich immunoassay for the detection of *Escherichia coli O157:H7* in buffer. The sensor was modified with protein A or protein G, and a mouse monoclonal or a rabbit polyclonal antibody, respectively, was then immobilized. Depending on the antibody used for detection, LODs in the range $5–7 \times 10^7$ cfu/mL have been achieved. *Vibrio cholerae O1* was also identified in buffer with an SPR biosensor functionalized with a protein G layer [64]. The sensor chip was modified with a self-assembled monolayer of a mixture of 11-mercaptoundecanoic acid and hexanethiol on which protein G was covalently bound and used to immobilize a monoclonal antibody specific to *V. cholera O1*. The LOD of the assay was 10^5 cfu/mL. The same antibody immobilization approach was employed to immobilize an antibody against *Legionella pneumophila* onto SPR chips, achieving an LOD of 10^5 cfu/mL in buffer [65]. *Salmonella enterica* serovar *Enteritidis* and *Escherichia coli* were detected in spiked skim milk by a direct binding assay on SPR chips modified with protein

G to which the anti-bacteria specific antibodies were then bound [66]. The LOD achieved after 1 h assay was 25 cfu/mL for *E. coli* and 23 cfu/mL for *Salmonella*.

The detection of *Salmonella* groups B, D, and E in buffer with SPR has also been reported, employing a sandwich assay format and using antibody pairs from different animals [67]. It was found that the LOD improved 200 times compared to direct detection [67]. The benefits of the sandwich immunoassay format for bacteria detection have also been demonstrated in a study for the detection of *Staphylococcus aureus* with SPR in buffer, where the LOD was improved from 1×10^7 to 1×10^5 cfu/mL when a sandwich immunoassay format was followed instead of direct detection [68]. In another report, where the direct binding assay was compared to a sandwich assay for the detection of *E. coli* O157:H7 in buffer, a 1000-fold improvement in sensitivity was reported, leading to an LOD of 10^3 cfu/mL [69]. The immobilization of the polyclonal antibody against *E. coli* was performed through covalent bonding to a monolayer of mixed thiol-terminated polyethylene glycol with thiol-terminated polyethylene-carboxylic acid. In another report, the free amine groups of protein A were converted to thiol groups, through a reaction with 2-iminothiolane, to facilitate its immobilization onto SPR chips, which are then modified with an antibody against *Salmonella paratyphi* [70]. An LOD of 10^2 cfu/mL was achieved in buffer employing the antibody-modified chip in a direct binding assay.

In another study, the SPR chip was modified with brushes of poly(carboxybetaine acrylamide) to reduce the non-specific binding of bacteria to its surface, while antibody-modified gold nanoparticles were used as labels to increase the detection sensitivity [71]. This immunosensor could detect *E. coli* O157:H7 in hamburger and cucumber samples at concentrations as low as 57 and 17 cfu/mL and *Salmonella* spp. at 7.4×10^3 and 11.7×10^3 cfu/mL, respectively [71]. Gold nanoparticles modified with an antibody against *Campylobacter jejuni* were also employed as labels in an SPR sandwich immunoassay that allowed for the detection of this bacterium in buffer at concentrations as low as 4×10^4 cfu/mL [72]. A gold-labeled secondary antibody was also employed to increase the detection sensitivity in a sandwich SPR for *L. monocytogenes* by two–four orders of magnitude compared to the direct binding assay, providing an LOD of 10^2 cfu/mL in buffer [73]. In another study, in order to achieve signal enhancement, a precipitate 3,3′,5,5′-tetramethylbenzidine (TMB) substrate was used in combination with an antibody labeled with horse radish peroxidase (HRP) to detect bacteria captured by an antibody attached to an SPR chip through covalent bonding to a self-assembled monolayer of mercaptoundecanoic acid [74]. Following this approach, 250% signal enhancement was achieved with respect to the assay not employing the HRP/TMB system, leading to an LOD of 10^4 cfu/mL for the detection of *E. coli* in spinach leaves [74].

SPR chips were modified in a plasma reactor in the presence of cyclopropylamine vapors to induce reactive moieties containing nitrogen which were in turn used to immobilize antibodies using glutaraldehyde activation [75]. The chips were employed to detect *Salmonella typhimurium* in buffer by a direct binding assay that provided an LOD of 10^5 cfu/mL.

Along with the optimization of the SPR assays for bacteria detection, efforts have been devoted to sample treatment methods aiming either to improve detection sensitivity or alleviate non-specific matrix effects. For example, the performance of an SPR sensor for the detection of live, heat-killed, or detergent-lysed *E. coli* O157:H7 cells was investigated, and LODs of 10^6, 10^5, and 10^4 cfu/mL, respectively, were reported [76]. The differences observed were ascribed to changes in cell size and morphology upon treatment with ethanol, whereas treatment with detergent probably led to fragmentation of cells to smaller particles that were recognized more efficiently by the antibody. The effect of the sample preparation method was also evident in another study for the detection of *E. coli* O157:H7 in different food samples [77]. In this study, milk, apple juice, and ground beef patties were spiked with *E. coli* O157:H7 at various concentrations and then analyzed with a portable SPR instrument commercialized by Texas Instruments Inc. under the tradename SPREETA[TM]. The sensor chip was modified with neutravidin to enable the immobilization of biotinylated antibodies against *E. coli* O157:H7. The spiked milk and apple juice samples were run without pretreatment, whereas the ground beef sample was extracted with buffer and homogenized prior to analysis. The LODs achieved ranged from 10^2 to 10^3 cfu/mL depending on the sample analyzed. In another report, where a sandwich SPR assay for *Salmonella* with an LOD of 1.25×10^5 cells/mL was developed [78], the authors claimed that the presence of milk did not affect the assay performance, alleviating the need for sample preparation or clean-up steps.

It has been suggested that the detection of whole bacteria using SPR generally results in lower sensitivity compared to other techniques, due to the limited interaction of bacteria with the evanescent wave electromagnetic field (the size of the bacteria is multiple times the penetration depth of the evanescent wave field) and the small difference in the refractive index between the bacterial cytoplasm and the surrounding aqueous medium [79]. Thus, instead of running over the sensor the bacteria-containing sample, it is first incubated with the pathogen-specific antibody, and after the separation of the free antibody from the bound antibody, the free antibody is quantified. This assay format is known as subtraction inhibition assay (SIA) and has been applied for the detection of *L. monocytogenes* [80], *E. coli* O157:H7 [81], and *B. anthracis* spores in buffer [82]. The LODs reported were 1×10^5, 3.0×10^4, and 10^4 cfu/mL, respectively, and were one order of magnitude lower than those achieved with the direct binding assay. The SIA format was also applied for the detection of fungal cells that are considerably larger than bacterial ones [83]. Thus, it has been applied for the detection of *Phytophthora infestans* [83] and the detection of *Puccinia striiformis* in buffer with LODs of 2.2×10^6 sporangia/mL and 3.1×10^5 urediniospores/mL, respectively [84]. A similar assay format was also applied for the SPR detection of *Cryptosporidium parvum* oocysts in water with an LOD of 1×10^2 oocysts/mL [85].

In a different approach for indirect bacteria detection by SPR, a polyclonal antibody against a cell extract enriched for the invasion-associated protein, internalin B, was used to develop an inhibition assay for *Listeria monocytogenes* [86]. After incubation of bacteria containing solutions with the antibody, the mixture was injected over an SRP chip modified with purified-recombinant internalin B, and the signal was inversely proportional to the *L. monocytogenes* concentration achieving an LOD of 2×10^5 cells/mL.

In addition to single bacteria detection, multiplexed bacteria detection with SPR systems has been also explored. Thus, an antibody microarray was developed on an SPR chip for the simultaneous detection of *S. typhimurium*, *E. coli* O157:H7, *Yersinia enterocolitica*, and *Legionella pneumophila* by modifying the chip with protein G to allow for the immobilization of each one of the anti-bacteria specific antibodies at different areas of the chip through spotting [87]. All bacteria were detected simultaneously in buffer samples each at a concentration of 10^5 cfu/mL. *E. coli* O157:H7, *L. monocytogenes*, *Campylobacter jejuni*, and *S. choleraesuis* were also simultaneously detected both in buffer and apple juice using a multi-channel SPR system [88]. The whole chip surface was modified with streptavidin, and the four bacteria antibodies were immobilized using an eight-channel fluidic device (two channels per antibody; one for the specific and the other for the non-specific signal monitoring). The LODs achieved were 1.4×10^4 cfu/mL for *E. coli*, 4.4×10^4 cfu/mL for *S. choleraesuis*, 1.1×10^5 cfu/mL for *C. jejuni*, and 3.5×10^3 cfu/mL for *L. monocytogenes*.

An SPR imaging device was combined with an array of antibodies specific to different serotypes of *L. monocytogenes* with the aim of monitoring the growth of live listeria cells in culture [89]. Emphasis was placed on the characterization of the antibodies rather than the analytical performance of the sensor. Similarly, the detection of *Salmonella* with an SPR imaging array was optimized, and LODs of 2.1×10^6 and 7.6×10^6 cfu/mL in buffer and chicken carcass rinse were demonstrated, respectively [90]. An SPR imaging sensor was applied for the simultaneous label-free detection of *Salmonella* spp., Shiga-toxin-producing *E. coli (STEC)*, and *L. monocytogenes* in chicken carcass rinse [91]. The antibodies specific to each bacterium were immobilized on the same chip, and an LOD for *Salmonella* of 10^6 cfu/mL was achieved. An SPR imaging sensor (Figure 6) was also implemented for the simultaneous detection of *Listeria monocytogenes* and *Listeria innocua*, achieving an LOD of 2×10^2 cfu/mL for both bacteria after 7 h incubation of the sample in the fluidic cell attached to the SPR chip [92].

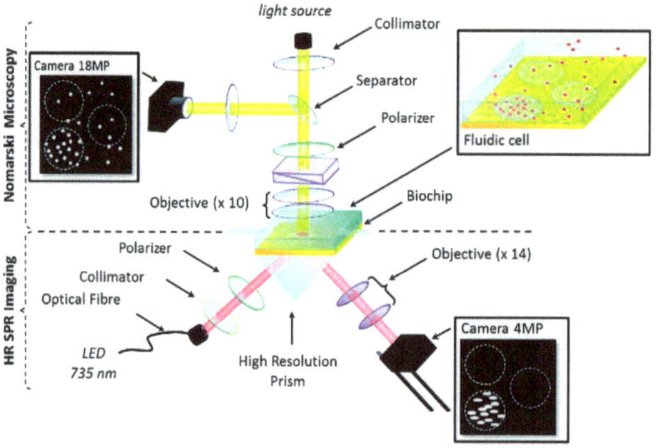

Figure 6. Schematic of a resolution-optimized prism-based SPR imaging apparatus [92]. Copyright 2019. Reproduced with permission from the Royal Society of Chemistry.

Despite the fact that SPR has found numerous applications in diverse fields and several companies have commercialized devices based on this transduction principle, the majority of these instruments are suitable for use in a lab. Thus, considerable efforts have been devoted to reducing the equipment size and complexity in order to build up systems appropriate for analysis at the point of need. The SPREETA™ SPR biosensor mentioned above was a successful outcome of such an effort. It included an AlGaAs light-emitting diode (LED, 840 nm), a polarizer, a temperature sensor, two photodiode arrays, and a reflecting mirror combined with a gold-coated glass slide and a silicone rubber gasket of two channels. The instrument was accompanied by software that provided all the information related to the analysis of the SPR curve, the real-time binding, the layer thickness, and the flow cell temperature. In addition to the determination of *E. coli* O157:H7 in various food samples [77], the SPREETA™ SPR biosensor has also been used for the detection of *Campylobacter jejuni* with an LOD of 10^3 cfu/mL in both buffer and spiked broiler meat samples [93]. SPREETA™ has also been employed to develop a sensor to detect *E. coli* O157:H7 in laboratory cultures [94]. The sensitivity and specificity of detection were determined. Thus, for an assay of 35 min, an LOD for *E. coli* O157:H7 of 8.7×10^6 cfu/mL was determined in single-bacteria culture, whereas in mixed cultures with non-target bacteria concentrations up to 10^6 cfu/mL or less, the LOD was 10^7 cfu/mL. For higher concentrations of non-target bacteria, the sensor's sensitivity was negatively affected. In another report, application of the SPREETA™ SPR sensor for *E. coli* detection in water with an LOD of 90 cfu/mL was reported for a direct binding assay that lasted less than 30 min [95]. The much lower LOD achieved in this report with respect to previous ones could be attributed to the fact that the specific antibody was immobilized onto the chip surface via streptavidin and not directly. Finally, SPREETA™ was applied to detect *Salmonella typhimurium* at concentrations equal to or higher than 1×10^6 cfu/mL in chicken [96]. To increase the detection sensitivity of a SPREETA™ sensor for the detection of *E. coli*, Au-coated magnetic nanoparticles were modified with an antibody against *E. coli* and used not only to concentrate *E. coli* cells from lake, river, puddle, and tap water samples but also as labels in a SPR sandwich immunoassay using SPREETA™ chips, which were modified with an anti-*E. coli* antibody, achieving a detection limit of 3 cfu/mL [97].

Apart from SPREETA™, other attempts to create portable instruments based on the SPR principle of detection have been reported in the literature. Thus, a portable instrument that combined microfluidic and SPR technologies on a single platform was applied for the determination of *E. coli* and *S. aureus* in buffer samples [98]. In this setup, an LED was used to illuminate a gold-covered rectangular prism, and the reflected light was captured by a CMOS sensor and then transferred for processing to a PC. The chip was modified with 11-mercaptoundecanoic acid to facilitate the covalent binding of protein G in which an antibody against the lipopolysaccharide (LPS) of *E. coli* was captured, enabling *E. coli* detection in buffer at a concentration of 3.2×10^5 cfu/mL. In another attempt, *Salmonella typhimurium* was detected in buffer at concentrations ranging from 10^7 to 10^9 cfu/mL within 1 h using an SPR biosensor in which the incident light from a diode laser (instead of an LED) was directed to the gold film by a rotating mirror, and the light reflected from the metal film was captured by a CMOS image sensor [99].

Another approach to surpass the portability limitations of standard SPR instruments is the implementation of a localized SPR or LSPR transduction approach. In LSPR, the continuous metal surface is replaced by noble-metal nanoparticles (nanospheres, nanorods, or nanodisks) of sub-wavelength size around which the surface plasmons are localized [100]. The light that strikes the nanostructures excites the surface plasmons and when resonance is achieved, certain wavelengths are scattered from the nanostructures. Thus, immunoreactions can be monitored in real time as shifts in the resonance wavelength [101]. The advent of LSPR opened up new horizons for the detection of pathogens, especially in the direction of portable systems. Nonetheless, the first report showed that LSPR was less sensitive than the classical SPR configuration [102] or more vulnerable to interferences from the matrix of the samples analyzed [103]. More recent reports, however, show improved detection sensitivity achieved mainly through optimization of the dimensions and stability of the nanoparticles [104]. Thus, an LSPR sensor was developed for the determination of *E. coli* O157:H7 in buffer employing spherical gold nanoparticles non-covalently modified with a specific anti-*E. coli* avian antibody. An LOD of 10 cfu/mL was achieved in less than 2 h, making the sensor suitable for *E. coli* O157:H7 determination at the point of need [104]. Instead of using non-continuous gold surfaces, structuring of the gold film through its deposition onto a nanostructured fluoropolymer enabled the development of an SPR sensor based on grating-coupled long-range surface plasmons, which were employed for the detection of *E. coli* O157:H7 in spiked buffer samples through a sandwich immunoassay implementing metal nanoparticles modified with another anti-*E. coli* O157:H7 antibody as labels to achieve an LOD of 50 cfu/mL [105].

The reports regarding bacteria detection in water and food samples with SPR-based immunosensors are summarized in Table 1.

Table 1. SPR-based immunosensors for bacteria detection.

Analyte	Detection Principle	Assay Type	Sample Type	Assay Duration	LOD	Ref. No.
E. coli	SPR	direct	milk	5–7 min	23 cfu/mL	[66]
S. enteritidis					25 cfu/mL	
E. coli O157:H7 *Salmonella* sp.	SPR	sandwich with biotinylated detection antibody and streptavidin labeled with gold nanoparticles	hamburger/ cucumber	~1 h	57/17 cfu/mL 7.4×10^3/ 11.7×10^3 cfu/mL	[71]
E. coli	SPR	sandwich with detection antibody labeled with peroxidase + TMB substrate	fresh spinach	~2 h	10^4 cfu/mL	[74]
E. coli O157:H7	SPR	direct	milk apple juice ground beef	30 min	10^2–10^3 cfu/mL	[77]
S. typhimurium	SPR	sandwich	milk	1 h	1.25×10^5 cfu/mL	[78]
C. parvum oocysts	SPR	subtractive inhibition	water	30 min	10^2 oocysts/mL	[85]
L. monocytogenes	SPR	inhibition	chocolate milk	30 min	2×10^5 cfu/mL	[86]

Table 1. Cont.

Analyte	Detection Principle	Assay Type	Sample Type	Assay Duration	LOD	Ref. No.
L. monocytogenes	SPR imaging	direct	apple juice	30 min	3.5×10^3 cfu/mL	[88]
E. coli O157:H7					1.4×10^4 cfu/mL	
C. jejuni					1.1×10^5 cfu/mL	
S. typhimurium					4.4×10^4 cfu/mL	
L. monocytogenes	SPR imaging	direct	lettuce	30 min	10^7 cfu/mL	[89]
S. typhimurium	SPR imaging	direct	chicken carcass rinse	~20 min	7.6×10^6 cfu/mL	[90]
Salmonella spp. Shiga-toxin producing *Escherichia coli* *L. monocytogenes*	SPR imaging	direct	chicken carcass rinse		10^6 cfu/mL	[91]
C. jejuni	Portable SPR	direct	broiler meat	<30 min	10^3 cfu/mL	[93]
E. coli	Portable SPR	direct	water	~17 min	90 cfu/mL	[95]
S. typhimurium	Portable SPR	direct	chicken carcass	3 min	10^6 cfu/mL	[96]
E. coli	SPR	direct	lake, river, puddle, and tap water	20 min	3 cfu/mL	[97]

3.1.2. Fiber Optic Immunosensors

Fiber optic immunosensors rely on the immobilization of immunoreagents onto a part of the optical fiber from which the cladding layer has been removed to allow interaction of the waveguided photons, through the evanescent wave field, with the analyte in the solution surrounding the fiber (Figure 7). In order to increase the evanescent field effect, fiber tapering is applied either in the form of a tapered tip or of a continuous tapered fiber [45,106]. Tapered tips are created by gradually reducing the diameter at the end of an optical fiber down to nanometers. In order to obtain the highest possible sensitivity due to reactions, the recognition biomolecules are immobilized on the tip region with the smallest diameter where the evanescent field is stronger. Continuous tapered fibers usually have a biconical taper, comprising a region of decreasing diameter, a region of constant diameter called the waist, and a region of increasing diameter [107]. Sensing takes place on the waist region where the evanescent field exhibits its higher intensity, whereas the emitted light is collected from the region of increasing diameter [108]. Fiber optic biosensors have been widely employed in the field of foodborne pathogen detection due to their convenience, small size, lack of electromagnetic interference, cost-effectiveness, high sensitivity, and accuracy [109].

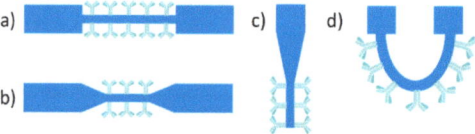

Figure 7. Main configurations of fiber optic sensors: (**a**) de-cladded optical fiber, (**b**) tapered optical fiber, (**c**) tapered tip, and (**d**) U-shaped optical fiber probe.

The first label-free approach for the detection of pathogens was realized using a U-bent optical fiber sensor [110]. Bending a de-cladded fiber into a U-shaped structure enhances the penetration depth of evanescent waves and, hence, the sensitivity of the probe. This system could detect *E. coli* in concentrations lower than 10^3 cfu/mL in buffer with an assay duration of 1 h. A similar approach employing a plastic fiber optic sensor with a U-shaped sensing probe functionalized with an antibody against *E. coli serotype O55* was employed for *E. coli* detection in buffer, resulting in an LOD of 10^3 cfu/mL for an assay duration of 10 min per sample [111]. Upon exposure of the sensor to bacteria solutions, the output signal decreased with time due to the attachment of the bacteria onto the probe surface, which resulted in an increase in the refractive index value close to the probe.

Several fiber optic immunosensors employing labels have also been reported for bacteria detection. Thus, tapered fiber tips have been used for the detection of *Salmonella* in culture medium [112] and *E. coli O157:H7* in ground beef [113] with LODs of 10^4 and 10^3 cfu/mL, respectively. The first employed silica fibers with tapered tips were modified with mercaptosilane to facilitate the covalent bonding of an anti-*Salmonella* antibody, while a second antibody labeled with a fluorescent dye was used as a detection antibody. In the second report [113], polystyrene fibers were first coated with biotinylated bovine serum albumin and then reacted with streptavidin and a biotinylated anti-*E. coli* antibody. A fluorescently labeled antibody was also used for detection in this case. Polystyrene fibers were integrated into a portable instrument commercialized under the name RAPTOR™ to develop an instrument that could be used for on-site determinations. This instrument was used for the detection of *S. typhimurium* in rinse water from sprouted alfalfa seeds through modification of the fibers first with streptavidin and then with a biotinylated antibody [114]. A second fluorescently labeled antibody was used for detection, achieving an LOD of 10^5 cfu/mL or 50 cfu/g of sprouted alfalfa seeds after 67 h of bacteria enrichment. The RAPTOR™ biosensor has also been used to detect *Enterococcus faecalis* in ambient water, with an LOD of 5.0×10^5 cells/mL [115], and *L. monocytogenes* in several food matrices, with LODs ranging from 10^3 to 4.3×10^3 cfu/mL [116–118]. In all cases, sandwich immunoassays were implemented with the exception of [117] where a fluorescently labeled aptamer (aptamer A8) specific for internalin A, an invasin protein of *L. monocytogenes*, was used for detection. The sensor was applied for the detection of *L. monocytogenes* in hot dog and bologna sausages at concentrations down to 10 cfu/g after 24 h of enrichment [116] as well as 10^2 cfu/g in ready-to-eat meat products (sliced beef, chicken, and turkey) after 18 h of enrichment [117]. A version of RAPTOR™ that supported multiplexed determinations was applied for the detection of *L. monocytogenes*, *E. coli O157:H7*, and *S. enterica* in several meat products [119]. The LODs achieved were 50 cfu/mL for *S. enterica* and 10^3 cfu/mL for *L. monocytogenes*, and the sensor could detect the three bacteria (*Salmonella*, *E. coli*, and *Listeria*) in ready-to-eat beef, chicken, and turkey meats inoculated with each pathogen at 100 cfu/25 g after enrichment for 18 h.

Fluorescence resonance energy transfer, i.e., the non-radiative energy transfer from a fluorescent donor molecule to an acceptor one when these two are in close proximity, has been also implemented for the detection of *S. typhimurium* with an optical fiber tip sensor in ground beef samples [120]. The anti-*Salmonella* antibody was labeled with the donor fluorophore (AlexaFluor 546) and protein G was labeled with the acceptor fluorophore (Alexa Fluor 594). Upon binding of *S. typhimurium* to the antibody, the induced conformation changes reduced the distance between the donor and acceptor molecules, resulting in an increase in emitted fluorescence and achieving an LOD of 10^5 cfu/g for the samples.

In recent years, fiber optic immunosensors based on surface modifications with nanomaterials have shown significant improvements compared to conventional fiber optic sensors regarding detection speed and sensitivity. For example, a fiber optic biosensor modified with zinc oxide (ZnO) nanorods for the detection of *E. coli* in water with an LOD of 10^3 cfu/mL was developed [121].

Fiber optic sensors in which the exposed fiber core is coated with a gold layer to take advantage of the SPR phenomenon have also been used for bacteria detection. Such a sensor has been employed for the detection of *Legionella pneumophila* by a direct assay after modification of the fiber gold-covered area with 11-mercaptoundecanoic to allow the covalent bonding of an anti-*L. pneumophila* antibody [122]. An LOD of 10 cfu/mL was achieved for a direct assay that lasted 1 h. In another report, a fiber optic SPR sensor was modified with MoS_2 nanosheets on which the specific antibodies were attached, and an LOD of 94 cfu/mL for *E. coli* was achieved compared to 391 cfu/mL received from fibers without MoS_2 nanosheets [123]. The reports regarding bacteria detection in water and food samples with fiber-optic-based immunosensors are summarized in Table 2.

Table 2. Fiber optic sensors for bacteria detection.

Analyte	Detection Principle	Assay Type	Sample Type	Assay Duration	LOD	Ref. No.
E. coli O157:H7	fiber optic fluorescence	sandwich	ground beef	2.5 h	10^3 cfu/mL	[113]
S. typhimurium	plastic fiber optic fluorescence	sandwich	sprouted alfalfa seeds rinse water	~30 min	10^5 cfu/mL 50 cfu/g (67 h) *	[114]
E. faecalis	plastic fiber optic fluorescence	indirect	ambient water	2.5 h	10^5 cfu/mL	[115]
L. monocytogenes	plastic fiber optic fluorescence	sandwich	hot dog bologna	2.5 h	4.3×10^3 cfu/mL 10 cfu/g (24 h) *	[116]
L. monocytogenes	plastic fiber optic fluorescence	sandwich	meat products	4.0 h	10^3 cfu/mL 10^2 cfu/g (18 h) *	[117]
L. monocytogenes	plastic fiber optic fluorescence	sandwich	frankfurter	12 min	5×10^5 cfu/mL	[118]
L. monocytogenes *E. coli* O157:H7 *S. enterica*	plastic fiber optic fluorescence	sandwich	beef chicken turkey	4.0 h	10^3 cfu/mL 4 cfu/g (18 h) *	[119]
S. typhimurium	FRET-based optical fiber	direct	buffer ground pork	5 min	10^3 cells/mL 10^5 cfu/g	[120]
E. coli	fiber optic with a gold layer and ZnO nanorods	direct	water	3 s	10^3 cfu/mL	[121]
E. coli O157:H7	fiber optic SPR	direct	drinking water	~15 min	94 cfu/mL	[123]

* Enrichment time required to achieve the relevant LOD.

3.1.3. Interferometric Immunosensors

Interferometric immunosensors are another category of devices that have been implemented for the detection of different bacteria. These sensors can detect refractive index changes down to 10^{-8} RIU and have demonstrated excellent analytical performance regarding the determination of analytes in complex matrices [124,125]. The most popular configurations of interferometric sensors are Mach–Zehnder (MZI) [125–128], Young (YI) [129], Hartman [130–132], and bimodal interferometers [133,134] (Figure 8).

In Mach–Zehnder interferometers (Figure 8a), a waveguide splits into two arms: one that can detect the variations in the refractive index over its surface through a window in the cladding layer (sensing arm) and the other which is fully covered by the cladding layer and operates as a reference (reference arm) [124,125]. The two arms combine again after some point to a single waveguide, and the output light intensity is monitored. Biomolecular reactions taking place in the sensing arm window change the refractive index and cause a phase difference between the light beams guided in the two arms. Thus, the output light is a cosine function of the input light. This means that the sensitivity to effective refractive index changes would be the maximum at the quadrature points and the minimum in the vicinity of the extrema. Regarding the geometrical characteristics of the transducer, most MZIs are symmetric, i.e., the sensing and the reference arms have equal lengths while asymmetric MZIs, i.e., MZIs with different lengths for the two arms, have been also explored. The majority of MZI-based detection systems implement monochromatic light sources, i.e., lasers, which complicate the instrument miniaturization and development of portable systems; therefore broad-band light sources have been explored instead of lasers. In this direction, external broad-band light sources have been coupled to MZIs integrated on the substrate [126,127], or silicon light-emitting diodes (LED) integrated onto the same silicon chip with planar silicon nitride waveguides have been implemented [125,127]. It should be noted that the detection sensitivities in terms of the refractive index achieved with these configurations were comparable to those of MZIs implementing lasers as light sources.

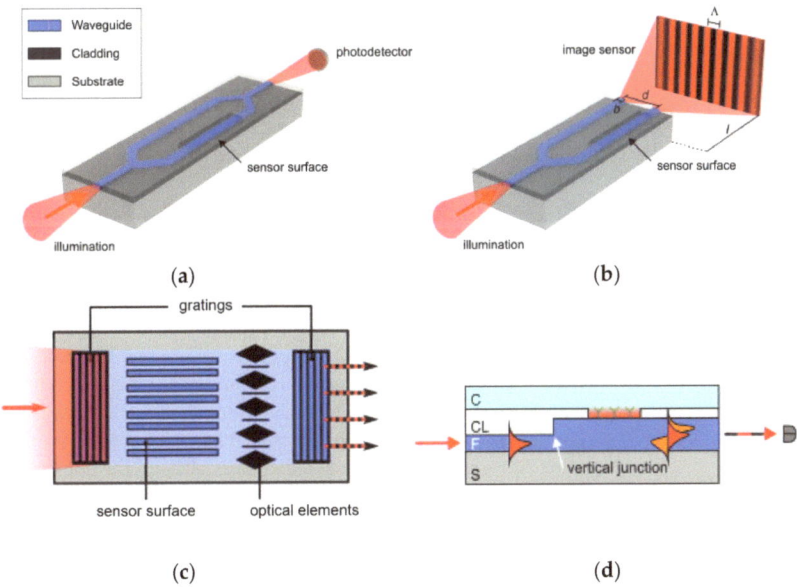

Figure 8. Schematic representation of (**a**) Mach–Zehnder, (**b**) Young, (**c**) Hartman, and (**d**) bimodal interferometer [124]. The arrows indicate the incoming and outgoing light beam. Copyright 2014. Reproduced with permission from the Elsevier B.V.

A Young interferometer (YI) also consists of a waveguide divided into two arms by means of a Y-junction (Figure 8b). The critical difference between MZIs and YIs is that in the second case, the two waveguides do not combine again but the output light interferes in air and the "interferogram" created is depicted on a CCD array. Thus, in YIs, the changes in the effective refractive index over the sensing arm due to binding reactions that cause the phase difference between the two interfering beams are recorded as a shift in the interference fringes [124]. There are considerably fewer reports of integrated YIs as compared to MZIs; nonetheless, it has been demonstrated that YI-based sensors can achieve an LOD 9×10^{-9} in terms of the bulk refractive index and therefore, they are supposed to be more sensitive than the other interferometric sensors as well as SPR.

Hartman interferometers are based on planar waveguides in which the two modes of light, TE and TM, propagate and interact with the adlayer on the same path (Figure 8c) [124]. Changes in the refractive index cause phase shifts of the two polarizations that are not equal because the sensitivity of the two modes to refractive index changes differs.

Bimodal interferometers are the fourth and most recently developed category of interferometric sensors (Figure 8d) [124]. In a bimodal interferometer, the transducer is a single waveguide with two different zones: the first supporting a single mode and the second supporting two modes (fundamental and first-order modes). Those two modes interfere and propagate until they reach the waveguide's output. When refractive index changes occur at the waveguide surface, due to binding reactions, the interference pattern at the waveguide output also changes since the velocity with which the two modes propagate depends on the refractive index of the waveguide adlayer.

The first report for bacteria detection based on interferometric sensors was the immunochemical detection of *S. typhimurium* with a Hartman interferometer in buffer [130]. The sensor could detect 5×10^8 cfu/mL of Salmonella in a direct assay that lasted 40 min. An integrated two-channel Hartman interferometer was also applied for the detection of *S. typhimurium* in spiked chicken rinse fluid [132]. Detection by direct binding of *Salmonella* cells to antibody-modified waveguides was compared to a sandwich assay. Both configurations provided an LOD of 10^4 cfu/mL for an assay duration of 10 min, which corresponded to 20 cfu/mL of chicken rinse fluid after 12 h of enrichment.

Mach–Zehnder interferometers (MZIs) have also been employed for bacteria detection. Thus, MZIs integrated onto silicon chips were modified with an antibody against *L. monocytogenes* following a chemical activation protocol specific to silicon nitride so as to limit antibody attachment to the sensing window areas [126]. The protocol consisted of chip treatment with HF for the creation of amine groups onto the silicon nitride followed by reaction with glutaraldehyde to enable the covalent bonding of antibody molecules through their free amine group. An LOD of 10^5 cfu/mL was achieved for a direct binding assay of 15 min. In another report, the MZIs' waveguide was formed by patterning a photoresist layer deposited on a glass coverslip [127]. The immobilization of an antibody against *E. coli* was performed after modification of the sensing arm with an aminosilane and subsequent activation with glutaraldehyde. The assay duration was 10 min and the LOD was 10^6 cfu/mL. In another report, a chip integrating ten MZIs along with the respective silicon light-emitting diodes was employed for the simultaneous detection of *S. typhimurium* and *E. coli* through a competitive immunoassay format [128]. The chip was activated with aminosilane, and then the liposaccharides of *S. typhimurium* and *E. coli* were spotted onto the sensing arm windows of different MZIs of the chip and immobilized through physical adsorption. MZIs spotted with the blocking protein (bovine serum albumin) were used as reference sensors. For the assay, mixtures of calibrators or samples with the bacteria-specific antibodies were run over the chip followed by reaction with biotinylated secondary antibodies and streptavidin for signal enhancement. Following this format, LODs of 40 cfu/mL for *S. typhimurium* and 110 cfu/mL for *E. coli* in both water and milk samples were achieved for a 10-min assay [128]. It was also calculated that 7.5 h enrichment was necessary to detect 1 cfu/25 mL of *Salmonella* spp. and *E. coli* in order to comply with the EU legislation (CE 1441/2007), which requires the absence of these bacteria (zero tolerance) in 25 g of milk or drinking water.

Regarding Young interferometers there are no reports for the detection of bacteria, although they have been employed for the immunochemical detection of viruses, and more specifically of herpes simplex virus type 1 (HSV-1), achieving an LOD of 850 particles/mL [129]. On the other hand, the label-free detection of *B. cereus* and *E. coli* with a bimodal interferometric immunosensor has also been reported [132]. The immobilization of antibodies was carried out either by physical adsorption on aminosilane-modified chips or by covalent binding to chips modified with a carboxysilane after conversion of surface carboxyl groups to active ester groups. The device could detect 70 cfu/mL of *B. cereus* in 12.5 min and 40 cfu/mL of *E. coli* in 25 min. The same bimodal interferometric sensor was also applied for the detection of multidrug-resistance bacteria genes without amplification through a DNA hybridization assay [133]. Table 3 summarizes the application and analytical performance of integrated interferometric sensors for the detection of bacteria in water and food samples.

3.1.4. Grating-Coupler-Based Immunosensors

Grating coupler sensors are based on planar waveguides with a grating incorporated into the waveguide to enable coupling and transmission of the incident light in a manner dependent on the refractive index of the medium over the waveguide surface [42]. Thus, binding reactions that occur onto the waveguide surface can be monitored by determining the incident light in-coupling angle. In addition, since the binding reactions affect the transmitted light (through interaction with the evanescent wave field), the light out-coupling angle is also altered and can also be used to monitor the binding reactions. The second configuration is advantageous compared to the first one since there is no need for precision alignment of the light, thus leading to simpler experimental setups. Over the years, grating coupler sensors have been upgraded through the introduction of two-dimensional grating structures [135] or new configurations such as wavelength-interrogated optical sensors (WIOSs), which implement two gratings for light in- and out-coupling to the waveguide [136], or angle interrogated optical sensors, which use a MEMS micro-mirror to scan the angle of the incident light of the gratings [137]. All these novel configurations have allowed for the development of systems for multiplexed determinations of analytes. Gratings have also been combined with other evanescent wave-based biosensors such as interferometric [138], SPR [139], SPR imaging [140], optical fiber [141], and silicon microring resonators [142] in order to create more flexible and sensitive biosensors.

Table 3. Bacteria detection with integrated interferometric immunosensors.

Analyte	Detection Principle	Assay Type	Sample Type	Assay Duration	LOD	Ref. No.
S. typhimurium	Hartman interferometer	sandwich and direct	buffer	10 min	10^4 cfu/mL	[132]
			chicken carcass		20 cfu/mL (12 h) *	
S. typhimurium E. coli	MZI	competitive with biotinylated secondary antibodies and streptavidin	water milk	10 min	40 cfu/mL 110 cfu/mL 1 cfu/25 g (7.5 h) *	[128]
E. coli	Bimodal interferometer	direct	buffer	25 min	40 cfu/mL	[133]
B. cereus				12.5 min	70 cfu/mL	

* Enrichment time required to achieve the relevant LOD.

A reverse symmetry waveguide sensor with an integrated grating coupler was applied for the detection of *E. coli K12* in buffer by monitoring the adhesion of bacteria cells onto the sensor surface, which was modified with poly-L-lysine, creating a positively charged layer that enabled bacteria binding through electrostatic interactions [143]. An LOD of 60 cells/mm^2 was achieved following this approach [143]. Optical fiber long-period grating (LPFG) sensors have also been used for the detection of *Staphylococcus aureus* in buffer through modification of the fiber surface with ionic self-assembled multilayers that facilitated the covalent bonding of antibodies specific to penicillin-binding-protein 2a of methicillin-resistant staphylococci. The sensor could discriminate between methicillin-resistant and methicillin-sensitive bacteria, with an LOD of 10^2 cfu/mL for methicillin-resistant bacteria [144]. Similarly, a long-period fiber grating sensor modified with nano-pitted polyelectrolyte coatings and an antibody was implemented for detection of *S. aureus* in buffer, with an LOD of 224 cfu/mL for a 30-min assay [145]. Another LPG immunosensing platform formed by two identical cascaded chirped long period gratings was applied for the detection of *E. coli* in buffer (Figure 9) [146]. The sensor worked like a Mach–Zehnder interferometer due to the space between the two gratings and could detect *E. coli* at concentrations as low as 7 cfu/mL. Finally, a grating-coupler-based biosensing platform known as Optical Waveguide Lightmode Spectroscopy System (OWLS), commercialized by Microvacuum Ltd. (Microvacuum Ltd.; URL: http://www.owls-sensors.com/, accessed on 18 June 2023), was applied for the label-free immunochemical detection of *S. typhimurium* [147] and *L. pneumophila* [148] in water following a direct binding assay format with LODs of 1.3×10^3 and 1.3×10^4 cfu/mL, respectively.

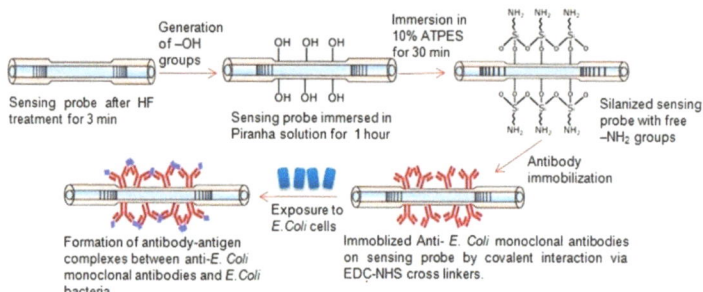

Figure 9. Schematic of antibody immobilization and *Escherichia coli* bacteria detection by a cascaded chirped long-period gratings immunosensor [146]. Copyright 2019. Reproduced with permission from AIP Publishing.

3.1.5. Ring-Resonator-Based Immunosensors

Ring resonators rely on the coupling of light propagating along a linear waveguide, through the evanescent wave field, to a circular one on which it propagates in the form of whispering-gallery modes. Any change in the refractive index in the proximity of the ring surface affects the spectral position of the whispering-gallery modes and changes the wavelength of the incident light for which resonance is achieved. As the light propagating in the ring can interact with the molecules on its surface multiple times, ring resonators are expected to provide the same performance (denoted by the Q factor) as that obtained from linear waveguides with many times longer lengths. Thus, by implementing ring resonators as transducers, smaller-sized devices as compared to linear waveguides and denser transducer arrays can be realized. Ring resonators can adopt not only the 2D format of a microdisk [149] or microring [150] but also the 3D format of a microtoroid [151]. Toroids claim higher Q factors than the planar resonators and therefore, higher detection sensitivity is expected. Sensors based on microring resonators have also been explored with regard to bacteria detection. The detection of *E. coli* in buffer with a microring resonator sensor has been reported and the relatively high LOD of 10^5 cfu/mL was ascribed to suboptimal functionalization of resonators with bacteria-binding antibodies (Figure 10) [152].

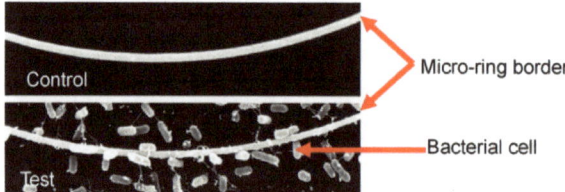

Figure 10. Scanning electron micrographs of a small section of the test and control micro-rings on a resonator chip showing specific bacterial binding at 2200× magnification [152]. Copyright 2007. Reproduced with permission from Elsevier B.V.

Finally, a whispering gallery mode optical microdisk resonator was modified with the phage protein LysK, an endolysin from the staphylococcal phage K that binds strongly to staphylococci and used to detect *S. aureus* in buffer with an LOD of 5×10^6 cfu/mL [153].

3.2. Reflectometric Immunosensors

Reflectometric sensors rely on monitoring shifts in the interference spectrum due to binding reactions taking place on a stack of materials with different refractive indices. The illumination sources usually employed are white light sources.

A commercially available RIfS-based biosensing device was used to detect *L. pneumophila* in buffer by monitoring either the direct capture of bacteria cells via electrostatic interaction onto the chip surface or through a sandwich assay [154]. The device's performance was compared to that of SPR, and an LOD of 1×10^5 cfu/mL was determined in both cases. A white-light-reflectance-spectroscopy-based immunosensor exhibited better performance regarding the rapid and sensitive detection of *S. typhimurium* in drinking water (Figure 11) [155]. The sensor chip consisted of a Si die with a thin SiO_2 layer on top, and *S. typhimurium* detection was performed through a competitive immunoassay format between the bacteria in the sample and the *Salmonella* liposaccharide immobilized onto the chip. An LOD of 320 cfu/mL in drinking water was achieved for an assay duration of 15 min.

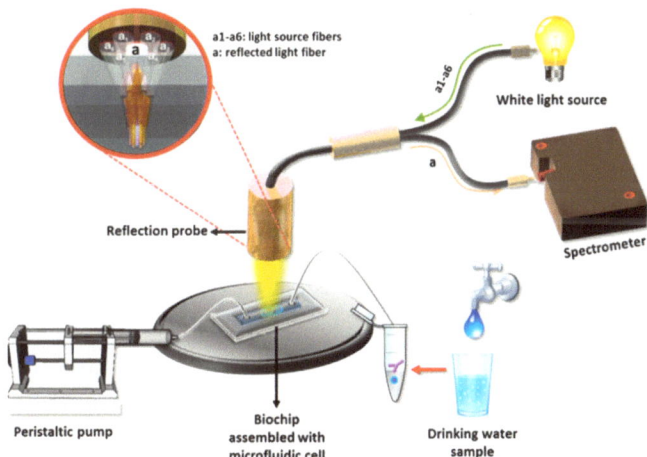

Figure 11. Illustration of the white light reflectance spectroscopy (WLRS) optical setup and sensing principle [155]. Copyright 2021. Reproduced with permission from MDPI.

Porous silicon has been also employed as substrate for detection of bacteria through reflectance spectroscopy. Thus, the label-free detection of *E. coli K12* in buffer was achieved using a sensor based on a nanostructured oxidized porous silicon thin film [156,157]. The sensor surface was functionalized with specific antibodies against *E. coli* through aminosilanization and activation of surface amine groups with bis(N-succinimidyl)carbonate for the coupling of antibodies via their free amine groups [156,157]. The LOD determined was about 10^4 cells/mL and the assay was completed in 30 min. Optimization of the same sensor, in which a different surface modification was followed (Figure 12), resulted in an LOD of 10^3 cells/mL in water for an assay of 45 min [158]. In this case, after aminosilanization, the surface was functionalized with glutaraldehyde to introduce aldehyde groups through which streptavidin was bound onto the surface to facilitate immobilization of biotinylated anti-bacteria-specific antibodies.

Another biosensor developed for *E. coli* detection on porous silicon substrates involved surface modification with a hydrogel made of polyacrylamide to which biotinylated specific monoclonal antibodies were immobilized onto the streptavidin covalently bound to the surface after appropriate chemical functionalization [159]. A detection limit in the range of 10^3–10^5 cell/mL was determined for a direct bacteria binding assay of 30 min. Furthermore, for the direct detection of *E. coli*, a biosensor based on the blockage of nanopores created by etching of a Si chip was presented, achieving a detection limit in buffer of 10^3 cfu/mL [160]. More specifically, when *E. coli* cells were trapped into the chip nanopores, a decrease in effective optical thickness was recorded. Thus, by monitoring the change of effective optical thickness value, it was possible to quantitatively determine the cells captured in the nanopores via indirect Fourier Transformed Reflectometric Interference Spectroscopy.

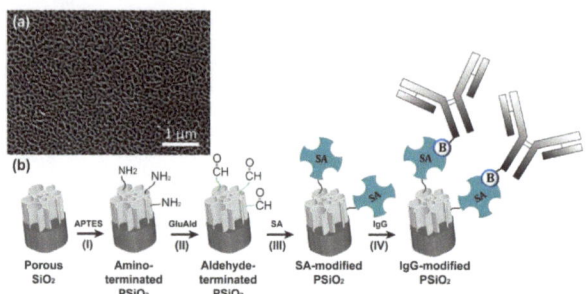

Figure 12. (a) A top-view high-resolution scanning electron microscope image of a porous SiO$_2$ film with pores in the range of 60–100 nm. (b) Schematic representation of the steps followed to bio-functionalize the porous SiO$_2$ surface with IgG, including: (I) 3-aminopropyl-triethoxysilane modification, (II) reaction of amine-terminated porous SiO$_2$ with one of the aldehyde groups of the cross-linker glutaraldehyde, (III) grafting of streptavidin onto the aldehyde-terminated surface, and (IV) conjugation of biotinylated-IgG (*E. coli*) via biotin-streptavidin binding [158]. Copyright 2016. Reproduced with permission from Springer Nature.

Finally, an interferometric reflectance imaging system (IRIS) was employed for the label-free detection of *E. coli* in tap water [161]. The bacteria specific antibody was spotted onto a SiO$_2$/Si chip modified with a polymer and after incubation for 2 h with the bacteria solutions the bound cells were counted using a low-magnification optical setup accompanied by an appropriate software. Based on experimental data, an extrapolated LOD of 2.2 cfu/mL was calculated which is the lowest, so far, reported for direct bacteria detection. Table 4 summarizes the analytical performance of reflectometric sensors for detection of bacteria in water and food samples.

Table 4. Bacteria detection with reflectrometric immunosensors.

Analyte	Detection Principle	Assay Type	Sample Type	Assay Duration	LOD	Ref. No
S. typhimurium	WLRS	competitive	drinking water	15 min	320 cfu/mL	[155]
E. coli	reflectance spectroscopy on porous silicon	direct	water	45 min	10^3 cells/mL	[158]
E. coli	reflectance spectroscopy imaging	direct	tap water	2 h	2.2 cfu/mL	[161]

3.3. Photoluminescence-Based Immunosensors

Photoluminescence (fluorescence and phosphorescence) is the phenomenon of light emission from a molecule that has been excited by absorption of photons in the visible or UV region [162]. Thus, the decrease in photoluminescence intensity of TiO$_2$ nanoparticles deposited on glass slides and modified with antibodies against *S. typhimurium* upon binding of bacteria was exploited in an immunosensor that could detect *S. typhimurium* in the range from 10^3 to 10^5 cells/mL [163]. In another report, a soda lime glass was employed as a waveguide and microarray substrate for the detection of *S. typhimurium* in several food samples as well as in chicken excretal samples through a sandwich immunoassay with fluorescently labeled detection antibodies [164]. The LOD achieved was 8×10^4 cfu/mL for an assay duration of 15 min and could be improved 10-times by increasing the assay duration to 1 h [164]. This array biosensor was also used to detect *Shigella dysenteriae* and *Campylobacter jejuni* in buffer and a variety of food and beverage samples (chicken carcass rinse, ground turkey, buffered milk, and lettuce leaf rinse) at concentrations of 4.9×10^4 and 9.7×10^2 cfu/mL, respectively, by applying sandwich immunoassays that lasted 25 min [165]. The same sensor was applied to detect the bacterium *Campylobacter jejuni* following a 25-min sandwich immunoassay in a number of different food matrices (ground turkey, sausage, ham, carnation non-fat dried milk, and vanilla fat-free yogurt) with an LOD of 500 cells/mL [166]. Finally, the sensor was applied to detect *Escherichia coli* in less than 30 min in various spiked food matrices (ground beef, turkey sausage, carcass wash, and apple juice) with LODs in the range $1-5 \times 10^4$ cells/mL [167].

A homogeneous FRET immunosensor using antibodies conjugated to graphene oxide quantum dots and graphene oxide sheets was designed for the detection of *C. jejuni* cells in food samples [168]. The antibody conjugated to graphene oxide quantum dots interacted with the graphene oxide sheets through π-π stacking, leading to fluorescence quenching. When *C. jejuni* was selectively captured by the antibody, this interaction was disrupted, and the fluorescence emission increased proportionally to the concentration of bacteria in the sample. The assay was completed in 1.5 h and the LOD for the detection of *C. jejuni* to poultry liver was 100 cfu/mL. In Table 5, the analytical performance of photoluminescence-based sensors for the detection of bacteria in water and food samples are presented.

Table 5. Photoluminescence-based immunosensors for bacteria detection.

Analyte	Detection Principle	Assay Type	Sample Type	Assay Duration	LOD	Ref. No.
S. typhimurium	fluorescence	sandwich	sausage cantaloupe whole liquid egg alfalfa sprouts chicken carcass rinse chicken excretal samples	15 min 1 h	8×10^4 cfu/mL 8×10^3 cfu/mL or cfu/g	[164]
Shigella dysenteriae C. jejuni	fluorescence	sandwich	buffer chicken carcass rinse ground turkey buffered milk lettuce leaf rinse	25 min	4.9×10^4 cfu/mL 9.7×10^2 cfu/mL	[165]
C. jejuni	fluorescence	sandwich	ground turkey sausage ham carnation non-fat dried milk vanilla fat-free yogurt	25 min	500 cells/mL	[166]
E. coli	fluorescence	sandwich	ground beef turkey sausage carcass wash apple juice	30 min	$1-5 \times 10^4$ cells/mL	[167]
C. jejuni	homogeneous FRET	direct	poultry liver	1.5 h	100 cfu/mL	[168]

3.4. Surface Enhanced Raman Scattering (SERS)-Based Immunosensors

SERS-based biosensors combine Raman inelastic scattering of incident light with signal enhancement provided by nanostructured noble metal substrates [169]. More specifically, the Raman signal of molecules adsorbed onto the SERS surfaces is enhanced by a factor of 10^4–10^8 due to the strong electromagnetic field generated on the surface of these substrates. This enhanced scattering phenomenon results in characteristic peaks due to resonance with the vibrational modes of the molecules, which are unique for each molecule. Thus, in SERS-based bioanalysis, the target molecules can either be detected directly after their attachment to the nanostructured surfaces (Figure 13a), through their binding to surface-anchored recognition elements, such as antibodies (Figure 13b), or through sandwich immunoassays employing SERS-active labels for signal enhancement (Figure 13c) [170–173]. SERS active labels or tags are prepared by conjugating the analyte-specific antibodies to nanoparticles along with Raman reporter molecules, which are low-molecular-weight moieties with strong and distinguishable Raman signals. All of these approaches have been implemented for bacteria detection through SERS.

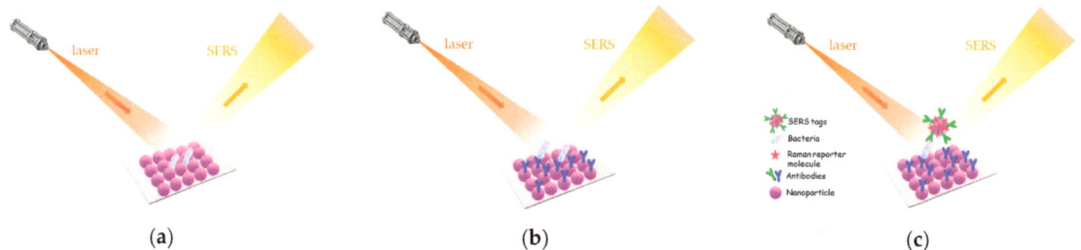

Figure 13. Schematic illustration of SERS-based approaches for detection (**a**) without any specific recognition molecule, (**b**) through binding to a surface-immobilized antibody, or (**c**) through sandwich immunoassays employing a surface-immobilized capture antibody and a detection antibody labeled with a SERS tag.

Knauer et al. developed direct label-free SERS-based immunochemical methods for the detection of *L. pneumophila* and *S. typhimurium* in a single run [174], and of *E. coli* separately [175]. Regarding the simultaneous determination of *S. typhimurium* and *L. pneumophila*, glass substrates were modified with an epoxy-silane and then reacted with a diamine-polyethylene glycol to introduce amine groups onto the surfaces, which were then implemented for the covalent bonding of antibodies specific against the two bacteria by spotting of the respective solutions at different areas of the substrate. After bacteria binding, the surfaces were incubated with an Ag colloid preparation that aggregated onto the immobilized bacteria, creating "hot spots" which provided a strong Raman signal. The assay lasted 65 min, and the LODs achieved were 10^6 and 10^8 cells/mL for *L. pneumophila* and *S. typhimurium*, respectively, in the water samples [174]. For *E. coli* detection, the antibody-modified substrates were combined with a flow-through system, enabling the detection of *E. coli* strains in water samples at concentrations down to 4.3×10^3 cfu/mL [175].

An LOD of 10 cfu/mL in spiked ground beef homogenates was achieved for *E. coli* O157:H7 with a SERS-based non-competitive immunoassay that combined antibody-modified magnetic particles and gold nanoparticles modified with an anti-bacterium antibody and a SERS tag molecule [176]. At first, the bacteria solution was incubated with the antibody-labeled magnetic nanoparticles, and after magnetic separation, the bound bacteria were incubated with the SERS-tagged antibody-modified gold nanoparticles. The immunocomplexes formed were separated from free antibodies using a membrane filter on which the Raman signal was determined [176].

A similar immunoassay format was applied to the simultaneous detection of *E. coli* O157:H7 and *S. aureus*, with LODs of 10 and 25 cfu/mL, respectively [177]. In this case, magnetic beads and SERS active gold nanoparticles functionalized with specific antibody pairs against each bacterium were used, and the simultaneous determination was based on implementation of a different SERS tag for each bacterium. After the formation of immunocomplexes in the liquid phase, a magnetic field was applied and the Raman signals from the two different tags were quantified [177]. The sensor was applied for the detection of the two bacteria in spiked samples of bottled water and milk.

In another report, the capture antibodies were immobilized on magnetite-gold nanoparticles to enable separation and concentration of the *E. coli O157* cells from the liquid, followed by reaction with gold nanoparticles modified with antibodies and Raman tags [178]. This method provided rapid separation and detection (less than one hour) of *E. coli* in apple juice, achieving an LOD of 10^2 cfu/mL. A similar approach was applied for the isolation and detection of multiple pathogenic bacteria through the implementation of lectin functionalized silver coated magnetic nanoparticles prior to reaction with SERS-tagged silver nanoparticles functionalized with antibodies specific to each one of the targeted bacteria. Thus, *E. coli, S. typhimurium,* and *methicillin-resistant S. aureus* isolation and detection in buffer were achieved at concentrations as low as 10 cfu/mL [179].

Gold-coated magnetic nanoparticles (gold-coated $MnFe_2O_4$ nanoparticles) were conjugated with antibodies against *S. aureus* and used as SERS tags, as well as for bacteria capture and separation from spiked buffer samples, resulting in an assay with an LOD of 10 cfu/mL [180]. Spherical and rod-shaped gold nanoparticles modified with Raman tags and an antibody against *E. coli* were compared as labels in a sandwich immunoassay with capture antibody immobilized onto a gold-coated glass slide [181]. The LODs determined in buffer using the gold nanorods and the spherical gold nanoparticles as labels were 4 and 5 cfu/mL, respectively.

Rod-shaped gold-covered magnetic nanoparticles modified with an antibody against *E. coli* have also been investigated as labels in a liquid phase assay for the detection of *E. coli* in buffer at concentrations as low as 35 cfu/mL [182]. The difference in the LOD of this report compared to the previous one that used the same label [181] was attributed to the lower capture efficiency of magnetic gold nanorod particles as compared to the gold-coated glass slide surface. Thus, when the same group implemented gold-coated magnetic spherical nanoparticles modified with an anti-*E. coli* antibody in combination with rod-shaped gold nanoparticles modified with Raman tags and an anti-*E. coli* antibody, an LOD for *E. coli* detection in water samples of 8 cfu/mL was achieved [183].

In another detection approach, SERS was combined with a microfluidic dielectrophoretic device to detect *Salmonella enterica serotype Choleraesuis* and *Neisseria lactamica* in buffer [184]. The SERS labels employed were silica-coated dye-induced aggregates of a small number of gold nanoparticles, denoted as nanoaggregate-embedded beads, and they were modified with antibodies specific to each bacterium to allow their online detection with an LOD of 70 cfu/mL.

SERS detection was also combined with lateral flow strip biosensors. Thus, *L. monocytogenes* and *S. typhimurium* were simultaneously detected with a lateral flow sandwich immunoassay employing gold nanoparticles modified with a Raman tag and specific antibodies against each one of the targeted bacteria [185]. The LODs achieved were 75 cfu/mL for both bacteria, and the strip assay was applied for bacteria detection in milk samples. Another lateral flow strip biosensor employing SERS labels combined with recombinase polymerase amplification (RPA) was applied for the simultaneous determination of *S. enteritidis* and *L. monocytogenes* [186]. The method made use of forward primers labeled with digoxin and fluorescein for *S. enteritidis* and *L. monocytogenes*, respectively, whereas the reverse primers were labeled with biotin. Thus, when the RPA product was applied to the sample pad, it was run along with gold nanoparticles modified with streptavidin and Raman tags toward the two test lines where antibodies against digoxin and fluorescein have been spotted, resulting in the creation of the respective colored lines (Figure 14). The LODs achieved in buffer were 27 and 19 cfu/mL for *S. enteritidis* and *L. monocytogenes*, respectively [186]. The method was applied for the detection of the two bacteria in milk, chicken breast, and beef samples with slightly increased LODs, which were 31, 35, and 35 cfu/mL for *S. Enteritidis* and 36, 29, and 22 cfu/mL for *L. monocytogenes*, respectively. The same approach was employed to detect *Escherichia coli O157:H7*, with an LOD of 5×10^4 cfu/mL in milk, chicken breast, and beef [187].

In Table 6, data regarding bacteria detection with SERS-based biosensors in food and water samples are presented.

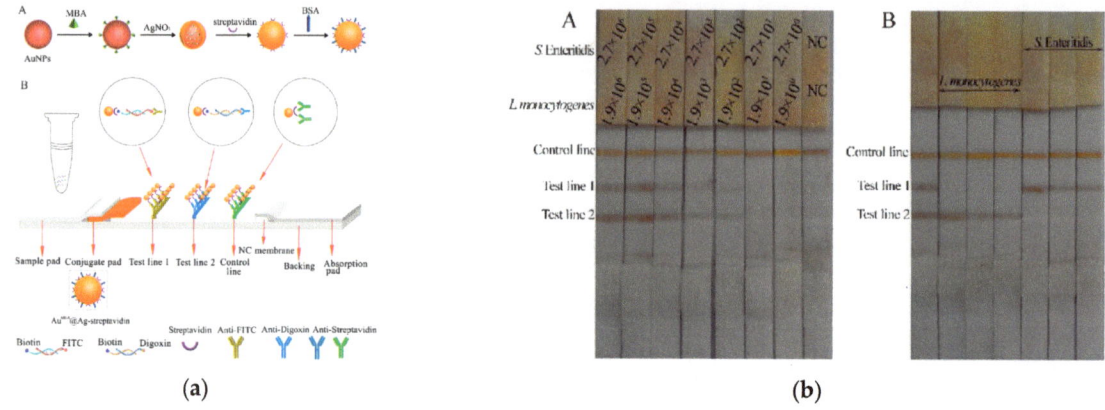

Figure 14. (**a**) Schematic illustration of: (A) preparation of gold nanoparticles modified with streptavidin and Raman tags (MBA), and (B) the multiplex lateral flow SERS assay. (**b**) (A) Results from lateral flow strips for the simultaneous detection of *S. enteritidis* (test line 1) and *L. monocytogenes* (test line 2). The test lines could be observed by the naked eye at a concentration of 1.9×10^2 cfu/mL of *L. monocytogenes* and 2.7×10^2 cfu/mL of *S. enteritidis*. The SERS signal of the test lines could be detected at a concentration of 1.9×10^1 cfu/mL of *L. monocytogenes* and 2.7×10^1 cfu/mL of *S. enteritidis*. (B) Specificity of the lateral flow strip assay against three *L. monocytogenes* and three *S. enteritidis* bacteria [186]. Copyright 2017. Reproduced with permission from the American Chemical Society.

Table 6. SERS-based immunosensors for bacteria detection.

Analyte	Detection Principle	Assay Type	Sample Type	Assay Duration	LOD	Ref. No
L. pneumophila *S. typhimurium*	SERS	direct	water	65 min	10^6 cfu/mL 10^8 cfu/mL	[174]
E. coli O157:H7	SERS	direct	water	>60 min	4.3×10^3 cfu/mL	[175]
E. coli O157:H7	SERS	sandwich	ground beef	~2 h	10 cfu/mL	[176]
E. coli *S. aureus*	SERS	sandwich	bottled water milk	~2 h	10 cfu/mL 25 cfu/mL	[177]
E. coli O157	SERS	sandwich	apple juice	<1 h	10^2 cfu/mL	[178]
E. coli	SERS	sandwich	water	70 min	8 cfu/mL	[183]
L. monocytogenes *S. typhimurium*	LFIA SERS	sandwich	milk	10 min	75 cfu/mL	[185]
S. enteritidis/ *L. monocytogenes*	LFIA SERS	sandwich	buffer chicken breast beef milk	30 min	27/19 cfu/mL 35/29 cfu/mL 35/22 cfu/mL 31/36 cfu/mL	[186]
*E. coli*O157:H7	LFIA SERS	sandwich	milk chicken breast beef	15 min	5×10^4 cfu/mL	[187]

4. Conclusions and Outlook

This review has outlined the principles and applications of antibody-based optical sensors in the detection of food pathogenic bacteria. Although there is an abundance of relative publications, it seems that very few of the available sensing principles have been exploited for the detection of bacteria in real food samples. An additional reason for that can be the fact that the LODs achieved in most cases cannot cover the requirements set by the existing regulations, according to which highly dangerous and contagious bacteria like *Salmonella* and *L. monocytogenes* should not be detected in 25 g of food samples (e.g., Commission Regulation (EC) No 2073/2005).

Thus, although fiber optic sensors employing fluorescent labels have been the first to be implemented for bacteria detection in food samples, the great majority of references, especially the most recent ones, rely on SPR sensors. The reason for that is probably due to the availability of different instruments based on the SPR detection principle, some of which are commercially available. Nevertheless, the LODs achieved in most of the cases ranged from 10^3–10^5 cfu/mL, which are considered high for direct application to food analysis, and they should be combined with a sample enrichment procedure that adds a few hours to the total time required from sampling to answer.

There are a few reports, however, that report LODs of a few or a few tens of cfu/mL [66,95,97,103–105], from which only one is performed in a standard benchtop SPR instrument [66], and two others with a commercially available portable SPR instrument (SPREETA™) [95,97], which is not currently in the market. The first two reports were based on direct binding assays for the label-free detection of targeted bacteria [66,95], whereas in the third, gold magnetic particles were employed as labels to drop the LOD to 3 cfu/mL [97]. The rest of the reports implement the localized SPR (LSPR) principle either in the form of gold nanoparticles [103,104] or in the form of nanostructured gold film [105]. In addition to higher detection sensitivity, LSPR is also considered more easily adaptative to portable low-cost devices, and it remains to be seen what would be carried out in this direction in the near future.

Regarding the majority of fiber-optic-based immunosensors, the LODs achieved for bacteria detection also ranged from 10^3–10^5 cfu/mL. LODs of less than 100 cfu/mL were achieved only when fibers were combined with SPR via covering their sensing area with a gold film or gold nanoparticles or films of other materials (e.g., MoS_2 nanosheets) [122,123]. Since fiber optic experimental setup can also be reduced in size and cost, the combination with SPR might also be proved a viable solution for portable devices in the future.

Integrated interferometers, particularly in the form of MZIs [128] and bimodal interferometers [133], have shown adequate sensitivity for bacteria detection (LODs in the range of a few tens of cfu/mL) combined with short assay times and great potential for multiplexed determinations. They also seem to be the most promising candidate in the direction of portable devices suitable for point-of-need determinations. On the other hand, microring resonators or microtoroids, despite their claimed high detection sensitivity, are the optical transducers less frequently employed for bacteria detection, whereas a few existing reports present LODs as non-competitive to other types of optical biosensors [152,153]. The same seems to be true for most of the immunosensors based on grating couplers for which there is only one report, with an LOD of a few cfu/mL in which an optical fiber grating device was implemented [144].

Immunosensors based on reflectance spectroscopy have also demonstrated potential for the detection of bacteria at concentrations as low as 2.2 cfu/mL [161], whereas the assay was performed in a 24-well plate with the interferometric reflectance imaging system (IRIS) chip placed at the bottom of each well and lasted over 2 h. A more compact system based on reflectance spectroscopy achieved an LOD of 320 cfu/mL with a 15-min assay, allowing for the detection of bacteria in food samples after a pre-enrichment step of approximately 8 h [155]. The most impressive performance regarding the percentage of reports that mention LODs of less than 100 cfu/mL present immunosensors based on SERS [176–182,184–186]. However, with the exception of two reports where Raman spectroscopy is combined with a lateral flow immunoassay [182,184], for which the assay duration was less than 30 min, most sensors required more than 1 h to complete the assay.

Regarding the assay type implemented for bacteria detection with optical immunosensors, although direct binding onto antibody-modified transducers has been widely used, it has rarely led to high detection sensitivity combined with a short assay time. Thus, competitive or non-competitive immunoassay formats have been employed to increase detection sensitivity and/or decrease assay duration. This might complicate the development of portable devices a bit since it requires the integration of optical sensors with microfluidic modules, pumps, and valves, which will provide for the automated execution of the assay steps. In addition, this integration should be performed in such a way that it will not increase either the device size or the cost.

In the direction of developing portable devices based on optical transducers, the tools offered by the continuously evolving smartphone gadgets, either as light sources or for the detection of optical signals or the ability to run the instrument software in such a device, process the data, and transfer them wirelessly to central facilities, are considered a significant asset.

Although the current technological limitations are more than certain to be surpassed in the near future, there are some aspects of optical immunosensors that need to be further addressed prior to their application for bacteria detection at the point of need. For example, the study of relevant literature reveals that sensitive bacteria detection with immunochemical techniques is often challenging due to lack of appropriate antibodies. Thus, in many instances, optical immunosensors have been developed using in-house produced antibodies that are not widely available. Even if antibodies with appropriate binding characteristics for the specific and sensitive detection of a particular bacterium are commercially available, their suitability for the detection of bacteria in processed food has to be investigated since the structure of bacteria epitopes could change dramatically when the food has been processed under certain temperature or pH conditions. In addition, in many instances, antibodies cannot discriminate viable from non-viable cells, providing false positive results.

Despite the above-mentioned limitations of optical immunosensors for bacteria detection, they remain one of our best hopes for sensitive portable devices that would be available at affordable prices with low operation costs but most importantly in a shorter time compared to the established techniques for bacteria detection in food. The long time required from sampling to testing by most of the established methods (especially the microbiological ones) is and will remain the driving force behind the development of biosensors for application in food safety. Even if the necessity for sample enrichment remains, suppression of the time required from days to a few hours will help to test a higher amount of raw food materials and ready-to-eat food by both the industry and the food-safety control departments.

Author Contributions: Conceptualization, D.K. and M.A.; writing—original draft preparation, D.K., M.A. and P.P.; writing—review and editing, M.A., P.P. and S.K.; funding acquisition, M.A., P.P. and S.K. All authors have read and agreed to the published version of the manuscript.

Funding: This research was co-financed by the European Regional Development Fund of the European Union and Greek national funds through the Operational Program Competitiveness, Entrepreneurship and Innovation, under the call RESEARCH–CREATE–INNOVATE (project code: T2EΔK-01934/FOODSENS), and the European Union's Horizon 2020 Research and Innovation program through the Marie Sklodowska-Curie grant agreement No. 101007299 (SAFEMILK).

Institutional Review Board Statement: Not applicable.

Informed Consent Statement: Not applicable.

Data Availability Statement: No new data were created or analyzed in this study. Data sharing is not applicable to this article.

Conflicts of Interest: The authors declare no conflict of interest.

References

1. World Health Organization. Foodborne Diseases. Available online: https://www.who.int/health-topics/foodborne-diseases#tab=tab_1 (accessed on 18 June 2023).
2. World Health Organization. The Top 10 Causes of Death. Available online: https://www.who.int/news-room/fact-sheets/detail/the-top-10-causes-of-death (accessed on 18 June 2023).
3. World Health Organization. *Estimating the Burden of Foodborne Diseases: A Practical Handbook for Countries*; World Health Organization: Geneva, Switzerland, 2021.
4. Addis, M.; Sisay, D. A review on major food borne bacterial illnesses. *J. Trop. Dis.* **2015**, *3*, 1000176.
5. Madigan, M.; Martinko, J.; Stahl, D.; Clark, D.P. *Brock Biology of Microorganisms*, 13th ed.; Benjamin Cummings: San Francisco, CA, USA, 2012; pp. 1022–1042.
6. Zourob, M.; Elwary, S.; Turner, A. *Principles of Bacterial Detection: Biosensors, Recognition Receptors and Microsystems*, 1st ed.; Springer Science+Business Media, LLC: New York, NY, USA, 2008.
7. Saravanan, A.; Kumar, P.S.; Hemavathy, R.V.; Jeevanantham, S.; Kamalesh, R.; Sneha, S.; Yaashikaa, P.R. Methods of detection of food-borne pathogens: A review. *Environ. Chem. Lett.* **2021**, *19*, 189–207. [CrossRef]
8. Gracias, K.S.; McKillip, J.L. A review of conventional detection and enumeration methods for pathogenic bacteria in food. *Can. J. Microbiol.* **2004**, *50*, 883–890. [CrossRef] [PubMed]
9. Zhao, X.; Lin, C.W.; Wang, J.; Oh, D.H. Advances in rapid detection methods for foodborne pathogens. *J. Microbiol. Biotechnol.* **2014**, *24*, 297–312. [CrossRef]
10. Wang, Y.; Salazar, J.K. Culture-independent rapid detection methods for bacterial pathogens and toxins in food matrices. *Compr. Rev. Food Sci. Food Saf.* **2016**, *15*, 183–205. [CrossRef] [PubMed]
11. Garrido-Maestu, A.; Tomás Fornés, D.; Prado Rodríguez, M. The use of multiplex real-time PCR for the simultaneous detection of foodborne bacterial pathogens. In *Foodborne Bacterial Pathogens: Methods and Protocols*; Bridier, A., Ed.; Springer Science+Business Media, LLC: New York, NY, USA, 2019; Volume 1918, pp. 35–45.
12. Liu, Y.; Cao, Y.; Wang, T.; Dong, Q.; Li, J.; Niu, C. Detection of 12 common food-borne bacterial pathogens by Taq Man real-time PCR using a single set of reaction conditions. *Front. Microbiol.* **2019**, *10*, 222. [CrossRef]
13. Moon, Y.-J.; Lee, S.-Y.; Oh, S.-W. A review of isothermal amplification methods and food-origin inhibitors against detecting food-borne pathogens. *Foods* **2022**, *11*, 322. [CrossRef]
14. Magliulo, M.; Simoni, P.; Guardigli, M.; Michelini, E.; Luciani, M.; Lelli, R.; Roda, A. A rapid multiplexed chemiluminescent immunoassay for the detection of *Escherichia coli* O157:H7, *Yersinia enterocolitica*, *Salmonella typhimurium*, and *Listeria monocytogenes* pathogen bacteria. *J. Agric. Food Chem.* **2007**, *55*, 4933–4939. [CrossRef]
15. Cavaiuolo, M.; Paramithiotis, S.; Drosinos, E.H.; Ferrante, A. Development and optimization of an ELISA based method to detect *Listeria monocytogenes* and *Escherichia coli* O157 in fresh vegetables. *Anal. Methods* **2013**, *5*, 4622–4627. [CrossRef]
16. Zhu, L.; He, J.; Cao, X.; Huang, K.; Luo, Y.; Xu, W. Development of a double-antibody sandwich ELISA for rapid detection of *Bacillus cereus* in food. *Sci. Rep.* **2016**, *6*, 16092. [CrossRef]
17. Hochel, I.; Slavíčková, D.; Viochna, D.; Škvor, J.; Steinhauserová, I. Detection of *Campylobacter* species in foods by indirect competitive ELISA using hen and rabbit antibodies. *Food Agric. Immunol.* **2007**, *18*, 151–167. [CrossRef]
18. Rohde, A.; Hammerl, J.A.; Boone, I.; Jansen, W.; Fohler, S.; Klein, G.; Dieckmann, R.; Al Dahouk, S. Overview of validated alternative methods for the detection of foodborne bacterial pathogens. *Trend Food Sci. Technol.* **2017**, *62*, 113–118. [CrossRef]
19. Law, J.W.F.; Mutalib, N.S.A.; Chan, K.G.; Lee, L.H. Rapid methods for the detection of foodborne bacterial pathogens: Principles, applications, advantages and limitations. *Front. Microbiol.* **2014**, *5*, 770. [CrossRef] [PubMed]
20. Khansili, N.; Rattu, G.; Krishna, P.M. Label-free optical biosensors for food and biological sensor applications. *Sens. Actuators B* **2018**, *265*, 35–49. [CrossRef]
21. Qiao, Z.; Fu, Y.; Lei, C.; Li, Y. Advances in antimicrobial peptides-based biosensing methods for detection of foodborne pathogens: A review. *Food Control* **2020**, *112*, 107116. [CrossRef]
22. Thévenot, D.R.; Toth, K.; Durst, R.A.; Wilson, G.S. Electrochemical biosensors: Recommended definitions and classification. *Biosens. Bioelectron.* **2001**, *16*, 121–131. [CrossRef]
23. Morales, M.A.; Halpern, J.M. Guide to selecting a biorecognition element for biosensors. *Bioconjugate Chem.* **2018**, *29*, 3231–3239. [CrossRef]
24. Wu, Q.; Zhang, Y.; Yang, Q.; Yuan, N.; Zhang, W. Review of electrochemical DNA biosensors for detecting food borne pathogens. *Sensors* **2019**, *19*, 4916. [CrossRef]
25. Ansari, N.; Yazdian-Robati, R.; Shahdordizadeh, M.; Wang, Z.; Ghazvini, K. Aptasensors for quantitative detection of *Salmonella typhimurium*. *Anal. Biochem.* **2017**, *533*, 18–25. [CrossRef]
26. Tawil, N.; Sacher, E.; Mandeville, R.; Meunier, M. Surface plasmon resonance detection of *E. coli* and methicillin-resistant S. aureus using bacteriophages. *Biosens. Bioelectron.* **2012**, *37*, 24–29. [CrossRef]
27. Wen, T.; Wang, R.; Sotero, A.; Li, Y. A portable impedance immunosensing system for rapid detection of *Salmonella typhimurium*. *Sensors* **2017**, *17*, 1973. [CrossRef]
28. Viswanathan, S.; Rani, C.; Ho, J.A.A. Electrochemical immunosensor for multiplexed detection of food-borne pathogens using nanocrystal bioconjugates and MWCNT screen-printed electrode. *Talanta* **2012**, *94*, 315–319. [CrossRef]

29. Fulgione, A.; Cimafonte, M.; Della Ventura, B.; Iannaccone, M.; Ambrosino, C.; Capuano, F.; Proroga, Y.T.R.; Velotta, R.; Capparelli, R. QCM-based immunosensor for rapid detection of *Salmonella typhimurium* in food. *Sci. Rep.* **2018**, *8*, 16137. [CrossRef] [PubMed]
30. Ye, Y.; Guo, H.; Sun, X. Recent progress on cell-based biosensors for analysis of food safety and quality control. *Biosens. Bioelectron.* **2019**, *126*, 389–404.
31. Templier, V.; Roux, A.; Roupioz, Y.; Livache, T. Ligands for label-free detection of whole bacteria on biosensors: A review. *TrAC Trend Anal. Chem.* **2016**, *79*, 71–79. [CrossRef]
32. Chakraborty, M.; Hashmi, M.S.J. An overview of biosensors and devices. *Ref. Mod. Mater. Sci. Mater. Eng.* **2017**, *1*, 1–23.
33. Dong, J.; Zhao, H.; Xu, M.; Maa, Q.; Ai, S. A label-free electrochemical impedance immunosensor based on AuNPs/PAMAM-MWCNT-Chi nanocomposite modified glassy carbon electrode for detection of *Salmonella typhimurium* in milk. *Food Chem.* **2013**, *141*, 1980–1986.
34. Lin, Y.H.; Chen, S.H.; Chuang, Y.C.; Lu, Y.C.; Shen, T.Y.; Chang, C.A.; Lin, C.S. Disposable amperometric immunosensing strips fabricated by Au nanoparticles-modified screen-printed carbon electrodes for the detection of foodborne pathogen *Escherichia coli* O157:H7. *Biosens. Bioelectron.* **2008**, *23*, 1832–1837. [CrossRef]
35. Mathelié-Guinlet, M.; Cohen-Bouhacina, T.; Gammoudi, I.; Martin, A.; Béven, L.; Delville, M.H.; Grauby-Heywang, C. Silica nanoparticles-assisted electrochemical biosensor for the rapid, sensitive and specific detection of *Escherichia coli*. *Sens. Actuators B* **2019**, *292*, 314–320.
36. Chen, S.H.; Wu, V.C.H.; Chuang, Y.C.; Lin, C.S. Using oligonucleotide-functionalized Au nanoparticles to rapidly detect foodborne pathogens on a piezoelectric biosensor. *J. Microbiol. Method* **2008**, *73*, 7–17. [CrossRef]
37. Kalograiaki, I.; Euba, B.; Proverbio, D.; Campanero-Rhodes, M.A.; Aastrup, T.; Garmendia, J.; Solís, D. Combined bacteria microarray and quartz crystal microbalance approach for exploring glycosignatures of nontypeable *Haemophilus influenzae* and recognition by host lectins. *Anal. Chem.* **2016**, *88*, 5950–5957. [CrossRef]
38. Yoo, S.M.; Lee, S.Y. Optical biosensors for the detection of pathogenic microorganisms. *Trend Biotechnol.* **2016**, *34*, 7–25. [CrossRef]
39. Sivakumar, R.; Lee, N.Y. Recent advances in airborne pathogen detection using optical and electrochemical biosensors. *Anal. Chim. Acta* **2022**, *1234*, 340297. [CrossRef] [PubMed]
40. Dudak, F.C.; Boyaci, I.H. Rapid and label-free bacteria detection by surface plasmon resonance (SPR) biosensors. *Biotechnol. J.* **2009**, *4*, 1003–1011. [CrossRef]
41. Bhunia, A.K. Biosensors and bio-based methods for the separation and detection of foodborne pathogens. *Adv. Food Nutr. Res.* **2008**, *54*, 1–44. [PubMed]
42. Chen, C.; Wang, J. Optical biosensors: An exhaustive and comprehensive review. *Analyst* **2020**, *145*, 1605–1628. [PubMed]
43. Angelopoulou, M.; Kakabakos, S.; Petrou, P. Label-free biosensors based onto monolithically integrated onto silicon optical transducers. *Chemosensors* **2018**, *6*, 52. [CrossRef]
44. Sanvicens, N.; Pastells, C.; Pascual, N.; Marco, M.P. Nanoparticle-based biosensors for detection of pathogenic bacteria. *TrAC Trend Anal. Chem.* **2009**, *28*, 1243–1252. [CrossRef]
45. Wang, X.D.; Wolfbeis, O.S. Fiber-optic chemical sensors and biosensors (2008–2012). *Anal. Chem.* **2013**, *85*, 487–508. [CrossRef]
46. Benito-Peña, E.; Valdés, M.G.; Glahn-Martínez, B.; Moreno-Bondi, M.C. Fluorescence based fiber optic and planar waveguide biosensors. A review. *Anal. Chim. Acta* **2016**, *943*, 17–40. [CrossRef]
47. Taitt, C.R.; Anderson, G.P.; Ligler, F.S. Evanescent wave fluorescence biosensors: Advances of the last decade. *Biosens. Bioelectron.* **2016**, *76*, 103–112. [CrossRef] [PubMed]
48. Vashist, S.K.; Luppa, P.B.; Yeo, L.Y.; Ozcan, A.; Luong, J.H.T. Emerging technologies for next-generation Point-of-Care testing. *Trend Biotechnol.* **2015**, *33*, 692–705. [CrossRef] [PubMed]
49. Gauglitz, G. Direct optical detection in bioanalysis: An update. *Anal. Bioanal. Chem.* **2010**, *398*, 2363–2372. [CrossRef]
50. Luan, E.; Shoman, H.; Ratner, D.M.; Cheung, K.C.; Chrostowski, L. Silicon photonic biosensors using label-free detection. *Sensors* **2018**, *18*, 3519. [CrossRef] [PubMed]
51. Lee, D.; Hwang, J.; Seo, Y.; Gilad, A.A.; Choi, J. Optical immunosensors for the efficient detection of target biomolecules. *Biotechnol. Bioproc. Eng.* **2018**, *23*, 123–133. [CrossRef]
52. Huertas, C.S.; Calvo-Lozano, O.; Mitchell, A.; Lechuga, L.M. Advanced evanescent-wave optical biosensors for the detection of nucleic acids: An analytic perspective. *Front. Chem.* **2019**, *7*, 724. [CrossRef]
53. Gauglitz, G.; Nahm, W. Observation of spectral interferences for the determination of volume and surface effects of thin films. *Fresenius J. Anal. Chem.* **1991**, *341*, 279–283. [CrossRef]
54. Rothmund, M.; Schütz, A.; Brecht, A.; Gauglitz, G.; Berthel, G.; Gräfe, D. Label free binding assay with spectroscopic detection for pharmaceutical screening. *Fresenius J. Anal. Chem.* **1997**, *359*, 15–22. [CrossRef]
55. Bleher, O.; Schindler, A.; Yin, M.X.; Holmes, A.B.; Luppa, P.B.; Gauglitz, G.; Proll, G. Development of a new parallelized, optical biosensor platform for label-free detection of autoimmunity-related antibodies multiplex platforms in diagnostics and bioanalytics. *Anal. Bioanal. Chem.* **2014**, *406*, 3305–3314. [CrossRef]
56. Schwartz, M.P.; Alvarez, S.D.; Sailor, M.J. A Porous SiO_2 interferometric biosensor for quantitative determination of protein interactions: Binding of protein A to immunoglobulins derived from different species. *Anal. Chem.* **2007**, *79*, 327–334. [CrossRef]
57. Tsang, C.K.; Kelly, T.L.; Sailor, M.J.; Li, Y.Y. Highly stable porous silicon-carbon composites as label-free optical biosensors. *ACS Nano* **2012**, *6*, 10546–10554. [CrossRef] [PubMed]

58. Mun, K.S.; Alvarez, S.D.; Choi, W.Y.; Sailor, M.J. A Stable, label-free optical interferometric biosensor based on TiO_2 nanotube arrays. *ACS Nano* **2010**, *4*, 2070–2076. [CrossRef] [PubMed]
59. Petrou, P.S.; Ricklin, D.; Zavali, M.; Raptis, I.; Kakabakos, S.E.; Misiakos, K.; Lambris, J.D. Real-time label-free detection of complement activation products in human serum by White Light Reflectance Spectroscopy. *Biosens. Bioelectron.* **2009**, *24*, 3359–3364. [CrossRef]
60. Tsounidi, D.; Koukouvinos, G.; Petrou, P.; Misiakos, K.; Zisis, G.; Goustouridis, D.; Raptis, I.; Kakabakos, S.E. Rapid and sensitive label-free determination of aflatoxin M1 levels in milk through a White Light Reflectance Spectroscopy immunosensor. *Sens. Actuators B* **2019**, *282*, 104–111. [CrossRef]
61. Koukouvinos, G.; Petrou, P.; Goustouridis, D.; Misiakos, K.; Kakabakos, S.; Raptis, I. Development and bioanalytical applications of a White Light Reflectance Spectroscopy label-free sensing platform. *Biosensors* **2017**, *7*, 46. [CrossRef]
62. Kurihara, Y.; Takama, M.; Masubuchi, M.; Ooya, T.; Takeuchi, T. Microfluidic reflectometric interference spectroscopy-based sensing for exploration of protein-protein interaction conditions. *Biosens. Bioelectron.* **2013**, *40*, 247–251. [CrossRef]
63. Fratamico, P.M.; Strobaugh, T.R.; Medina, M.B.; Gehring, A.G. Detection of *Escherichia coli* O157:H7 using a surface plasmon resonance biosensor. *Biotechnol. Technique* **1998**, *12*, 571–576. [CrossRef]
64. Jyoung, J.-Y.; Hong, S.H.; Lee, W.; Choi, J.W. Immunosensor for the detection of *Vibrio cholerae O1* using surface plasmon resonance. *Biosens. Bioelectron.* **2006**, *21*, 2315–2319. [CrossRef]
65. Oh, B.K.; Kim, Y.K.; Lee, W.; Bae, Y.M.; Lee, W.H.; Choi, J.W. Immunosensor for detection of *Legionella pneumophila* using surface plasmon resonance. *Biosens. Bioelectron.* **2003**, *18*, 605–611. [CrossRef]
66. Waswa, J.W.; Debroy, C.; Irudayaraj, J. Rapid detection of *Salmonella Enteritidis* and *Escherichia Coli* using surface plasmon resonance biosensor. *J. Food Proc. Eng.* **2006**, *29*, 373–385. [CrossRef]
67. Bokken, G.C.A.M.; Corbee, R.J.; Van Knapen, F.; Bergwerff, A.A. Immunochemical detection of *Salmonella* group B, D and E using an optical surface plasmon resonance biosensor. *FEMS Microbiol. Lett.* **2003**, *222*, 75–82. [CrossRef]
68. Subramanian, A.; Irudayaraj, J.; Ryan, T. Mono and dithiol surfaces on surface plasmon resonance biosensors for detection of *Staphylococcus aureus*. *Sens. Actuators B* **2006**, *114*, 192–198. [CrossRef]
69. Subramanian, A.; Irudayaraj, J.; Ryan, T. A Mixed self-assembled monolayer-based surface plasmon immunosensor for detection of *E. coli* O157:H7. *Biosens. Bioelectron.* **2006**, *21*, 998–1006. [CrossRef] [PubMed]
70. Oh, B.K.; Lee, W.; Kim, Y.K.; Lee, W.H.; Choi, J.W. Surface plasmon resonance immunosensor using self-assembled protein G for the detection of *Salmonella paratyphi*. *J. Biotechnol.* **2004**, *111*, 1–8. [CrossRef] [PubMed]
71. Vaisocherová-Lísalová, H.; Víšová, I.; Ermini, M.L.; Špringer, T.; Song, X.C.; Mrázek, J.; Lamačová, J.; Scott Lynn, N.; Šedivák, P.; Homola, J. Low-fouling surface plasmon resonance biosensor for multi-step detection of foodborne bacterial pathogens in complex food samples. *Biosens. Bioelectron.* **2016**, *80*, 84–90. [CrossRef]
72. Masdor, N.A.; Altintas, Z.; Tothill, I.E. Surface plasmon resonance immunosensor for the detection of *Campylobacter jejuni*. *Chemosensors* **2017**, *5*, 16. [CrossRef]
73. Poltronieri, P.; De Blasi, M.D.; D'Urso, O.F. Detection of *Listeria monocytogenes* through real-time PCR and biosensor methods. *Plant Soil Environ.* **2009**, *55*, 363–369. [CrossRef]
74. Linman, M.J.; Sugerman, K.; Cheng, Q. Detection of low levels of *Escherichia coli* in fresh spinach by surface plasmon resonance spectroscopy with a TMB-based enzymatic signal enhancement method. *Sens. Actuators B* **2010**, *145*, 613–619. [CrossRef]
75. Makhneva, E.; Farka, Z.; Skládal, P.; Zajíčková, L. Cyclopropylamine plasma polymer surfaces for label-free SPR and QCM immunosensing of *Salmonella*. *Sens. Actuators B* **2018**, *276*, 447–455. [CrossRef]
76. Taylor, A.D.; Yu, Q.; Chen, S.; Homola, J.; Jiang, S. Comparison of *E. coli* O157:H7 preparation methods used for detection with surface plasmon resonance sensor. *Sens. Actuators B* **2005**, *107*, 202–208. [CrossRef]
77. Waswa, J.; Irudayaraj, J.; DebRoy, C. Direct detection of *E. coli* O157:H7 in selected food systems by a surface plasmon resonance biosensor. *LWT Food Sci. Technol.* **2007**, *40*, 187–192. [CrossRef]
78. Mazumdar, S.D.; Hartmann, M.; Kämpfer, P.; Keusgen, M. Rapid method for detection of *Salmonella* in milk by surface plasmon resonance (SPR). *Biosens. Bioelectron.* **2007**, *22*, 2040–2046. [CrossRef] [PubMed]
79. Leonard, P.; Hearty, S.; Quinn, J.; O'Kennedy, R. A generic approach for the detection of whole *Listeria monocytogenes* cells in contaminated samples using surface plasmon resonance. *Biosens. Bioelectron.* **2004**, *19*, 1331–1335. [CrossRef] [PubMed]
80. Wang, Y.; Ye, Z.; Si, C.; Ying, Y. Subtractive inhibition assay for the detection of *E. coli* O157:H7 using surface plasmon resonance. *Sensors* **2011**, *11*, 2728–2739. [CrossRef] [PubMed]
81. Wang, D.B.; Bi, L.J.; Zhang, Z.P.; Chen, Y.Y.; Yang, R.F.; Wei, H.P.; Zhou, Y.F.; Zhang, X.E. Label-free detection of *B. anthracis* spores using a surface plasmon resonance biosensor. *Analyst* **2009**, *134*, 738–742. [CrossRef] [PubMed]
82. Skottrup, P.D.; Nicolaisen, M.; Justesen, A.F. Towards on-site pathogen detection using antibody-based sensors. *Biosens. Bioelectron.* **2008**, *24*, 339–348. [CrossRef]
83. Skottrup, P.; Nicolaisen, M.; Justesen, A.F. Rapid determination of *Phytophthora infestans* sporangia using a surface plasmon resonance immunosensor. *J. Microbiol. Method* **2007**, *68*, 507–515. [CrossRef]
84. Skottrup, P.; Hearty, S.; Frøkiær, H.; Leonard, P.; Hejgaard, J.; O'Kennedy, R.; Nicolaisen, M.; Justesen, A.F. Detection of fungal spores using a generic surface plasmon resonance immunoassay. *Biosens. Bioelectron.* **2007**, *22*, 2724–2729. [CrossRef]
85. Kang, C.D.; Cao, C.; Lee, J.; Choi, I.S.; Kim, B.W.; Sim, S.J. Surface plasmon resonance-based inhibition assay for real-time detection of *Cryptosporidium parvum* oocyst. *Water Res.* **2008**, *42*, 1693–1699. [CrossRef]

86. Leonard, P.; Hearty, S.; Wyatt, G.; Quinn, J.; O'Kennedy, R. Development of a surface plasmon resonance-based immunoassay for *Listeria monocytogenes*. *J. Food Protect.* **2005**, *68*, 728–735. [CrossRef]
87. Oh, B.K.; Lee, W.; Chun, B.S.; Bae, Y.M.; Lee, W.H.; Choi, J.W. The fabrication of protein chip based on surface plasmon resonance for detection of pathogens. *Biosens. Bioelectron.* **2005**, *20*, 1847–1850. [CrossRef] [PubMed]
88. Taylor, A.D.; Ladd, J.; Yu, Q.; Chen, S.; Homola, J.; Jiang, S. Quantitative and simultaneous detection of four foodborne bacterial pathogens with a multi-channel SPR sensor. *Biosens. Bioelectron.* **2006**, *22*, 752–758. [CrossRef] [PubMed]
89. Morlay, A.; Duquenoy, A.; Piat, F.; Calemczuk, R.; Mercey, T.; Livache, T.; Roupioz, Y. Label-free immuno-sensors for the fast detection of *Listeria* in Food. *Measurement* **2017**, *98*, 305–310. [CrossRef]
90. Chen, J.; Park, B. Label-free screening of foodborne *Salmonella* using surface plasmon resonance imaging. *Anal. Bioanal. Chem.* **2018**, *410*, 5455–5464. [CrossRef] [PubMed]
91. Park, B.; Wang, B.; Chen, J. Label-free immunoassay for multiplex detections of foodborne bacteria in chicken carcass rinse with surface plasmon resonance imaging. *Foodborne Pathog. Dis.* **2021**, *18*, 202–209. [CrossRef] [PubMed]
92. Boulade, M.; Morlay, A.; Piat, F.; Roupioz, Y.; Livache, T.; Charette, P.G.; Canva, M.; Leroy, L. Early detection of bacteria using SPR imaging and event counting: Experiments with *Listeria monocytogenes* and *Listeria innocua*. *RSC Adv.* **2019**, *9*, 15554–15560. [CrossRef]
93. Wei, D.; Oyarzabal, O.A.; Huang, T.S.; Balasubramanian, S.; Sista, S.; Simonian, A.L. Development of a surface plasmon resonance biosensor for the identification of *Campylobacter jejuni*. *J. Microbiol. Method* **2007**, *69*, 78–85. [CrossRef]
94. Meeusen, C.A.; Alocilja, E.C.; Osburn, W.N. Detection of *E. coli* O157:H7 using a miniaturized surface plasmon resonance biosensor. *Transact. ASAE* **2005**, *48*, 2409–2416. [CrossRef]
95. Dudak, F.C.; Boyaci, I.H. Development of an immunosensor based on surface plasmon resonance for enumeration of *Escherichia coli* in water samples. *Food Res. Int.* **2007**, *40*, 803–807. [CrossRef]
96. Lan, Y.-B.; Wang, S.-Z.; Yin, Y.-G.; Hoffmann, W.C.; Zheng, X.-Z. Using a surface plasmon resonance biosensor for rapid detection of *Salmonella typhimurium* in chicken carcass. *J. Bionic Eng.* **2008**, *5*, 239–246. [CrossRef]
97. Torun, Ö.; Hakki Boyaci, I.; Temür, E.; Tamer, U. Comparison of sensing strategies in SPR biosensor for rapid and sensitive enumeration of bacteria. *Biosens. Bioelectron.* **2012**, *37*, 53–60. [CrossRef]
98. Tokel, O.; Yildiz, U.H.; Inci, F.; Durmus, N.G.; Ekiz, O.O.; Turker, B.; Cetin, C.; Rao, S.; Sridhar, K.; Natarajan, N.; et al. Portable microfluidic integrated plasmonic platform for pathogen detection. *Sci. Rep.* **2015**, *5*, 9154. [CrossRef]
99. Nguyen, H.H.; Yi, S.Y.; Woubit, A.; Kim, M. A portable surface plasmon resonance biosensor for rapid detection of *Salmonella typhimurium*. *Appl. Sci. Converg. Technol.* **2016**, *25*, 61–65. [CrossRef]
100. Choi, J.; Lee, J.; Son, J.; Choi, J. Noble metal-assisted surface plasmon resonance immunosensors. *Sensors* **2020**, *20*, 1003. [CrossRef] [PubMed]
101. Lopez, G.A.; Estevez, M.C.; Soler, M.; Lechuga, L.M. Recent advances in nanoplasmonic biosensors: Applications and lab-on-a-chip integration. *Nanophotonics* **2017**, *6*, 123–136. [CrossRef]
102. Fu, J.; Park, B.; Zhao, Y. Limitation of a localized surface plasmon resonance sensor for *Salmonella* detection. *Sens. Actuators B* **2009**, *141*, 276–283. [CrossRef]
103. Song, L.; Zhang, L.; Huang, Y.; Chen, L.; Zhang, G.; Shen, Z.; Zhang, J.; Xiao, Z.; Chen, T. Amplifying the signal of localized surface plasmon resonance sensing for the sensitive detection of *Escherichia coli* O157:H7. *Sci. Rep.* **2017**, *7*, 3288. [CrossRef]
104. Yaghubi, F.; Zeinoddini, M.; Saeedinia, A.R.; Azizi, A.; Samimi Nemati, A. Design of localized surface plasmon resonance (LSPR) biosensor for immunodiagnostic of *E. coli* O157:H7 using gold nanoparticles conjugated to the chicken antibody. *Plasmonics* **2020**, *15*, 1481–1487. [CrossRef]
105. Wang, Y.; Knoll, W.; Dostalek, J. Bacterial pathogen surface plasmon resonance biosensor advanced by long range surface plasmons and magnetic nanoparticle assays. *Anal. Chem.* **2012**, *84*, 8345–8350. [CrossRef]
106. Sharma, H.; Mutharasan, R. Review of biosensors for foodborne pathogens and toxins. *Sens. Actuators B* **2013**, *183*, 535–549. [CrossRef]
107. Rijal, K.; Leung, A.; Shankar, P.M.; Mutharasan, R. Detection of pathogen *Escherichia coli* O157:H7 at 70 cells/mL using antibody-immobilized biconical tapered fiber sensors. *Biosens. Bioelectron.* **2005**, *21*, 871–880. [CrossRef]
108. Wiejata, P.J.; Shankar, P.M.; Mutharasan, R. Fluorescent sensing using biconical tapers. *Sens. Actuators B* **2003**, *96*, 315–320. [CrossRef]
109. Yu, W.; Lang, T.; Bian, J.; Kong, W. Label-free fiber optic biosensor based on thin-core modal interferometer. *Sens. Actuators B* **2016**, *228*, 322–329. [CrossRef]
110. Bharadwaj, R.; Sai, V.V.R.; Thakare, K.; Dhawangale, A.; Kundu, T.; Titus, S.; Verma, P.K.; Mukherji, S. Evanescent wave absorbance based fiber optic biosensor for label-free detection of *E. coli* at 280 nm wavelength. *Biosens. Bioelectron.* **2011**, *26*, 3367–3370. [CrossRef]
111. Wandemur, G.; Rodrigues, D.; Allil, R.; Queiroz, V.; Peixoto, R.; Werneck, M.; Miguel, M. Plastic optical fiber-based biosensor platform for rapid cell detection. *Biosens. Bioelectron.* **2014**, *54*, 661–666. [CrossRef]
112. Zhou, C.; Pivarnik, P.; Auger, S.; Rand, A.; Letcher, S. A compact fiber-optic immunosensor for *Salmonella* based on evanescent wave excitation. *Sens. Actuators B* **1997**, *42*, 169–175. [CrossRef]
113. Geng, T.; Uknalis, J.; Tu, S.I.; Bhunia, A.K. Fiber-optic biosensor employing Alexa-Fluor conjugated antibody for detection of *Escherichia coli* O157:H7 from ground beef in four hours. *Sensors* **2006**, *6*, 796–807. [CrossRef]

114. Kramer, M.F.; Lim, D.V. A rapid and automated fiber optic-based biosensor assay for the detection of *Salmonella* in spent irrigation water used in the sprouting of sprout seeds. *J. Food Protect.* **2004**, *67*, 46–52. [CrossRef]
115. Leskinen, S.D.; Lim, D.V. Rapid ultrafiltration concentration and biosensor detection of *Enterococci* from large volumes of Florida recreational water. *Appl. Environ. Microbiol.* **2008**, *74*, 4792–4798. [CrossRef]
116. Geng, T.; Morgan, M.T.; Bhunia, A.K. Detection of low levels of *Listeria monocytogenes* cells by using a fiber-optic immunosensor. *Appl. Environ. Microbiol.* **2004**, *70*, 6138–6146. [CrossRef]
117. Ohk, S.H.; Koo, O.K.; Sen, T.; Yamamoto, C.M.; Bhunia, A.K. Antibody-aptamer functionalized fibre-optic biosensor for specific detection of *Listeria monocytogenes* from food. *J. Appl. Microbiol.* **2010**, *109*, 808–817. [CrossRef] [PubMed]
118. Nanduri, V.; Kim, G.; Morgan, M.T.; Ess, D.; Hahm, B.K.; Kothapalli, A.; Valadez, A.; Geng, T.; Bhunia, A.K. Antibody immobilization on waveguides using a flow–through system shows improved *Listeria monocytogenes* detection in an automated fiber optic biosensor: RAPTOR^{TM}. *Sensors* **2006**, *6*, 808–822. [CrossRef]
119. Ohk, S.H.; Bhunia, A.K. Multiplex fiber optic biosensor for detection of *Listeria monocytogenes*, *Escherichia coli* O157: H7 and *Salmonella enterica* from ready-to-eat meat samples. *Food Microbiol.* **2013**, *33*, 166–171. [CrossRef]
120. Ko, S.; Grant, S.A. A Novel FRET-based optical fiber biosensor for rapid detection of *Salmonella typhimurium*. *Biosens. Bioelectron.* **2006**, *21*, 1283–1290. [CrossRef]
121. Fallah, H.; Asadishad, T.; Parsanasab, G.M.; Harun, S.W.; Mohammed, W.S.; Yasin, M. Optical fiber biosensor toward *E. coli* bacterial detection on the pollutant water. *Eng. J.* **2021**, *25*, 1–8. [CrossRef]
122. Lin, H.Y.; Tsao, Y.C.; Tsai, W.H.; Yang, Y.W.; Yan, T.R.; Sheu, B.C. Development and application of side-polished fiber immunosensor based on surface plasmon resonance for the detection of *Legionella pneumophila* with halogens light and 850 nm-LED. *Sens. Actuators A* **2007**, *138*, 299–305. [CrossRef]
123. Kaushik, S.; Tiwari, U.K.; Pal, S.S.; Sinha, R.K. Rapid detection of *Escherichia coli* using fiber optic surface plasmon resonance immunosensor based on biofunctionalized molybdenum disulfide (MoS_2) nanosheets. *Biosens. Bioelectron.* **2019**, *126*, 501–509. [CrossRef]
124. Kozma, P.; Kehl, F.; Ehrentreich-Förster, E.; Stamm, C.; Bier, F.F. Integrated planar optical waveguide interferometer biosensors: A comparative review. *Biosens. Bioelectron.* **2014**, *58*, 287–307. [CrossRef]
125. Makarona, E.; Petrou, P.; Kakabakos, S.; Misiakos, K.; Raptis, I. Point-of-need bioanalytics based on planar optical interferometry. *Biotechnol. Adv.* **2016**, *34*, 209–233. [CrossRef]
126. Sarkar, D.; Gunda, N.S.K.; Jamal, I.; Mitra, S.K. Optical biosensors with an integrated Mach-Zehnder interferometer for detection of *Listeria monocytogenes*. *Biomed. Microdev.* **2014**, *16*, 509–520. [CrossRef]
127. Mathesz, A.; Valkai, S.; Újvárosy, A.; Aekbote, B.; Sipos, O.; Stercz, B.; Kocsis, B.; Szabó, D.; Dér, A. Integrated optical biosensor for rapid detection of bacteria. *Optofluid. Microfluid. Nanofluid.* **2016**, *2*, 15–21. [CrossRef]
128. Angelopoulou, M.; Petrou, P.; Misiakos, K.; Raptis, I.; Kakabakos, S. Simultaneous detection of *Salmonella typhimurium* and *Escherichia coli* O157:H7 in drinking water and milk with Mach–Zehnder interferometers monolithically integrated on silicon chips. *Biosensors* **2022**, *12*, 507. [CrossRef]
129. Ymeti, A.; Greve, J.; Lambeck, P.V.; Wink, T.; Van Novell, S.W.F.M.; Beumer, T.A.M.; Wijn, R.R.; Heideman, R.G.; Subramaniam, V.; Kanger, J.S. Fast, ultrasensitive virus detection using a Young interferometer sensor. *Nano Lett.* **2007**, *7*, 394–397. [CrossRef] [PubMed]
130. Schneider, B.H.; Edwards, J.G.; Hartman, N.F. Hartman interferometer: Versatile integrated optic sensor for label- free, real-time quantification of nucleic acids, proteins, and pathogens. *Clin. Chem.* **1997**, *43*, 1757–1763. [CrossRef]
131. Seo, K.H.; Brackett, R.E.; Hartman, N.F.; Campbell, D.P. Development of a rapid response biosensor for detection of *Salmonella typhimurium*. *J. Food Protect.* **1999**, *62*, 431–437. [CrossRef]
132. Nagel, T.; Ehrentreich-Förster, E.; Singh, M.; Schmitt, K.; Brandenburg, A.; Berka, A.; Bier, F.F. Direct detection of Tuberculosis infection in blood serum using three optical label-free approaches. *Sens. Actuators B* **2008**, *129*, 934–940. [CrossRef]
133. Maldonado, J.; González-Guerrero, A.B.; Domínguez, C.; Lechuga, L.M. Label-free bimodal waveguide immunosensor for rapid diagnosis of bacterial infections in cirrhotic patients. *Biosens. Bioelectron.* **2016**, *85*, 310–316. [CrossRef]
134. Maldonado, J.; González-Guerrero, A.B.; Fernández-Gavela, A.; González-López, J.J.; Lechuga, L.M. Ultrasensitive label-free detection of unamplified multidrug-resistance bacteria genes with a bimodal waveguide interferometric biosensor. *Diagnostics* **2020**, *10*, 845. [CrossRef]
135. Grego, S.; McDaniel, J.R.; Stoner, B.R. Wavelength interrogation of grating-based optical biosensors in the input coupler configuration. *Sens. Actuators B* **2008**, *131*, 347–355. [CrossRef]
136. Adrian, J.; Pasche, S.; Diserens, J.M.; Sánchez-Baeza, F.; Gao, H.; Marco, M.P.; Voirin, G. Waveguide interrogated optical immunosensor (WIOS) for detection of sulfonamide antibiotics in milk. *Biosens. Bioelectron.* **2009**, *24*, 3340–3346. [CrossRef]
137. Kehl, F.; Etlinger, G.; Gartmann, T.E.; Tscharner, N.S.R.U.; Heub, S.; Follonier, S. Introduction of an angle interrogated, MEMS-based, optical waveguide grating system for label-free biosensing. *Sens. Actuators B* **2016**, *226*, 135–143. [CrossRef]
138. Kozma, P.; Hamori, A.; Cottier, K.; Kurunczi, S.; Horvath, R. Grating coupled interferometry for optical sensing. *Appl. Phys. B* **2009**, *97*, 5–8. [CrossRef]
139. Teotia, P.K.; Kaler, R.S. Multilayer with Periodic Grating Based High Performance SPR Waveguide Sensor. *Opt. Commun.* **2017**, *395*, 154–158. [CrossRef]

140. Marusov, G.; Sweatt, A.; Pietrosimone, K.; Benson, D.; Geary, S.J.; Silbart, L.K.; Challa, S.; Lagoy, J.; Lawrence, D.A.; Lynes, M.A. A microarray biosensor for multiplexed detection of microbes using grating-coupled surface plasmon resonance imaging. *Environ. Sci. Technol.* **2012**, *46*, 348–359. [CrossRef]
141. Davies, E.; Viitala, R.; Salomäki, M.; James, S.W.; Tatam, R.P. Optical fibre long-period grating sensors: Characteristics and application. *Meas. Sci. Technol.* **2003**, *14*, R49.
142. Shi, W.; Wang, X.; Zhang, W.; Yun, H.; Lin, C.; Chrostowski, L.; Jaeger, N.A.F. Grating-coupled silicon microring resonators. *Appl. Phys. Lett.* **2012**, *100*, 121118. [CrossRef]
143. Horváth, R.; Pedersen, H.C.; Skivesen, N.; Selmeczi, D.; Larsen, N.B. Optical waveguide sensor for on-line monitoring of bacteria. *Opt. Lett.* **2003**, *28*, 1233. [CrossRef]
144. Bandara, A.B.; Zuo, Z.; Ramachandran, S.; Ritter, A.; Heflin, J.R.; Inzana, T.J. Detection of methicillin-resistant *Staphylococci* by biosensor assay consisting of nanoscale films on optical fiber long-period gratings. *Biosens. Bioelectron.* **2015**, *70*, 433–440. [CrossRef]
145. Yang, F.; Chang, T.L.; Liu, T.; Wu, D.; Du, H.; Liang, J.; Tian, F. Label-free detection of *Staphylococcus aureus* bacteria using long-period fiber gratings with functional polyelectrolyte coatings. *Biosens. Bioelectron.* **2019**, *133*, 147–153. [CrossRef]
146. Kaushik, S.; Tiwari, U.; Nilima; Prashar, S.; Das, B.; Sinha, R.K. Label-free detection of *Escherichia coli* bacteria by cascaded chirped long period gratings immunosensor. *Rev. Sci. Instr.* **2019**, *90*, 25003. [CrossRef] [PubMed]
147. Kim, N.; Park, I.S.; Kim, W.Y. *Salmonella* detection with a direct-binding optical grating coupler immunosensor. *Sens. Actuators B* **2007**, *121*, 606–615. [CrossRef]
148. Cooper, I.R.; Meikle, S.T.; Standen, G.; Hanlon, G.W.; Santin, M. The rapid and specific real-time detection of *Legionella pneumophila* in water samples using optical waveguide lightmode spectroscopy. *J. Microbiol. Method* **2009**, *78*, 40–44. [CrossRef] [PubMed]
149. Lee, S.; Chan Eom, S.; Soo Chang, J.; Huh, C.; Yong Sung, G.; Shin, J.H. A silicon nitride microdisk resonator with a 40-nm-thin horizontal air slot. *Opt. Exp.* **2010**, *18*, 11209–11215. [CrossRef] [PubMed]
150. De Vos, K.; Bartolozzi, I.; Schacht, E.; Bienstman, P.; Baets, R. Silicon-on-insulator microring resonator for sensitive and label-free biosensing. *Opt. Exp.* **2007**, *15*, 7610–7615. [CrossRef] [PubMed]
151. Armani, D.K.; Kippenberg, T.J.; Spillane, S.M.; Vahala, K.J. Ultra-high-Q toroid microcavity on a chip. *Nature* **2003**, *421*, 925–928. [CrossRef]
152. Ramachandran, A.; Wang, S.; Clarke, J.; Ja, S.J.; Goad, D.; Wald, L.; Flood, E.M.; Knobbe, E.; Hryniewicz, J.V.; Chu, S.T.; et al. A universal biosensing platform based on optical micro-ring resonators. *Biosens. Bioelectron.* **2008**, *23*, 939–944. [CrossRef]
153. Ghali, H.; Chibli, H.; Nadeau, J.L.; Bianucci, P.; Peter, Y.A. Real-time detection of *Staphylococcus aureus* using whispering gallery mode optical microdisks. *Biosensors* **2016**, *6*, 20. [CrossRef]
154. Merkl, S.; Vornicescu, D.; Dassinger, N.; Keusgen, M. Detection of whole cells using reflectometric interference spectroscopy. *Phys. Status Solidi A* **2014**, *211*, 1416–1422. [CrossRef]
155. Angelopoulou, M.; Tzialla, K.; Voulgari, A.; Dikeoulia, M.; Raptis, I.; Kakabakos, S.E.; Petrou, P. Rapid detection of *Salmonella typhimurium* in drinking water by a White Light Reflectance Spectroscopy immunosensor. *Sensors* **2021**, *21*, 2683. [CrossRef]
156. Massad-Ivanir, N.; Shtenberg, G.; Tzur, A.; Krepker, M.A.; Segal, E. Engineering nanostructured porous SiO_2 surfaces for bacteria detection via "direct cell capture". *Anal. Chem.* **2011**, *83*, 3282–3289. [CrossRef]
157. Massad-Ivanir, N.; Shtenberg, G.; Segal, E. Optical detection of *E. coli* Bacteria by Mesoporous Silicon Biosensors. *J. Vis. Exp.* **2013**, *81*, e50805.
158. Massad-Ivanir, N.; Shtenberg, G.; Raz, N.; Gazenbeek, C.; Budding, D.; Bos, M.P.; Segal, E. Porous silicon-based biosensors: Towards real-time optical detection of target bacteria in the food industry. *Sci. Rep.* **2016**, *6*, 38099. [CrossRef] [PubMed]
159. Massad-Ivanir, N.; Shtenberg, G.; Segal, E. Advancing nanostructured porous Si-based optical transducers for label free bacteria detection. *Adv. Exp. Med. Biol.* **2012**, *733*, 37–45. [PubMed]
160. Tang, Y.; Li, Z.; Luo, Q.; Liu, J.; Wu, J. Bacteria detection based on its blockage effect on silicon nanopore array. *Biosens. Bioelectron.* **2016**, *79*, 715–720. [CrossRef] [PubMed]
161. Zaraee, N.; Kanik, F.E.; Bhuiya, A.M.; Gong, E.S.; Geib, M.T.; Lortlar Ünlü, N.; Ozkumur, A.Y.; Dupuis, J.R.; Ünlü, M.S. Highly sensitive and label-free digital detection of whole cell *E. coli* with interferometric reflectance imaging. *Biosens. Bioelectron.* **2020**, *162*, 112258. [CrossRef]
162. Reardon, K.F.; Zhong, Z.; Lear, K.L. Environmental applications of photoluminescence-based biosensors. In *Optical Sensor Systems in Biotechnology*; Rao, G., Ed.; Springer: Berlin/Heidelberg, Germany, 2009; Volume 116, pp. 99–123.
163. Viter, R.; Tereshchenko, A.; Smyntyna, V.; Ogorodniichuk, J.; Starodub, N.; Yakimova, R.; Khranovskyy, V.; Ramanavicius, A. Toward development of optical biosensors based on photoluminescence of TiO_2 nanoparticles for the detection of *Salmonella*. *Sens. Actuators B* **2017**, *252*, 95–102. [CrossRef]
164. Taitt, C.R.; Shubin, Y.S.; Angel, R.; Ligler, F.S. Detection of *Salmonella enterica serovar typhimurium* by using a rapid, array-based immunosensor. *Appl. Environ. Microbiol.* **2004**, *70*, 152–158. [CrossRef]
165. Sapsford, K.E.; Rasooly, A.; Taitt, C.R.; Ligler, F.S. Detection of *Campylobacter* and *Shigella* Species in Food Samples Using an Array Biosensor. *Anal. Chem.* **2004**, *76*, 433–440. [CrossRef]
166. Sapsford, K.E.; Ngundi, M.M.; Moore, M.H.; Lassman, M.E.; Shriver-Lake, L.C.; Taitt, C.R.; Ligler, F.S. Rapid detection of foodborne contaminants using an array biosensor. *Sens. Actuators B* **2006**, *113*, 599–607. [CrossRef]

167. Shriver-Lake, L.C.; Turner, S.; Taitt, C.R. Rapid detection of *Escherichia coli* O157:H7 spiked into food matrices. *Anal. Chim. Acta* **2007**, *584*, 66–71. [CrossRef]
168. Dehghani, Z.; Mohammadnejad, J.; Hosseini, M.; Bakhshi, B.; Rezayan, A.H. Whole cell FRET immunosensor based on graphene oxide and graphene dot for *Campylobacter jejuni* detection. *Food Chem.* **2020**, *309*, 125690. [CrossRef]
169. Wang, Z.; Zong, S.; Wu, L.; Zhu, D.; Cui, Y. SERS-activated platforms for immunoassay: Probes, encoding methods, and applications. *Chem. Rev.* **2017**, *117*, 7910–7963. [CrossRef]
170. Mosier-Boss, P.A. Review on SERS of bacteria. *Biosensors* **2017**, *7*, 51. [CrossRef]
171. Zhou, X.; Hu, Z.; Yang, D.; Xie, S.; Jiang, Z.; Niessner, R.; Haisch, C.; Zhou, H.; Sun, P. Bacteria detection: From powerful SERS to its advanced compatible techniques. *Adv. Sci.* **2020**, *7*, 2001739. [CrossRef]
172. Qu, L.L.; Ying, Y.L.; Yu, R.J.; Long, Y.T. In situ food-borne pathogen sensors in a nanoconfined space by surface enhanced Raman scattering. *Microchim. Acta* **2021**, *188*, 201.
173. Xie, X.; Pu, H.; Sun, D.W. Recent advances in nanofabrication techniques for SERS substrates and their applications in food safety analysis. *Crit. Rev. Food Sci. Nutr.* **2018**, *58*, 2800–2813. [CrossRef]
174. Knauer, M.; Ivleva, N.P.; Liu, X.; Niessner, R.; Haisch, C. Surface-enhanced Raman scattering-based label-free microarray readout for the detection of microorganisms. *Anal. Chem.* **2010**, *82*, 2766–2772. [CrossRef] [PubMed]
175. Knauer, M.; Ivleva, N.P.; Niessner, R.; Haisch, C. A flow-through microarray cell for the online SERS detection of antibody-captured *E. coli* bacteria. *Anal. Bioanal. Chem.* **2012**, *402*, 2663–2667. [CrossRef] [PubMed]
176. Cho, I.H.; Bhandari, P.; Patel, P.; Irudayaraj, J. Membrane filter-assisted surface enhanced Raman spectroscopy for the rapid detection of *E. coli* O157:H7 in ground beef. *Biosens. Bioelectron.* **2015**, *64*, 171–176. [CrossRef]
177. Bai, X.; Shen, A.; Hu, J. A sensitive SERS-based sandwich immunoassay platform for simultaneous multiple detection of foodborne pathogens without interference. *Anal. Method* **2020**, *12*, 4885–4891. [CrossRef]
178. Najafi, R.; Mukherjee, S.; Hudson, J.; Sharma, A.; Banerjee, P. Development of a rapid capture-cum-detection method for *Escherichia coli* O157 from apple juice comprising nano-immunomagnetic separation in tandem with surface enhanced Raman scattering. *Int. J. Food Microbiol.* **2014**, *189*, 89–97. [CrossRef]
179. Kearns, H.; Goodacre, R.; Jamieson, L.E.; Graham, D.; Faulds, K. SERS detection of multiple antimicrobial-resistant pathogens using nanosensors. *Anal. Chem.* **2017**, *89*, 12666–12673. [CrossRef]
180. Wang, J.; Wu, X.; Wang, C.; Rong, Z.; Ding, H.; Li, H.; Li, S.; Shao, N.; Dong, P.; Xiao, R.; et al. Facile synthesis of Au-coated magnetic nanoparticles and their application in bacteria detection via a SERS method. *ACS Appl. Mater. Interface* **2016**, *8*, 19958–19967. [CrossRef] [PubMed]
181. Tamer, U.; Boyaci, I.H.; Temur, E.; Zengin, A.; Dincer, I.; Elerman, Y. Fabrication of magnetic gold nanorod particles for immunomagnetic separation and SERS application. *J. Nanopart. Res.* **2011**, *13*, 3167–3176. [CrossRef]
182. Temur, E.; Boyaci, I.H.; Tamer, U.; Unsal, H.; Aydogan, N. A Highly Sensitive detection platform based on surface-enhanced Raman scattering for *Escherichia coli* enumeration. *Anal. Bioanal. Chem.* **2010**, *397*, 1595–1604. [CrossRef] [PubMed]
183. Guven, B.; Basaran-Akgul, N.; Temur, E.; Tamer, U.; Boyaci, I.H. SERS-based sandwich immunoassay using antibody coated magnetic nanoparticles for *Escherichia coli* enumeration. *Analyst* **2011**, *136*, 740–748.
184. Lin, H.Y.; Huang, C.H.; Hsieh, W.H.; Liu, L.H.; Lin, Y.C.; Chu, C.C.; Wang, S.T.; Kuo, I.T.; Chau, L.K.; Yang, C.Y. On-line SERS detection of single bacterium using novel SERS nanoprobes and a microfluidic dielectrophoresis device. *Small* **2014**, *10*, 4700–4710. [CrossRef]
185. Wu, Z. Simultaneous detection of *Listeria monocytogenes* and *Salmonella typhimurium* by a SERS-based lateral flow immunochromatographic assay. *Food Anal. Method* **2019**, *12*, 1086–1091. [CrossRef]
186. Liu, H.-B.; Du, X.-J.; Zang, Y.-X.; Li, P.; Wang, S. SERS-based lateral flow strip biosensor for simultaneous detection of *Listeria monocytogenes* and *Salmonella enterica* serotype *enteritidis*. *J. Agric. Food Chem.* **2017**, *65*, 10290–10299. [CrossRef]
187. Liu, H.-B.; Chen, C.-Y.; Zhang, C.-N.; Du, X.-J.; Li, P.; Wang, S. Functionalized AuMBA@Ag nanoparticles as an optical and SERS dual probe in a lateral flow strip for the quantitative detection of *Escherichia coli* O157:H7. *J. Food Sci.* **2019**, *84*, 2916–2924. [CrossRef]

Disclaimer/Publisher's Note: The statements, opinions and data contained in all publications are solely those of the individual author(s) and contributor(s) and not of MDPI and/or the editor(s). MDPI and/or the editor(s) disclaim responsibility for any injury to people or property resulting from any ideas, methods, instructions or products referred to in the content.

Article

Utilising Portable Laser-Induced Breakdown Spectroscopy for Quantitative Inorganic Water Testing

Nils Schlatter [1,*], Bernd G. Lottermoser [1], Simon Illgner [2] and Stefanie Schmidt [3]

1. Institute of Mineral Resources Engineering, RWTH Aachen University, 52062 Aachen, Germany; lottermoser@mre.rwth-aachen.de
2. MHI Gruppe, 63456 Hanau, Germany; simon.illgner@gmail.com
3. Institute of Applied Geosciences, Hydrogeology, Technical University of Darmstadt, 64287 Darmstadt, Germany; schmidt@geo.tu-darmstadt.de
* Correspondence: schlatter@mre.rwth-aachen.de

Abstract: At present, the majority of water testing is carried out in the laboratory, and portable field methods for the quantification of elements in natural waters remain to be established. In contrast, portable instruments like portable X-ray fluorescence (pXRF) analysis and portable laser-induced breakdown spectroscopy (pLIBS) have become routine analytical methods for the quantification of elements in solids. This study aims to show that pLIBS can also be used for chemical compositional measurements of natural waters. Bottled mineral waters were selected as sample materials. A surface-enhanced liquid-to-solid conversion technique was used to improve the detection limits and circumvent the physical limitations in liquid analysis. The results show that low to medium mineralised waters can be analysed quantitatively for their ions using the documented method. For more highly concentrated samples, typically above an electrical conductivity (EC) of 1000 µS/cm, further adjustment is required in the form of self-absorption correction. However, water with a conductivity up to this limit can be analysed for the main cations (Li^+, Na^+, Mg^{2+}, K^+, Ca^{2+}, and Sr^{2+}) as well as the main anions (SO_4^{2-} and Cl^-) using the documented method. This study demonstrates that there is significant potential for developing field-based pLIBS as a tool for quantitative water analysis.

Keywords: portable laser-induced breakdown spectroscopy; pre-screening; handheld; inorganic water analysis; hydrochemistry; in-field water analysis; SciAps Z-300; water-quality testing

1. Introduction

An understanding of water quality is traditionally obtained using laboratory-based analyses and continuous monitoring techniques [1–3]. This is despite the need for fast, reliable, and, if possible, inexpensive in-field measurement methods, particularly in remote regions. Although hand-held instruments like portable X-ray fluorescence (pXRF) and portable laser-induced breakdown spectroscopy (pLIBS) have been established in geochemical analysis of solid samples for many years [4–7], inorganic water analyses are still performed almost exclusively in the laboratory [2]. Typical equipment used includes ion chromatography (IC), atomic absorption spectroscopy (AAS), and inductively coupled plasma-mass spectroscopy (ICP-MS). These techniques require trained personnel and proper sample transport, storage, and handling prior to analysis; are expensive to maintain; and are time consuming [8]. This often prevents quick action, as it can take more than a week from the time the sample is taken to the actual analysis of samples and data generation. In addition, in less developed countries, analyses are less likely to be carried out due to the cost and expertise required. Therefore, reliable field instruments are needed to quantify as many elements and compounds as possible. In on-site analysis, lower sensitivity and higher detection limits are usually accepted, if immediate results and higher data density are feasible, especially if pre-screening is performed [4,9].

Typical field-ready measuring equipment for inorganic water analysis available on the market includes photometers, test kits, and ion selective electrodes. The main reason why field methods have not yet been widely adopted for measuring the inorganic chemistry of waters is presumably due to the fact that reliable methods for simultaneous determination of a range of elements have not been developed to date.

Laser-induced breakdown spectroscopy (LIBS) is an atomic emission spectroscopy technique capable of simultaneously determining the complete elemental chemistry of a sample. A focused, pulsed laser beam is directed at a sample to form a plasma containing the elements of the small sample volume that is being ablated. By spectral analysis of the emitted light, it is possible to obtain qualitative and quantitative data on the elements present, provided a suitable calibration is used [4,6]. Although LIBS currently plays rather a niche role in water analysis, several studies have shown that laboratory-based LIBS systems can be used to simultaneously quantify almost any element in water with very low detection limits [10–23]. For example, Na has been quantitatively analysed in aqueous solutions with a detection limit of 0.57 µg/L [20]. Mg, Ca, Sr, and Ba have been detected down to 0.3, 0.6, 1.0, and 0.7 ppm, respectively [16], and Mg, Cr, Mn, and Re have been detected down to 0.1, 0.4, 0.7, and 8 mg/L, respectively [11].

However, many of these laboratory applications described use complicated experimental setups, such as measurement in a liquid jet [11,20] or liquid to aerosol [16]. This is because direct bulk liquid analysis by LIBS is prone to low sensitivity and accuracy due to energy losses as a result of liquid evaporation, plasma cooling, and intense splashing [24]. A simpler sample preparation method, which is also feasible in the field and adaptable to a portable laser-induced breakdown spectroscopy (pLIBS) instrument, is liquid-to-solid conversion (LSC). At the same time, this method offers the advantage that the detection limits are lowered by pre-concentration. Therefore, in a previous work, a surface-enhanced (SE) liquid-to-solid conversion (LSC) method was adapted to a pLIBS for quantitative analysis of Li, Na, and K in standard solutions containing nearly no other cations [25]. Instead of directly shooting the liquid with the laser, the evaporation residue (EvR) was analysed on a commercially available aluminium foil, which was SE with a thin layer of graphite pencil powder. Low detection limits could be achieved by LSC while avoiding negative physical effects such as splashing and cooling of the plasma that occur when analysing liquid samples. Moreover, by preparing the aluminium foil with pencil powder, the surface became more hydrophobic, and therefore, the EvR were distributed more homogenously, leading to better reproducibility. A self-designed template that fits on the nose of the SciAps Z-300 guaranteed fixation during the analysis of 100 positions in a fixed grid on and around the EvR. Results of the study showed that a portable LIBS analyser is well suited for the quantitative analysis of light alkali elements in standard solutions up to 160 mg/L [25]. To our knowledge, that was the first time that a handheld LIBS instrument had been used to quantify dissolved elements in aqueous solutions using an LSC technique. The portability of the method opens up new possibilities for on-site screening and quantitative analysis of inorganic water chemistry.

However, to date, the method has only been applied to single-element standard solutions. In order to identify possible matrix effects and to further adapt the method for field use, bottled mineral waters from different manufacturers with as diverse a chemistry as possible were chosen as examples in this study (see Table 1 and Figure 1). When using bottled mineral waters from grocery stores, it is possible to choose from a wide range of different mineralised waters, as the manufacturers in the European Union are obliged to print analysis results on their bottles. These may often not be particularly up to date, but they provide a rough guide on the likely chemical composition of bottled waters. There has also been a lot of research into the testing of bottled mineral water for mineralisation [26–29]. As can be seen in Figure 1, the choice of bottled mineral water allowed very different types of water to be selected. This is important in order to have a diverse test series to study possible matrix effects.

Table 1. List of bottled mineral waters analysed by this study.

Abbr.	Name/Brand	Spring	Location	State	Country	Bottle	TDS (mg/L)	EC (µS/cm)
Adhz	Adelholzener	Adelholzener Alpen Quell	Bergen	BY	De	PET	511	598
AqMi	Aqua Mia	Geotaler	Löhne	NW	De	PET	1284	1575
BeWa	Bergische Waldquelle	Bergische Waldquelle	Haan	NW	De	PET	159	257
Blk	Blank sample	-	laboratory	-	-	HDPE	-	1.3
Extl	Extaler-Mineralquell	Extaler-Mineralquell	Rinteln-Exten	NI	De	TP	1603	1708
Fach	Staatl. Fachingen	Staatl. Fachingen	Fachingen	RP	De	glass	4306	2726
Gest	Gerolsteiner Naturell	Naturell	Gerolstein	RP	De	PET	807	878
Laur	Lauretana	Lauretana	Graglia	21	It	PET	17	20.5
Löng	K-Classic	Quelle Löningen	Löningen	NI	De	PET	162	275
Mar	Marius-Quelle	Marius-Quelle	Sachsenheim	BW	De	PET	2435	2650
Nat	Naturalis still	Urstromquelle	Wolfhagen	HE	De	PET	126	212
Odwq	Odenwald-Quelle	Odenwald Quelle	Heppenheim	HE	De	PET	760	727
Qubr	Quellbrunn Werretaler	Werretaler	Löhne	NW	De	PET	729	971
Rosb	Rosbacher Naturell	Rosbacher Naturell	Rosbach v.d. Höhe	HE	De	PET	1235	1363
Saw	Sawell	Genussquelle 3	Emsdetten	NW	De	PET	377	589
Vit	Vittel	Vittel Bonne Source	Vittel	88	Fr	PET	490	444

BY = Bavaria, NW = North Rhine-Westphalia, NI = Lower Saxony, RP = Rhineland-Palatinate, BW = Baden-Württemberg, HE = Hesse, 21 = Piemont, 88 = Département des Vosges, De = Germany, It = Italy, Fr = France, PET = polyethylene terephthalate, HDPE = high-density polyethylene, TP = Tetra Pak, TDS = total dissolved solids (as indicated), EC = electrical conductivity (as measured).

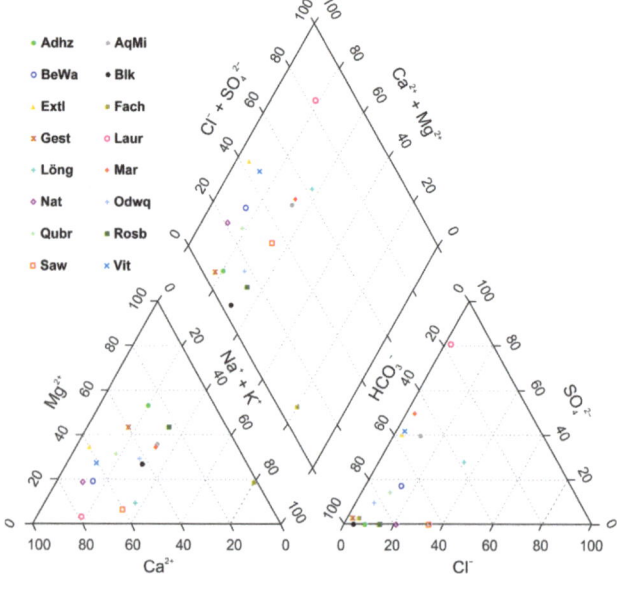

Figure 1. Piper plot of the bottled mineral waters. The chemical data were obtained by IC and field photometer (hydrogen carbonate) analyses.

In this study, the analytical approach of the former study [25] was extended to include elements and compounds (Ca^{2+}, Mg^{2+}, Sr^{2+}, Cl^-, SO_4^{2-}, and NO_3^-) to cover the main cations and anions in natural waters and documented on bottled mineral water. The results of this study therefore contribute to the ongoing development of hydrochemical field testing tools.

2. Materials and Methods

2.1. Sample Preparation

Water samples were taken from commercially available bottled mineral waters purchased from supermarkets. Fifteen different brands were chosen, and waters with the most diverse chemistry according to their information labels were selected to obtain a diverse test series (Table 1 and Figure 1). Only non-carbonated bottled water was selected to compensate for precipitation by degassing. With two exceptions (Fach., glass, and Extl., Tetra Pak), all mineral waters were bottled in PET. Two 50 mL samples of each mineral water were prepared by filling centrifuge tubes. One was acidified with HNO_3 to prevent precipitation (cation sample). The anion sample was not prepared further. An additional blank sample was prepared from distilled deionised water (18 MΩ) in the same way as the mineral waters. A portion of each anion sample was used for the analysis of physico-chemical parameters (temperature (T), electrical conductivity (EC), pH, and HCO_3^-) and pLIBS analysis before all samples were sealed with Parafilm® and sent directly to the laboratory for subsequent ion exchange chromatography (IC) and inductively coupled plasma–mass spectrometry (ICP-MS) analysis (Technical University of Darmstadt, Darmstadt, Germany).

For subsequent measurements, aqueous single-element AAS standard solutions on a 2% nitric acid basis with a concentration of 1000 mg/L were used for the calibration of the cations (ROTI®Star, Carl Roth, Karlsruhe, Germany). From each of the standard solutions, 16 different concentrations ranging from 0.1 to 1000 mg/L were prepared by diluting with 2% nitric acid. The latter was made by diluting 65% nitric acid (ROTIPURAN® ≥ 65%, Carl Roth, Karlsruhe, Germany) with distilled deionised water (18 MΩ). Li^+, Na^+, and K^+ were prepared as single-element standard dilution series. Mg^{2+} and Ca^{2+} were mixed with each other as paired standard dilution series. Concentrations above 500 mg/L were prepared as single-element standards. The same procedure was used for Zn^{2+} and Sr^{2+}. In addition, a water-based multi-element anion IC standard solution containing Cl^-, SO_4^{2-}, and NO_3^- was used to calibrate the main anions by dilution with distilled deionised water at concentrations ranging from 0.5 to 1000 mg/L.

Furthermore, mixed solutions were prepared by using the single-element standard solutions in equal amounts and diluting with 2% nitric acid. Six mixed solutions containing Li, Na, and K (ranging from 2 to 250 mg/L) and six containing Li^+, Na^+, K^+, Mg^{2+}, Sr^{2+}, and Ca^{2+} (ranging from 1 to 125 mg/L) were prepared. Cl^-, SO_4^{2-}, and NO_3^- were not included in the mixed standards, as the cationic single element standards already contained nitrate in different quantities.

2.2. Instrumentation

As in the previous study [25], the commercially available handheld LIBS analyser SciAps Z-300 (SciAps, Woburn, MA, USA) was used. It contains a class 3B Nd:YAG laser that produces laser light with a wavelength of 1064 nm and an energy of 5–6 mJ/pulse with a duration of 1 ns and an adjustable firing rate of between 1 and 50 Hz [30]. Three spectrometers consisting of time-gated charge-coupled diodes (CCD) detect the emitted light in the spectral range of 190–950 nm by three spectrometers [30].

The acquisition settings of the pLIBS analyses have been detailed by a previous study [25]. Due to the small size of the EvR after LSC and the possibility of laying a raster over an area with the SciAps Z-300, it was possible to analyse the whole EvR and a small area around it. One hundred locations were fired at four times each in order to obtain data. For each location, the four individual analyses were averaged into a spectrum, resulting in 100 spectra per sample. Since fresh samples were applied to the SE aluminium foil, no cleaning shots were needed, and this setting was set to 0. The use of gating can improve the signal-to-noise ratio because the continuum radiation, which contains no useful information and occurs mainly at the beginning of a measurement, has a lower proportion with a slightly delayed measurement [23]. In the previous study, the gate delay for Li, Na, and K was optimised [25]. In order to obtain comparable data, the gate delay was not further adjusted in this work. An internal integration delay (IID) of 84 was used, corresponding to a delay time (t_d) of approximately 2 µs. In order to achieve signal enhancement [30,31], an Ar atmosphere was used for the measurements. The coordinates were set to start at 100, 100, 70 and end at 350, 350. Finally, the test rate was set to 50 Hz. A custom stencil designed for the SciAps Z-300 described in [25] was used to facilitate and speed up handling by fixing the substrate and the device itself.

2.3. Liquid Analysis

First, the pH and EC of each bottled mineral water were determined. For this purpose, the bottled waters were poured into centrifuge tubes large enough to fit the probes of the pH and conductivity electrode. This method has already been described by [32] and helps to reduce the required sample volume. In this study, a slightly larger sample volume of about 20–25 mL was used. Conductivity and temperature were measured first (CA 10141 with conductivity probe XCP4ST1, Chauvin Arnoux, Asnières-Sur-Seine, France), making sure that the probe was thoroughly cleaned with distilled water and then dried before measurement. Afterwards, the pH value was analysed with a likewise cleaned probe (HI 991002 with pH/ORP probe HI1297, Hanna Instruments, Vöhringen, Germany). Hydrogen carbonate was measured with a field-ready photometer (HI775 Checker HC, Hanna Instruments, Vöhringen, Germany) using a sample of maximum 15 mL.

For the comparative measurements in the laboratory, one cation and one anion sample were prepared, the former being acidified with 10 droplets of 10% nitric acid (diluted ROTIPURAN® ≥ 65%, Carl Roth, Karlsruhe, Germany) to prevent precipitation. The samples were analysed for the main cations and anions with IC at the Institute of Applied Geosciences of Technical University of Darmstadt (Metrohm 882 Compact IC plus, Metrohm, Herisau, Switzerland).

Sr and Zn were analysed using ICP-MS (Analytik Jena Plasma Quant MS Elite®, Jena, Germany). Instrumental conditions were optimised using a 1 µg/L tuning solution (diluted 10 mg/L Analytik Jena Tuning solution, Jena, Germany), leading to high sensitivities of the containing elements Be, In, Pb, and Th in the tuning solution with simultaneous low oxide and a doubly charged ion ratio. Helium was used as a collision gas in the integrated collision–reaction cell (iCRC) for the minimisation of potential interferences. A 10 µg/L Y solution (diluted ROTI®Star 1000 mg/l Y, Carl Roth, Karlsruhe, Germany) was added online via a peristaltic pump to all samples and standards in order to compensate for drifts of the ICP-MS system. Standard deviations of laboratory analyses are provided in Table A2.

The ion balances were calculated from the equivalent concentrations of the cations and anions according to [33]. High ion balance deviations are an indication that certain ions have not been recorded or have been recorded incorrectly. The algebraic sign is an indicator of whether the error could be on the anion or cation side. Another plausibility check is the calculation of the EC of a water sample from the main cations and anions [34]. This method was used to evaluate whether there was any inconsistency between the measured and calculated conductivity, indicating that certain cations or anions had not been detected or had been incorrectly detected.

The underlying method for the analysis of cations and anions in water with pLIBS has been presented in detail by [25] for the three alkali elements Li, Na, and K. Here, the SE aluminium foil, prepared with a thin layer of pencil powder, was placed between the base and the stencil and the sample solutions were applied through the recesses using a micropipette (cf. Figure 2). A total of 0.75 µL of the sample solutions was applied to the aluminium foil in the same way each time with the help of an auxiliary line. In this way, 15 droplets can be placed on one sample foil. After application, the aluminium foil was removed and heated on the heating table to achieve LSC until only the EvR remained. The aluminium foil was then placed back between the base and the stencil and the EvR was analysed with the pLIBS. Standard deviations and relative standard deviations of replicate pLIBS analyses are provided in Table A3. By fixing the nose of the device, it is possible to quickly switch between the individual samples without additional focusing. In addition, the measuring device does not slip during the measurement of the 100 spots per EvR.

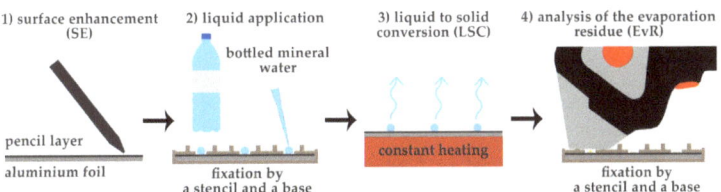

Figure 2. Summary of the method involving surface enhancement, fixed liquid application, liquid-to-solid conversion (LSC), and fixed analysis by portable LIBS.

2.4. Calibration Settings

With the calibrations in this work, it should later be possible to examine as wide a range of differently mineralised waters as possible. It can thus be assumed that the matrices in the EvR will also vary strongly. However, even small changes in the matrix, e.g., due to different concentrations of the analytes, alter the physical and chemical properties in the plasma to such an extent that the emitted signal is no longer proportional to the concentration. This leads to so-called matrix effects [35]. To compensate for matrix effects [35], multivariate calibration was performed for all elements of interest (EoI) using multiple linear regression (MLR). Using the intensities of Al in the denominator to normalise the numerator (analyte) adds an internal standard [35,36], as Al foil was used as a substrate. Intensity ratios (IR) were calculated by using the intensities of the EoI in the numerator (analyte) and Al intensities in the denominator. The intensities chosen for both the EoIs and Al were selected to avoid possible overlap of nearby peaks of other EoIs due to peak broadening and to maintain an equal sequence of intensities as the concentration of the analytes increases and constant intensity at the intensities of the standard. For this reason, small changes were also made to the lines used for Li, Na, and K compared to [25]. The spectral lines used for calibration in this study are listed in Table 2. However, the integration was performed in the same way in SciAps' Profile Builder, applying the zeroth order derivative and Savitzky–Golay smoothing with 7 and 9 as input values for the analytes and Al, respectively, to filter out noise [25].

Table 2. Statistical assessment of the calibrations for single-element standards (Li, Na, K) and paired-element standards (Mg/Ca, Zn/Sr, NO$_3$/SO$_4$/Cl).

z	EoI	State	λ nm	LoD mg/L	Range mg/L	y	R^2	RMSE mg/L	S mg/L
3	Li	I	497.1		0.1–2.5	28.133x − 0.3725	0.918	0.17	0.18
		I	610.4	0.006	2.5–100	37.973x − 2.3069	0.995	6.65	6.71
		I	670.8		100–1000	11.632$x^{2.1134}$	0.954	202.18	204.19
		I	812.9						
11	Na	II	330.2		0.1–2.5	45.219x + 0.0149	0.971	2.00	2.03
		I	589.0	0.014	2.5–100	184.47x − 11.45	0.960	3.54	4.01
		I	589.6		100–1000	8152.3$x^{8.359}$	0.538	1131.01	1141.34
		I	818.3						
		I	819.5						

Table 2. Cont.

z	EoI	State	λ nm	LoD mg/L	Range mg/L	y	R^2	RMSE mg/L	S mg/L
12	Mg	II	279.6		0.1–10	78.378x − 0.0589	0.979	0.00	0.00
		II	279.8		10–100	173.46x − 14.439	0.973	6.83	6.93
		II	280.3		100–1000	-	-	-	-
		I	285.2						
		I	293.6						
		I	382.9	0.008					
		I	383.2						
		I	383.8						
		I	516.7						
		I	517.3						
		I	518.4						
19	K	I	691.2		0.1–10	343.08 + 9.4034	0.987	0.49	0.49
		I	693.9	0.006	10–160	292.55x + 9.7589	0.973	10.70	10.82
		I	766.5		160–1000	$62.902e^{1.6945x}$	0.913	266.10	269.77
		I	769.8						
20	Ca	II	315.9		0.1–2.5	23.014x − 0.0512	0.990	0.03	0.03
		II	317.9		2.5–100	103.34x − 12.964	0.893	3.79	3.82
		II	318.1		100–1000	$-527.31x^2 + 2456.7x - 1766.7$	0.889	155.42	157.59
		II	370.6						
		II	393.3						
		II	396.8						
		I	422.6						
		I	430.2	0.021					
		I	430.8						
		I	443.5						
		I	445.5						
		I	526.5						
		I	527.0						
		I	558.9						
30	Zn	II	202.6	0.0005	0.1–2.5	621.9x − 0.1649	0.988	0.07	0.07
		I	213.9		2.5–50	544.01 − 0.4883	0.998	1.38	1.40
		I	334.5		50–1000	$2449.2x^2 + 1115.7x - 94.322$	0.989	102.14	103.02
		I	468.0						
		I	481.1						
		I	636.2						
38	Sr	II	215.3	0.0008	0.1–5	38.764x + 0.0607	0.999	0.14	0.14
		II	216.6		5–75	$136.68x^2 - 0.3885x + 3.2393$	0.998	7.64	7.74
		II	338.1		75–1000	$6.1616e^{3.7x}$	0.851	4708.79	4754.73
		II	407.7						
		II	416.2						
		II	421.6						
		II	430.6						
		I	460.7						
		I	496.2						
		I	525.7						
		I	548.1						
7	N	II	567.6	0.0017	0.5–160	2014.3x	0.853	105.13	105.97
		II	568.6		160–1000	$100.07e^{7.0029x}$	0.872	195.30	197.96

Table 2. Cont.

z	EoI	State	λ nm	LoD mg/L	Range mg/L	y	R^2	RMSE mg/L	S mg/L
16	S	I	921.2		0.5–160	175098x + 106.97	0.647	93.42	94.12
		I	922.8	0.0002	160–1000	$109.42e^{2043.1x}$	0.829	621.23	629.69
		I	923.7						
17	Cl	I	833.3		0.5–160	33588x	0.976	97.18	98.04
		I	837.6		160–1000	$18.869e^{561.95x}$	0.912	1113.22	1126.09
		I	857.5	0.0004					
		I	858.6						
		I	894.8						
13	Al *	I	236.7						
		I	237.3						
		I	308.2						
		I	394.4						
		I	396.1						

z = atomic number, EoI = element of interest (EoI), λ = wavelength of the lines used for calibration (* Al used as internal standard), LoD = limit of detection: calculated according to the 3σ-IUPAC criterion (LoD = 3* σB/k) [37], y = formula used for calibration, R^2 = coefficient of determination, RMSE = root mean square error, S = residual standard deviation, σB = standard deviation of the background signal at the lowest solution concentration, k = slope of the calibration line.

For more extensive possibilities in calibration, the IRs were calculated with these settings, exported as a .csv file, and used in a spreadsheet for calibration. Subsequently, outliers were eliminated by 1.5 inter-quartile range (IQR) method. For most concentrations, 15 IR values were used for calibration, of which at least 12 remained after outlier elimination. Some higher concentrations were tested with up to 27 IR values. The mean of the respective IR values per element for blanks was then subtracted from all IR values. Due to effects such as self-absorption at higher analyte concentrations [35], the slope of a single calibration curve over the entire concentration range (0.1–1000 mg/L) changes strongly. Ref. [21] encountered the same problem when analysing K, Na, Ca, and Mg in liquid solutions dried on filter paper. They had to use two calibration lines over the concentration range of 10–1000 mg/L [21]. Therefore, in this work, three concentration ranges were defined for all cationic species, which were chosen differently due to the different slopes for the individual elements. Only two concentration ranges were defined for the anionic species. Wherever possible, linear calibration lines were used. For higher concentration ranges, exponential or quadratic calibrations were often required. The different concentration ranges are listed in Table 2. According to the 3σ-IUPAC criterion [37], LoDs for the EoI were calculated and are also listed in Table 2. Within the concentration ranges, other statistical parameters are given, such as the coefficient of determination (R^2) of the calibration lines, the root mean square error (RMSE) of the regression, and the residual standard deviation based on the regression (S).

3. Results

In Table 2, the statistical evaluation of the calibrations for single-element standards (Li, Na, K) and paired-element standards (Mg/Ca, Zn/Sr, NO_3/SO_4/Cl) is shown. The calculated LoDs were quite low (<0.03 mg/L) and therefore notably lower than the lowest concentration used for calibration (0.1 mg/L). In general, high coefficients of determination (R^2) were obtained for the low and medium concentration ranges. The third concentration range generally suffered from lower R^2 and higher RMSE and S (e.g., Na, Sr). The analysis of anionic species was generally less sensitive than that of cationic species, as indicated by the high to very high RMSE and S. It was not possible to establish a calibration line for Mg above 100 mg/L, as the IR did not increase with further increases in concentration. The calibration curves for Sr are shown in Figure 3. In order not to go beyond the scope of this paper, the other calibration curves are not shown here.

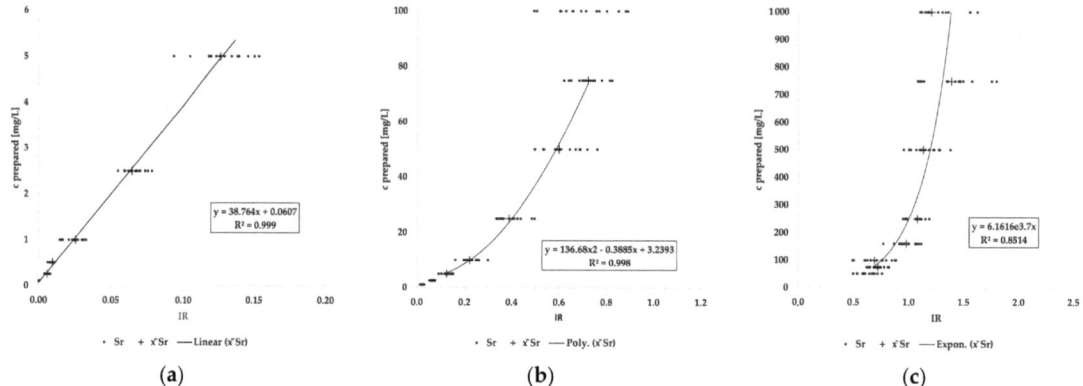

Figure 3. Calibration curves for Sr between the prepared concentration and the calculated intensity ratio (IR) for the concentration ranges (**a**) 0.1–5 mg/L, (**b**) 5–75 mg/L, and (**c**) 75–1000 mg/L.

By applying the calibration curves provided in Table 2, subtracting the respective IR blank values (see Table A1) and using the respective threshold values from Table A1 to select the correct concentration range, the results of the analysis of the bottled mineral waters were obtained and are given in Table 3. The values were compared with the laboratory analyses (IC and ICP-MS). The absolute and relative deviations are provided. In addition, ion balances of all waters are presented and were compared between the laboratory and the pLIBS analysis. The electrical conductivities calculated from the laboratory and pLIBS analyses are shown and the latter were compared with values measured before the analysis. As highly mineralised bottled mineral waters, typically above 1000 µS/cm, showed very low accuracy, these were excluded from evaluation in order to increase clarity. In addition, Zn and NO_3 were excluded, as they showed low analytical performance (low accuracy—in particular, inconsistent overestimation).

Table 3. Results of the analysis with the pLIBS compared to laboratory analysis (lab), excluding SO_4 and Zn. pLIBS values are presented as the mean of five measurements (Adhz: 15). All bottled mineral waters with a conductivity above 1000 µS/cm were omitted. Standard deviations are given in Tables A2 and A3. Laboratory analyses were performed using ion exchange chromatography (IC) except for Sr (inductively coupled plasma–mass spectrometry (ICP-MS)). The absolute deviation (dev) and the relative deviation (r-dev) of the pLIBS from the laboratory value are given for each value. The ionic balance (IB) was calculated using additional hydrogen carbonate values measured with a field photometer. Electrical conductivity was calculated according to [34] for both laboratory and LIBS analysis.

Abbr.		Li	Na	Mg	K	Ca	Sr	SO_4	Cl	Unit	IB eq-%	EC µS/cm	
Adhz	lab	<LoD	11.95	30.63	1.26	18.40	1.891	28.46	18.95	mg/L	−41.3	496	lab
	pLIBS	<LoD	16.11	22.69	0.32	18.74	1.460	<LoD	16.71	mg/L	−44.9	434	pLIBS
	dev	0.022	4.16	7.94	0.95	0.34	0.431	28.46	2.24	mg/L		598	meas
	r-dev	78.6	34.8	25.9	74.8	1.9	22.8	100	11.8	%	−3.7	−164	dev
BeWa	lab	<LoD	6.19	6.96	0.97	30.16	0.103	21.81	15.46	mg/L	−0.7	252	lab
	pLIBS	<LoD	9.41	7.06	1.86	40.15	<LoD	17.98	11.08	mg/L	30.2	277	pLIBS
	dev	0.022	3.21	0.10	0.88	9.99	0.1022	3.83	4.37	mg/L		257	meas
	r-dev	78.6	51.9	1.4	91.1	33.1	99.2	17.5	28.3	%	30.9	20	dev

Table 3. Cont.

Abbr.		Li	Na	Mg	K	Ca	Sr	SO$_4$	Cl	Unit	IB eq-%	EC µS/cm	
Blk	lab	<LoD	<LoD	<LoD	<LoD	0.24	0.004	1.68	2.01	mg/L	−171.2	17	lab
	pLIBS	<LoD	<LoD	<LoD	0.01	<LoD	<LoD	<LoD	0.28	mg/L	−193.6	9	pLIBS
	dev	0.022	0.005	0.051	0.047	0.077	0.003	1.68	1.73	mg/L		1	meas
	r-dev	78.6	26.3	86.4	88.7	78.6	79.4	100	86.2	%	−22.4	8	dev
Gest	lab	<LoD	11.86	39.42	4.45	31.57	0.529	24.23	20.61	mg/L	−45.8	684	lab
	pLIBS	<LoD	13.71	22.08	4.96	33.25	0.363	10.98	6.36	mg/L	−61.8	573	pLIBS
	dev	0.022	1.85	17.34	0.52	1.69	0.166	13.25	14.25	mg/L		878	meas
	r-dev	78.6	15.6	44.0	11.6	5.3	31.4	54.7	69.2	%	−15.9	−305	dev
Laur	lab	<LoD	0.97	0.34	0.36	1.38	0.010	3.21	2.13	mg/L	−73.2	26	lab
	pLIBS	<LoD	0.46	0.04	0.01	1.74	<LoD	42.50	0.95	mg/L	−163.0	81	pLIBS
	dev	0.022	0.51	0.29	0.35	0.36	0.009	39.29	1.18	mg/L		21	meas
	r-dev	78.6	52.1	86.9	98.3	26.2	91.8	1225	55.3	%	−89.9	60	dev
Löng	lab	<LoD	15.05	3.23	1.70	30.33	0.066	40.08	28.79	mg/L	4.8	281	lab
	pLIBS	<LoD	21.31	3.23	4.17	30.05	0.001	24.99	22.56	mg/L	29.1	278	pLIBS
	dev	0.022	6.25	0.00	2.47	0.28	0.065	15.09	6.22	mg/L		275	meas
	r-dev	78.6	41.5	0.0	145	0.9	98.8	37.6	21.6	%	24.3	3	dev
Nat	lab	<LoD	6.28	7.01	3.09	22.48	0.057	12.02	9.29	mg/L	46.8	177	lab
	pLIBS	<LoD	5.21	6.05	2.60	37.28	0.001	0.00	6.36	mg/L	98.8	182	pLIBS
	dev	0.022	1.07	0.96	0.49	14.80	0.056	12.02	2.94	mg/L		212	meas
	r-dev	78.6	17.0	13.7	15.8	65.9	98.6	100	31.6	%	52.1	−30	dev
Odwq	lab	0.036	13.04	23.87	5.83	41.52	0.834	17.56	24.54	mg/L	−42.7	592	lab
	pLIBS	<LoD	22.15	15.25	10.54	36.16	0.694	35.49	19.86	mg/L	−57.1	602	pLIBS
	dev	0.030	9.11	8.62	4.71	5.37	0.140	17.94	4.68	mg/L		727	meas
	r-dev	83.3	69.9	36.1	80.7	12.9	16.8	102	19.1	%	−14.4	−125	dev
Rosb	lab	<LoD	59.91	65.13	3.09	30.59	0.284	18.15	91.57	mg/L	−9.2	1016	lab
	pLIBS	0.057	57.49	42.53	7.67	37.75	0.001	0.0002	44.85	mg/L	−16.4	876	pLIBS
	dev	0.029	2.42	22.60	4.58	7.17	0.284	18.15	46.72	mg/L		1363	meas
	r-dev	103	4.0	34.7	148	23.4	100	100	51.0	%	−7.3	−487	dev
Saw	lab	<LoD	18.73	3.52	1.17	71.03	0.722	58.00	52.78	mg/L	3.7	524	lab
	pLIBS	<LoD	41.18	4.29	0.62	65.17	0.296	0.0002	32.02	mg/L	60.5	439	pLIBS
	dev	0.022	22.45	0.78	0.55	5.86	0.427	58.00	20.76	mg/L		589	meas
	r-dev	78.6	120	22.1	47.1	8.2	59.0	100	39.3	%	56.8	−150	dev
Vit	lab	0.055	6.02	19.45	5.07	54.26	0.877	113.17	6.87	mg/L	5.0	511	lab
	pLIBS	<LoD	5.93	11.94	7.16	43.63	0.764	70.51	4.33	mg/L	1.5	390	pLIBS
	dev	0.049	0.09	7.51	2.09	10.63	0.113	42.65	2.54	mg/L		444	meas
	r-dev	89.1	1.5	38.6	41.4	19.6	12.9	37.7	37.0	%	−3.6	−54	dev
Median r-dev		78.6	34.8	34.7	80.7	19.6	79.4	100.0	37.0	%			

Adhz = Adelholzner, BeWa = Bergische Waldquelle, Blk = blank sample, Gest = Gerolsteiner, Laur = Lauretana, Löng = Löningen, Nat = Naturalis, Odwq = Odenwaldquelle, Rosb = Rosbacher, Saw = Sawell, Vit = Vittel.

Figures 4–6 show correlations between pLIBS predicted concentrations and laboratory concentrations for all EoIs, excluding bottled mineral waters with a conductivity greater than 1000 µS/cm. Of these, Figure 4 shows only the singly charged cations (alkali metals) and Figure 5 the doubly charged cations (alkaline earth elements). The correlations for the anionic species, excluding NO$_3$, are illustrated in Figure 6.

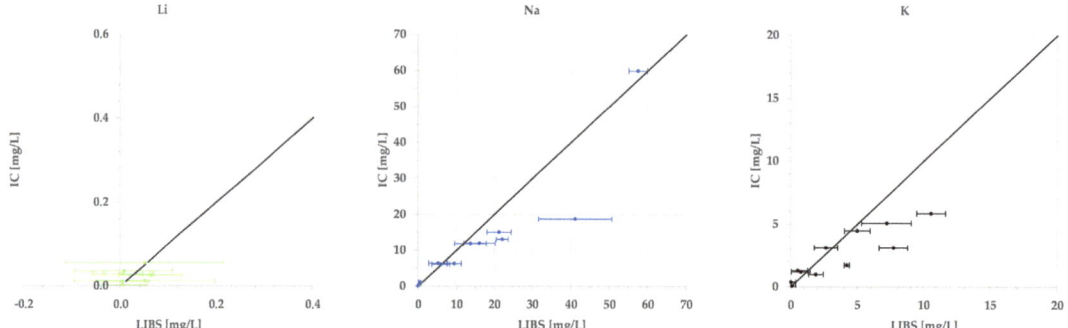

Figure 4. Correlations of LIBS predicted concentrations versus IC for the light alkali metals. Li and K values were adjusted by subtracting a slightly higher value than the blank IR value, otherwise the results would be overestimated. An optimal correlation is indicated by the black line. Results for Aqua Mia, Extaler, Fachinger, Marius, and Quellbrunn are not included.

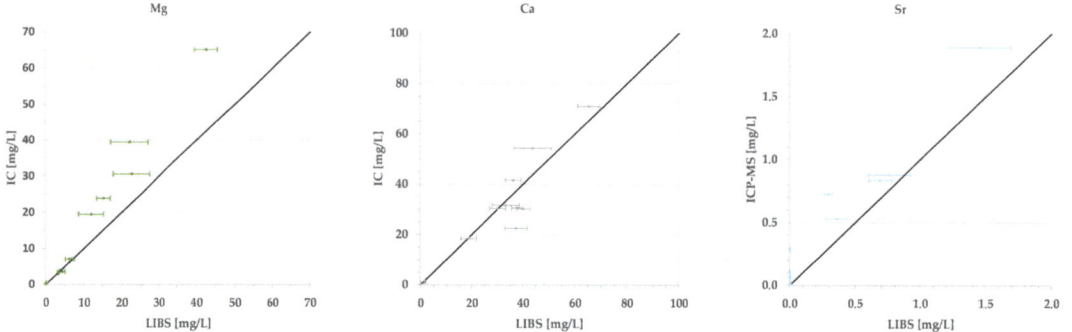

Figure 5. Correlations of LIBS predicted concentrations versus IC or ICP-MS analyses (Sr) for the alkaline earth metals. An optimal correlation is indicated by the black line. Results for Aqua Mia, Extaler, Fachinger, Marius, and Quellbrunn are not included.

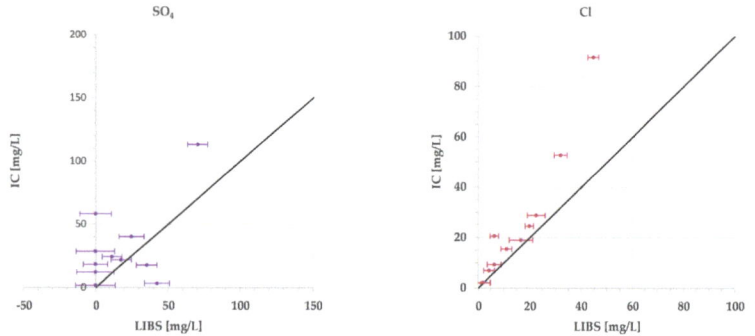

Figure 6. Correlations of LIBS predicted concentrations versus IC for the anionic species, excluding SO_4. An optimal correlation is indicated by the black line. Results for Aqua Mia, Extaler, Fachinger, Marius, and Quellbrunn are not included.

Figure 7 shows combined Stiff diagrams for all selected bottled mineral waters. Stiff plots simplify the comparison of waters [38] and are usually applied to compare different waters—for example, to illustrate spatial or temporal differences in water chemistry. Here, combined Stiff diagrams were used to compare the same water in different analyses (pLIBS and laboratory). For each water sample, a Stiff diagram is shown for the laboratory and for the pLIBS analysis results. Perfectly matching analyses should produce exactly the same polygon for both analyses. Since an additional photometer was used for the HCO_3 concentrations, the results for laboratory and LIBS analysis are identical for HCO_3.

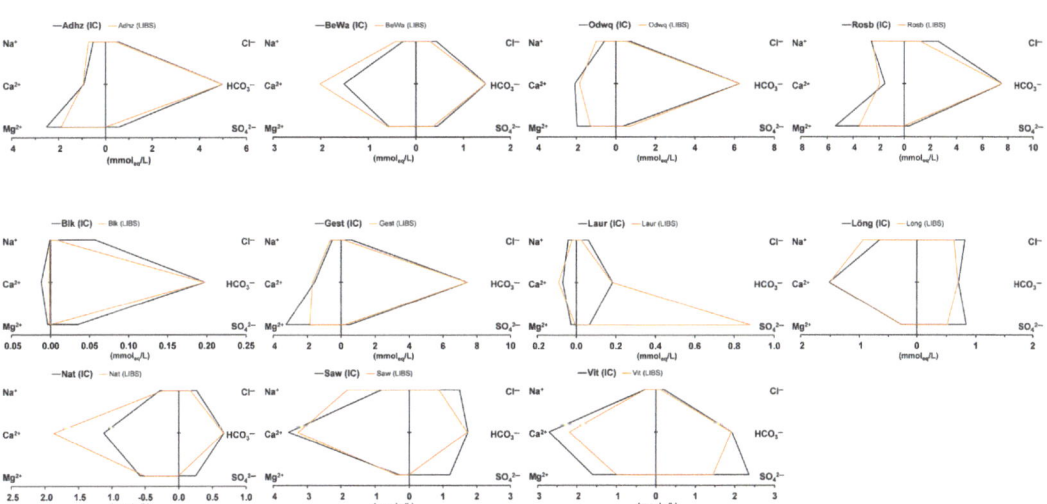

Figure 7. Combined Stiff diagrams of the selected bottled mineral waters. One shows the diagram for the IC and one for the pLIBS results. HCO_3 concentrations were measured with a field-ready photometer and were therefore identical for both analyses.

Figures 8 and 9 show the results of applying the calibrations used for single-element or paired-element standard solutions (Table 2) to mixed standard solutions. The first of the two figures shows the results of the mixed test series containing all three singly charged cations simultaneously (cf. Figure 8). K concentrations appear to have been slightly underestimated for prepared concentrations below 10 mg/L and overestimated for prepared concentrations between 40 mg/L and 75 mg/L. For higher concentrations, the prepared concentrations were clearly underestimated. Li concentrations were slightly overestimated for concentrations up to 30 mg/L, fit relatively well for concentrations up to 80 mg/L, and were underestimated for even higher concentrations. Na concentrations up to 20 mg/L seem to have fit well, but at higher prepared concentrations the predicted concentrations also seem to have been underestimated.

Figure 9 shows the results of the test series with mixed standards containing all six cations simultaneously. Compared to the mixed standards with less different cations, all three alkali elements seem to have behaved differently. The overestimation at low concentrations and underestimation at high concentrations was even more pronounced for Li, Na, and K in the second series of tests. Li, in particular, changed and ended up behaving very similarly to Na. It is noticeable that the alkaline earth metals (doubly charged cations) Mg, Ca, and Sr behaved similarly to each other but quite differently to the alkali elements (singly charged cations). They were more clearly underestimated at higher concentrations but not overestimated at low concentrations. For all elements in both test series, there appears to have been a plateau at higher concentrations where even higher concentrations did not produce more signal and therefore a predicted concentration. Attenuation at higher concentrations appears to have been greater for alkaline earth elements (divalent cations) than for alkali elements (monovalent cations). A series of attenuations can be formed from low to high: K < Li < Na < Mg < Sr < Ca.

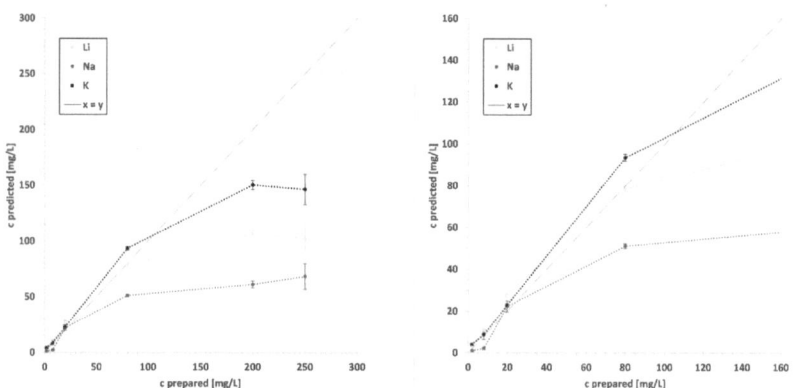

Figure 8. Results of applying the calibrations to a series of tests using mixed solutions of known concentration containing equal concentrations of Li, Na, and K. The (**right**) graph shows a section of the (**left**) graph for better comparison with Figure 9.

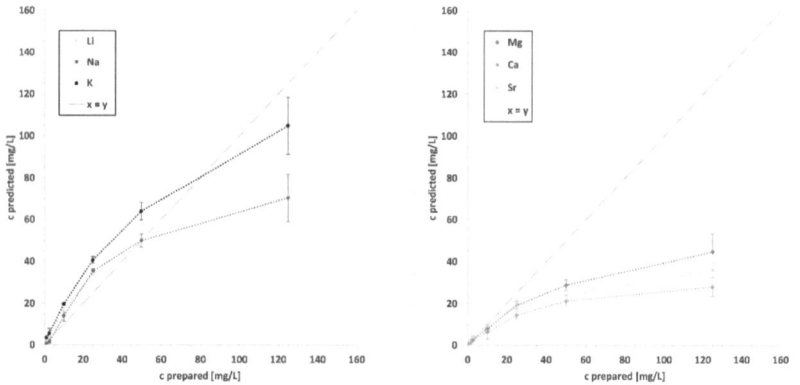

Figure 9. Results of applying the calibrations to a series of tests using mixed solutions of known concentration containing equal concentrations of Li, Na, K, Mg, Ca, and Sr. The light alkali elements are shown on the (**left**) and the alkaline earth metals on the (**right**).

4. Discussion

All calculated LoDs were quite low (in the ppb range). However, these low detection limits are deceptive. The lowest concentrated standard used was 0.1 mg/L for the cations and 0.5 mg/L for the anions. Nevertheless, the highest LoD calculated was 0.021 mg/L (Ca). It is particularly striking that the detection limits for the anions were particularly low. It has been shown in earlier research that the quantitative analysis of sulphur and chlorine with LIBS is subject to some difficulties and that indirect determination is often necessary to achieve low detection limits [39,40]. This is due to the low excitation in the plasma caused by the high ionisation energy of Cl and S [39]. However, [41] pointed out that the determination of LoDs with the outdated IUPAC formula used in this work is not particularly appropriate for multivariate LIBS analysis [41]. Yet, since this formula is currently the most widely used calculation of the LoD and comparability with [25] should be ensured, a different calculation was not used. Although the absolute detection limits may be somewhat higher in reality, it has been shown that very low concentrations (<0.1 mg/L) can be detected in standard solutions with pLIBS.

It may seem cumbersome to have several calibration curves within one EoI for different concentration ranges, but by using THVs (cf. Table A1) the selection of the correct range and therefore the formula can be completed automatically in a spreadsheet. The advantage of having multiple calibration curves for an EoI at different concentration ranges is that the changing slope of a single curve can be better represented. If self-absorption correction is applied in the future, it may not be necessary to have several concentration ranges but rather only one calibration curve, as the slope will no longer change as much.

The high determination coefficients for Li, Na, and K in [41] could also be achieved for other elements and compounds, especially at low to medium concentrations (cf. Table 2). At higher concentrations (generally > 100 mg/L), the determination coefficients were higher, as expected. For Mg, no calibration line could be established, as the IR did not increase with increasing concentration. This is an indication of strong self-absorption [35].

Zn, and NO_3 were excluded from the evaluation, as they showed low analytical performance. For Zn, on the one hand, the test series was not diverse enough to make statements on the applicability to natural waters and, on the other hand, the measured values were clearly overestimated. NO_3 concentrations showed no clear correlation between pLIBS and IC analysis, with a strong tendency of overestimation. For the pLIBS analysis of NO_3, it cannot be excluded that the results were partly falsified. It is possible that some cation samples were used for the pLIBS measurements instead of the unaltered anion samples. In contrast to the anion samples, these were acidified with HNO_3 (see liquid analyses) to prevent the cations from precipitating during transport to the laboratory for IC and ICP-MS analyses. Since only very small amounts of diluted high-purity nitric acid were used, this should only have had an influence on the nitrate concentrations. However, this would explain why there was no correlation between pLIBS and IC analyses for nitrate.

The median of the relative deviation for all Ca analyses with pLIBS compared to laboratory analyses was fairly good, at 19.6% (cf. Table 3). Figure 5 shows a fairly good correlation between pLIBS and IC data for Ca up to 75 mg/L.

The median of the relative deviation for all Na and Mg analyses with pLIBS compared to laboratory analyses was reasonable, at 34.8% and 34.7%, respectively (cf. Table 3). Figure 4 shows a fairly good correlation between pLIBS and IC data for Na with only one conspicuous outlier with a very high standard deviation (9.52 mg/L). Figure 5 shows a correlation between pLIBS and IC data for Mg, with a tendency for higher concentrations to be underestimated. This trend could be interpreted as a progressive exponential function, which could indicate an increase in self-absorption with increasing concentration.

The median of the relative deviation for all Cl analyses with pLIBS compared to laboratory analyses was still reasonable, at 37.0% (cf. Table 3). Figure 6 shows a fairly good correlation between pLIBS and IC up to 30 mg/L, with a tendency to underestimate higher concentrations, similar to Mg (cf. Figure 5), from which the same conclusions can be drawn.

The median of the relative deviation with pLIBS compared to laboratory analyses was quite high for Li, Sr, and K, at 78.6, 79.4, and 80.7%, respectively (cf. Table 3). However, the test series was not very diverse for Li, with most values close to or below the LoD of the IC analyses (0.027 mg/L). It is therefore hardly surprising that most of the values for the pLIBS Li analysis were also close to or below the LoD of the pLIBS analysis. A large part of the relative deviation for Li thus resulted from the different LoDs between pLIBS and IC analysis. As the test series was not diverse enough for Li concentrations (cf. Figure 4), it is also difficult to say whether there was a good correlation between pLIBS and IC analysis. For Sr, the deviation also mainly came from very low concentrations. Many pLIBS results were below the LoD of 0.0008 mg/L. However, there was a correlation between pLIBS and IC data for concentrations up to 2 mg/L (cf. Figure 5). For K, Figure 4 shows a fairly good correlation between pLIBS and IC data, with a tendency for all concentrations to be slightly overestimated.

The median of the relative deviation for all SO_4 analyses with pLIBS compared to laboratory analyses was quite high, at 100.0% (cf. Table 3). In addition, the correlation between pLIBS and IC data was quite poor (cf. Figure 6).

The IB can help to identify possible analytical discrepancies between cationic and anionic species concentrations. Therefore, a negative IB indicates excessive findings of anionic species concentration or underestimation of cationic species concentration. A positive IB indicates too low an analysed anionic species concentration or too high a cationic species concentration. Seven out of 11 results of IB calculated with pLIBS had a negative IB, which indicates that mostly anions were overestimated and or cations underestimated.

Ideally, the calculated ECs for both analyses (pLIBS and laboratory) should match the measured EC value. A deviation from the measured value is a clear indication of non-analysed or incorrectly analysed ions. If the calculated EC value of the laboratory measurement differs from the measured value, it can be assumed that either ions precipitated, samples were contaminated, or they were measured incorrectly. Looking at the values in Table 3, one water stands out as having had a deviation of more than 30% for the laboratory EC measurements: the blank one. This was mainly due to the low mineralisation of the deionised water, where minimal absolute differences in the analysis result in large percentage deviations. For the pLIBS analysis, 4 of the 11 waters showed a deviation of more than 30% (Blk, Gest, Laur, Löng, Saw). This is an indication that the total of all determined ions for these waters differs from the real solution content.

Furthermore, precipitation of $CaCO_3$ prior to both pLIBS and IC analysis can be observed by comparing the analysis results with the values indicated on the bottles (Adhz, Gest, Odwq, Rosb, Vit). However, other cations or anions do not appear to have been affected and precipitation occurred prior to analysis, as shown by comparative measurements with newly purchased bottles and re-measurements of the original samples.

The Stiff diagrams perfectly illustrate the differences between pLIBS and laboratory analyses (cf. Figure 7). Based on the agreement between the analyses, the plots can be grouped into two categories: good correlation between laboratory and pLIBS analysis (first row) and moderate correlation (second and third rows). A third category with poor correlation would have been needed for waters with a conductivity greater than 1000 µS/cm or concentrations of several ions >> 6 $mmol_{eq}/L$.

For clarity, the uncertainty and the precision of the pLIBS analysis of the bottled waters are reported separately in Table A3 in the Appendix A. Standard deviations (SD) for replicates on one sample in the range of 0.003 to 14.01 mg/L for all selected samples and elements are quite acceptable for a portable instrument, taking into account the diverse chemistry, with up to approximately 120 mg/L solution content per element (cf. Table A3, highly mineralised waters excluded). Looking at the median relative standard deviation (RSD) for the different elements, the values appear quite high. The lowest RSD was 11% (Ca) and the highest 39% (SO_4). However, these values are comparable to the RSDs reported by other authors who analysed aqueous samples by laboratory LIBS. For example, a precision of 2–6% RSD was achieved for aerosol LIBS and a precision of 13–22% for microdrop LIBS [16]. For more similar sample preparation techniques using LSC, 11–17% RSD was achieved for geometric constraint LSC and 25–36% RSD for unconstrained direct LSC [19]. Precision in LIBS analysis is typically low (5–20%) due to shot-to-shot variability and matrix effects [42]. Other effects, such as the slightly different distribution of the EvR, may also occur, resulting in lower precision. It is therefore advisable to perform multiple measurements per sample. At least three or, better, five measurements per sample are recommended for the presented method. Due to the small sample volume required (0.75 µL) and the short measurement time, this can be achieved quickly and easily.

Compared to testing single-element standard solutions [25], it is to be expected that there are more effects affecting the results of an analysis of mixed solutions or natural waters with an even more complex matrix. As LIBS analysis is highly susceptible to the so-called matrix effect [35], small changes in the matrix can cause the emitted signal to be no longer proportional to the concentration. There are several indications of matrix effects in the results of the mineral water analyses. These are particularly evident in the fact that more highly mineralised waters such as Aqua Mia, Extaler, and Marius generally showed very poor analytical results and were omitted from the evaluation. If a suitable self-absorption correction is used in the future, these more highly mineralised waters should also be analysable. In addition, especially for Mg and Cl concentrations above 15 mg/L, there is a systematic underestimation, which can even be seen as a recognisable (progressive) exponential function in the correlation plots (cf. Figures 5 and 6). For the other cations, this effect might also occur if samples with higher concentrations had been analysed. The discarded data for Ca and Sr confirm this assumption. However, the self-absorption effect is difficult to investigate in complex natural waters.

In order to gain a better understanding of this effect, mixed standard solutions were analysed in addition to the bottled mineral waters, and the calibrations developed were used for analysis. A small mixed standard containing Li, Na, and K and a more comprehensive one containing Li, Na, K, Ca, Mg, and Sr were analysed. The results of the two test series show clearly that there were both amplifying and attenuating effects that cancelled out the proportionality (cf. Figures 8 and 9). Nevertheless, a clear correlation is recognisable. This can be described as degressive proportionality, in which the measured concentration increases less and less as the real concentration increases. Typically, with low concentrations attenuating effects can be visible, especially for the light alkali elements Li, Na, and K. This effect was less pronounced in the test series without the doubly charged cations. As with all elements in both test series, a plateau was reached at higher concentrations, where even higher concentrations did not produce significantly more signal and therefore a predicted concentration, the linearity, was cancelled out. This is a clear hint of self-absorption [41]. When using single-element standards, these problems were encountered with concentrations typically above 160 mg/L [25]. This limit seems to have dropped significantly for more complex waters and was more pronounced for the alkaline earth metals than for the alkali elements (cf. Figures 8 and 9).

Typically, self-absorption has several effects on the line shape, so these should be visible in the lines used for calibration. Self-reversal can occur in LIBS analysis when there are spatial gradients in plasma temperature and electron number density. This can lead to a dip at the centre of an emission peak, which can be strong enough to erroneously identify two peaks [43]. In this work, no line showed typical self-reversal effects such as a dip at any maximum. However, this does not mean that there was no self-absorption [44]. Self-absorption was visible in several lines, as the IR did not grow proportionally with increasing concentration (typically above 100 mg/L) and the curve saturated (cf. Figures 8 and 9). This can also be seen in the broadening of the peaks, which resulted in a higher full width half maximum (FWHM) (cf. Figure 10a,b). It can clearly be seen that for both K 769.8 and Li 670.7 the lines not only increased in height with increasing concentration but also became wider. Between 10 and 25 mg/L there was still a large difference in peak height for K 769.8 (cf. Figure 10b). Between 50 and 125 mg/L the difference was already smaller and the height variation at the same concentration was greater. In addition, the lines at 10 mg/L were only a little more than half a nm wide at the base. At 125 mg/L, it was already more than 2 nm. This made integration more difficult. If the integration range is too large, there may be overlap with other peaks. If it is too small, the area under the peaks will be underestimated for higher concentrations. The effect of peak broadening also occurred with all other lines of the other elements and compounds investigated. However, it was particularly pronounced for the higher-intensity peaks.

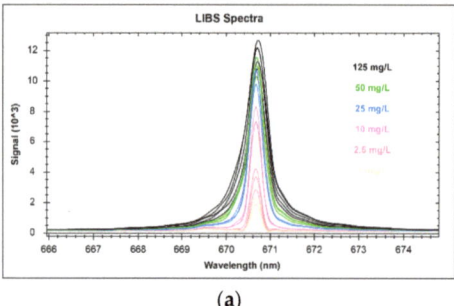
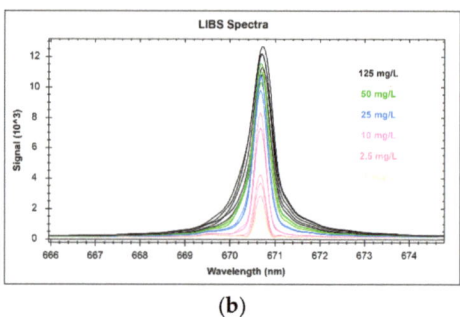

Figure 10. Line broadening with increasing concentration (1–125 mg/L). (**a**) Li 670.7 nm; (**b**) K 769.8 nm. Experimental parameters: 5–6 mJ/pulse, 50 Hz, 2 µs gate delay, Ar atmosphere.

Ref. [43] did not observe any self-reversal or self-absorption when analysing the liquids directly, even at a concentration of 40,000 mg/L. In contrast to [43], self-absorption played a significant role with increasing concentration when using LSC. However, [43] also found both self-reversal and self-absorption effects when analysing solids, and the main difference in this study is that by analysing the evaporation residue, solids were analysed instead of liquids.

This difference was also highlighted by [43] and attributed to the fact that the atomic densities of analytes in plasma are approximately 1000 times greater for a pure solid than for liquid solutions and are therefore optically thicker.

Ref. [21], who used LSC on filter paper in the concentration range of 0–1000 mg/L, also experienced self-absorption and therefore had to apply two calibration lines per element to fit the data. At lower concentrations, a steeper straight line could be applied than at higher concentrations [21]. This clearly reduced the sensitivity at higher concentrations, as in this work.

The same effect was observed by [17], who also used an SE LSC method. They explained the increased effect of self-absorption by the fact that analytes and standards are concentrated in a very small area after drying [17]. It can therefore be assumed that the effect is even stronger with SE methods without filter paper, since the evaporation residue is confined to a smaller area than when a filter paper is used.

In this work, a relatively long gate delay of 2 µs was used. The gate delay was initially optimised for Li, Na, and K [25] and not further adjusted in this work to obtain comparable data. Ref. [45] showed that sensitivity is not significantly affected by increasing the gate delay but that precision is increased and self-absorption reduced by a longer gate delay. However, they used significantly shorter gate delays of 250 ns and 500 ns. Ref. [46] also recommend mitigating self-absorption by recording the signal with a longer gate delay, since this effect tends be more prominent in the early stages of laser-generated plasma.

Looking at the spectra of the gate delay investigations in [25], which were recorded similarly to the data within this study, it is noticeable that not only was the peak height affected by a change in gate delay at the same concentration (cf. Figure 11).There is also a clear broadening of the line at shorter gate delays (more intense grey values). With a longer gate delay, the peaks become significantly narrower and the effect of self-absorption decreases. This effect was observed for the three tested elements of Li, Na, and K.

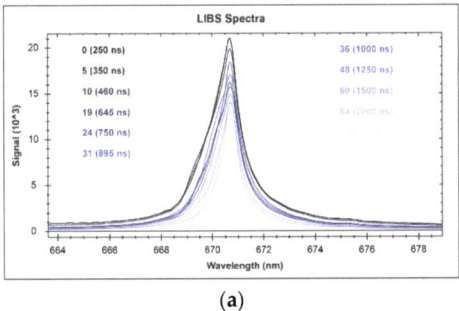 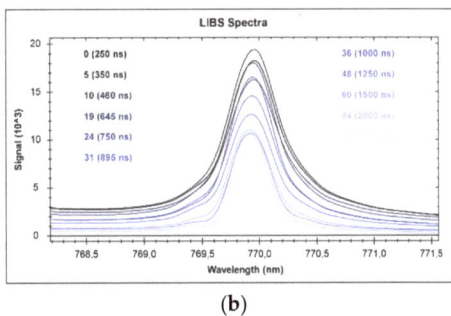

(a) (b)

Figure 11. Line broadening with increasing gate delay (0–132 IID). (**a**) Li 670.7 nm; (**b**) K 769.8 nm. Experimental parameters: stable concentration of evaporated droplets (1000 mg/L), 5–6 mJ/pulse, 50 Hz, variable gate delay, Ar atmosphere. Grey value increase with increasing gate delay.

Due to the strong self-absorption effects, future work will focus on the improvement of the method by the addition of a self-absorption correction to improve accuracy and compensate for the underestimation of higher predicted concentrations. When applying self-absorption correction, it may also be possible to have only one calibration curve instead of two or three for different concentration ranges. However, in this work, it was important to show that the method is basically applicable to natural waters and to determine the influencing factors. These seem to be determined less by the number of different elements than by self-absorption. In addition to filters to remove any undissolved components prior to analysis, a mobile hot plate to evaporate the micro droplets is required for future field application of the method.

The possibility to set up calibration curves for Zn with high coefficients of determination in standard solutions (cf. Table 3) proves that, in principle, it is also possible to analyse environmentally relevant elements in aqueous solutions with portable LIBS. In this work, no correlation between pLIBS and ICP-MS results for Zn could be found with the bottled mineral waters used. However, the test series was not very diverse for Zn, with all but one value below 0.1 mg/L in ICP-MS analyses. With an appropriate self-absorption correction and a diverse test series, Zn and possibly other problematic elements such as Pb and As should theoretically also be quantifiable.

It is clear that the documented analytical approach is not only applicable to single-element standard solutions but also to low mineralised natural waters with complex matrices. By adding a self-absorption correction, it should also be possible to quantitatively analyse more highly mineralised waters and improve the precision. As demonstrated, there is significant potential for developing field-based pLIBS for quantitative water analysis.

5. Conclusions

In our previous study, pLIBS was evaluated for the quantitative analysis of dissolved alkali metals in single-element standard solutions. The aim of this work was to show whether pLIBS can also be used for chemical compositional measurements of natural waters. The results of this study demonstrate that it is possible to quantitatively analyse low to medium mineralised bottled mineral waters with pLIBS for some of the main cations and anions. In higher mineralised waters with an EC above approximately 1000 µS/cm, the concentrations of the main cations and anions were mostly underestimated. This effect was mainly due to self-absorption, which was clearly visible in strong line broadening with increasing concentration. The effect of self-absorption was quite strong, despite a long gate delay being used, which should have compensated for high self-absorption. However, no self-reversal could be detected in any peak, which should have made self-absorption correction easier. There have been differences observed for singly ionised cations (alkali elements) compared to doubly ionised cations (alkaline earth metals). The self-absorption seems to have been pronounced for alkaline earth metals. Therefore, the analysis of alkali elements is currently more reliable than for alkaline earth metals, and the analysis of higher concentrations is more reliable for alkali elements. Analysis of anions is less reliable, even though LoDs may be calculated lower than for the other investigated ions. Of the anions, only Cl showed reasonably reliable results in natural waters. In general, the low detection limits are deceptive and do not reflect how well an EoI can be analysed. An analysable concentration range for natural mineral waters using the method described is approximately between 0.1 and 100 mg/L per element.

The next step in the development of the method is the introduction of an adapted self-absorption correction. It is quite promising that more elements will be calibratable (e.g., Zn). It should therefore be possible to analyse environmentally significant elements in the future. In any case, the ability to analyse natural mineral waters with complex matrices for their main ions opens up many new possibilities for pre-screening and on-site water analysis.

Author Contributions: Conceptualization, N.S.; methodology, N.S.; software, N.S.; validation, N.S.; formal analysis, N.S., S.I. and S.S.; investigation, N.S.; resources, N.S.; data curation, N.S.; writing—original draft preparation, N.S.; writing—review and editing, N.S., B.G.L. and S.S.; visualization, N.S.; supervision, B.G.L.; project administration, N.S.; funding acquisition, B.G.L. All authors have read and agreed to the published version of the manuscript.

Funding: This research received no external funding.

Institutional Review Board Statement: Not applicable.

Informed Consent Statement: Not applicable.

Data Availability Statement: The datasets generated during and/or analysed during the current study are not publicly available due to the fact that the data are part of a PhD thesis but are available from the corresponding author on reasonable request.

Acknowledgments: We thank the three anonymous reviewers for their constructive comments, which greatly improved this manuscript.

Conflicts of Interest: The authors declare no conflict of interest.

Appendix A

Table A1. Threshold values (THV) for selecting the correct formula according to the concentration range and the mean values of the blank IR (mean b), which were subtracted from the calculated IR values. THV I = transition of the first to the second concentration range. THV II = transition of the second to the third concentration range. Unitless IR values are given.

	Li	Na	Mg	K	Ca	Sr	SO_4	Cl
THV I	0.100	0.070	0.130	0.020	0.175	0.140	0.0002	0.0028
THV II	3.000	0.600	0.700	0.550	1.000	0.850		
mean b	0.015	0.0141	0.0222	0.022	0.0093	0.0019	0.0003	0.0001

Table A2. Relative standard deviations (RSDs) and detection limits (LoDs) of IC and ICP-MS* analysis.

Abbr.	Li	Na	Mg	K	Ca	Sr*	SO_4	Cl	Unit
RSD	0.611	0.476	0.449	0.666	0.355		0.149	0.072	%
LoD	0.028	0.019	0.059	0.053	0.098	0.0000214	0.218	0.075	mg/L

Table A3. Standard deviations (SDs) calculated from five pLIBS (Adhz: 15) measurements per water sample and median values of the relative standard deviations (RSDs) for the elements investigated. Readings below the detection limit were not included in the calculation of the RSD. For this reason, no RSD could be calculated for Li.

SD	Li	Na	Mg	K	Ca	Sr	SO$_4$	Cl	Unit
Adhz	0.06	4.21	4.86	0.73	3.08	0.23	13.51	4.56	mg/L
BeWa	0.06	1.89	0.51	0.54	2.62	0.00	7.00	2.14	
Blk	0.03	0.13	0.05	0.29	0.13	0.01	14.01	3.31	
Gest	0.06	4.23	4.94	0.97	5.28	0.08	7.00	1.65	
Laur	0.04	0.29	0.10	0.34	0.18	0.01	8.58	3.02	
Löng	0.05	3.21	0.21	0.20	3.23	0.01	8.58	3.44	
Nat	0.05	2.44	0.96	0.88	4.33	0.01	13.10	2.70	
Odwq	0.10	1.60	1.78	1.08	3.03	0.09	7.00	1.65	
Rosb	0.07	2.37	3.05	1.08	2.26	0.01	8.58	2.14	
Saw	0.15	9.52	0.74	0.71	4.24	0.03	11.07	2.53	
Vit	0.16	2.34	3.37	1.88	7.22	0.16	7.00	2.14	
Median RSD	-	25	17	29	11	22	39	26	%

References

1. Zulkifli, S.N.; Rahim, H.A.; Lau, W.-J. Detection of Contaminants in Water Supply: A Review on State-of-the-Art Monitoring Technologies and Their Applications. *Sens. Actuators B Chem.* **2018**, *255*, 2657–2689. [CrossRef] [PubMed]
2. Yaroshenko, I.; Kirsanov, D.; Marjanovic, M.; Lieberzeit, P.A.; Korostynska, O.; Mason, A.; Frau, I.; Legin, A. Real-Time Water Quality Monitoring with Chemical Sensors. *Sensors* **2020**, *20*, 3432. [CrossRef]
3. Jan, F.; Min Allah, N.; Düştegör, D. IoT Based Smart Water Quality Monitoring: Recent Techniques, Trends and Challenges for Domestic Applications. *Water* **2021**, *13*, 1729. [CrossRef]
4. Lemière, B.; Uvarova, Y.A. New Developments in Field-Portable Geochemical Techniques and on-Site Technologies and Their Place in Mineral Exploration. *Geochem. Explor. Environ. Anal.* **2020**, *20*, 205–216. [CrossRef]
5. Lemière, B.; Harmon, R.S. XRF and LIBS for Field Geology. In *Portable Spectroscopy and Spectrometry*; Crocombe, R., Leary, P., Kammrath, B., Eds.; Wiley: Hoboken, NJ, USA, 2021; pp. 455–497, ISBN 978-1-119-83557-8.
6. Harmon, R.S.; Senesi, G.S. Laser-Induced Breakdown Spectroscopy—A Geochemical Tool for the 21st Century. *Appl. Geochem.* **2021**, *128*, 104929. [CrossRef]
7. Schlatter, N.; Freutel, G.; Lottermoser, B.G. Evaluation of the Use of field-portable LIBS Analysers for on-site chemical Analysis in the Mineral Resources Sector. *GeoResources* **2022**, *2*, 32–38.
8. Tiihonen, T.E.; Nissinen, T.J.; Turhanen, P.A.; Vepsäläinen, J.J.; Riikonen, J.; Lehto, V.-P. Real-Time on-Site Multielement Analysis of Environmental Waters with a Portable X-ray Fluorescence (PXRF) System. *Anal. Chem.* **2022**, *94*, 11739–11744. [CrossRef]
9. Gałuszka, A.; Migaszewski, Z.M.; Namieśnik, J. Moving Your Laboratories to the Field—Advantages and Limitations of the Use of Field Portable Instruments in Environmental Sample Analysis. *Environ. Res.* **2015**, *140*, 593–603. [CrossRef]
10. Cremers, D.A.; Radziemski, L.J.; Loree, T.R. Spectrochemical Analysis of Liquids Using the Laser Spark. *Appl. Spectrosc.* **1984**, *38*, 721–729. [CrossRef]
11. Yueh, F.-Y.; Sharma, R.C.; Singh, J.P.; Zhang, H.; Spencer, W.A. Evaluation of the Potential of Laser-Induced Breakdown Spectroscopy for Detection of Trace Element in Liquid. *J. Air Waste Manag. Assoc.* **2002**, *52*, 1307–1315. [CrossRef]
12. Zhao, F.; Chen, Z.; Zhang, F.; Li, R.; Zhou, J. Ultra-Sensitive Detection of Heavy Metal Ions in Tap Water by Laser-Induced Breakdown Spectroscopy with the Assistance of Electrical-Deposition. *Anal. Methods* **2010**, *2*, 408. [CrossRef]
13. Lee, D.-H.; Han, S.-C.; Kim, T.-H.; Yun, J.-I. Highly Sensitive Analysis of Boron and Lithium in Aqueous Solution Using Dual-Pulse Laser-Induced Breakdown Spectroscopy. *Anal. Chem.* **2011**, *83*, 9456–9461. [CrossRef] [PubMed]
14. Lee, Y.; Oh, S.-W.; Han, S.-H. Laser-Induced Breakdown Spectroscopy (LIBS) of Heavy Metal Ions at the Sub-Parts per Million Level in Water. *Appl. Spectrosc.* **2012**, *66*, 1385–1396. [CrossRef]
15. Aguirre, M.A.; Legnaioli, S.; Almodóvar, F.; Hidalgo, M.; Palleschi, V.; Canals, A. Elemental Analysis by Surface-Enhanced Laser-Induced Breakdown Spectroscopy Combined with Liquid–Liquid Microextraction. *Spectrochim. Acta Part B At. Spectrosc.* **2013**, *79–80*, 88–93. [CrossRef]
16. Cahoon, E.M.; Almirall, J.R. Quantitative Analysis of Liquids from Aerosols and Microdrops Using Laser Induced Breakdown Spectroscopy. *Anal. Chem.* **2012**, *84*, 2239–2244. [CrossRef]
17. Bae, D.; Nam, S.-H.; Han, S.-H.; Yoo, J.; Lee, Y. Spreading a Water Droplet on the Laser-Patterned Silicon Wafer Substrate for Surface-Enhanced Laser-Induced Breakdown Spectroscopy. *Spectrochim. Acta Part B At. Spectrosc.* **2015**, *113*, 70–78. [CrossRef]
18. Yang, X.; Yi, R.; Li, X.; Cui, Z.; Lu, Y.; Hao, Z.; Huang, J.; Zhou, Z.; Yao, G.; Huang, W. Spreading a Water Droplet through Filter Paper on the Metal Substrate for Surface-Enhanced Laser-Induced Breakdown Spectroscopy. *Opt. Express* **2018**, *26*, 30456. [CrossRef]

19. Ma, S.; Tang, Y.; Ma, Y.; Chen, F.; Zhang, D.; Dong, D.; Wang, Z.; Guo, L. Stability and Accuracy Improvement of Elements in Water Using LIBS with Geometric Constraint Liquid-to-Solid Conversion. *J. Anal. At. Spectrom.* **2020**, *35*, 967–971. [CrossRef]
20. Nakanishi, R.; Ohba, H.; Saeki, M.; Wakaida, I.; Tanabe-Yamagishi, R.; Ito, Y. Highly Sensitive Detection of Sodium in Aqueous Solutions Using Laser-Induced Breakdown Spectroscopy with Liquid Sheet Jets. *Opt. Express* **2021**, *29*, 5205. [CrossRef]
21. Skrzeczanowski, W.; Długaszek, M. Application of Laser-Induced Breakdown Spectroscopy in the Quantitative Analysis of Elements—K, Na, Ca, and Mg in Liquid Solutions. *Materials* **2022**, *15*, 3736. [CrossRef] [PubMed]
22. Tian, H.; Li, C.; Jiao, L.; Zhao, X.; Dong, D. Study on Rapid Detection Method of Water Heavy Metals by Laser-Induced Breakdown Spectroscopy Coupled with Liquid-Solid Conversion and Morphological Constraints. In Proceedings of the International Conference on Optoelectronic Materials and Devices (ICOMD 2021), Guangzhou, China, 10–12 December 2021; Lu, Y., Gu, Y., Chen, S., Eds.; SPIE: Guangzhou, China, 2022; p. 44.
23. Zhang, Z.; Jia, W.; Shan, Q.; Hei, D.; Wang, Z.; Wang, Y.; Ling, Y. Determining Metal Elements in Liquid Samples Using Laser-Induced Breakdown Spectroscopy and Phase Conversion Technology. *Anal. Methods* **2022**, *14*, 147–155. [CrossRef] [PubMed]
24. Bhatt, C.R.; Goueguel, C.L.; Jain, J.C.; McIntyre, D.L.; Singh, J.P. LIBS Application to Liquid Samples. In *Laser-Induced Breakdown Spectroscopy*; Elsevier: Amsterdam, The Netherlands, 2020; pp. 231–246, ISBN 978-0-12-818829-3.
25. Schlatter, N.; Lottermoser, B.G. Quantitative Analysis of Li, Na, and K in Single Element Standard Solutions Using Portable Laser-Induced Breakdown Spectroscopy (pLIBS). *Geochem. Explor. Environ. Anal.* **2023**, *23*, geochem2023-019. [CrossRef]
26. Birke, M.; Rauch, U.; Harazim, B.; Lorenz, H.; Glatte, W. Major and Trace Elements in German Bottled Water, Their Regional Distribution, and Accordance with National and International Standards. *J. Geochem. Explor.* **2010**, *107*, 245–271. [CrossRef]
27. Birke, M.; Reimann, C.; Demetriades, A.; Rauch, U.; Lorenz, H.; Harazim, B.; Glatte, W. Determination of Major and Trace Elements in European Bottled Mineral Water—Analytical Methods. *J. Geochem. Explor.* **2010**, *107*, 217–226. [CrossRef]
28. Reimann, C.; Birke, M. *Geochemistry of European Bottled Water*; Gebrüder Borntraeger: Stuttgart, Germany, 2010; ISBN 978-3-443-01067-6.
29. Demetriades, A.; Reimann, C.; Birke, M. The Eurogeosurveys Geochemistry EGG Team European Ground Water Geochemistry Using Bottled Water as a Sampling Medium. In *Clean Soil and Safe Water*; Quercia, F.F., Vidojevic, D., Eds.; NATO Science for Peace and Security Series C: Environmental Security; Springer: Dordrecht, The Netherlands, 2012; pp. 115–139, ISBN 978-94-007-2239-2.
30. Wise, M.A.; Harmon, R.S.; Curry, A.; Jennings, M.; Grimac, Z.; Khashchevskaya, D. Handheld LIBS for Li Exploration: An Example from the Carolina Tin-Spodumene Belt, USA. *Minerals* **2022**, *12*, 77. [CrossRef]
31. Scott, J.R.; Effenberger, A.J.; Hatch, J.J. Influence of Atmospheric Pressure and Composition on LIBS. In *Laser-Induced Breakdown Spectroscopy*; Musazzi, S., Perini, U., Eds.; Springer Series in Optical Sciences; Springer: Berlin/Heidelberg, Germany, 2014; pp. 91–116, ISBN 978-3-642-45084-6.
32. Schäffer, R.; Götz, E.; Schlatter, N.; Schubert, G.; Weinert, S.; Schmidt, S.; Kolb, U.; Sass, I. Fluid–Rock Interactions in Geothermal Reservoirs, Germany: Thermal Autoclave Experiments Using Sandstones and Natural Hydrothermal Brines. *Aquat. Geochem.* **2022**, *28*, 63–110. [CrossRef]
33. DIN 38402-62:2014-12; German Standard Methods for the Examination of Water, Waste Water and Sludge—Part 62: Plausibility Check of Analytical Data by Performing an Ion Balance. Beuth Verlag GmbH: Berlin, Germany, 2014.
34. Rossum, J.R. Conductance Method for Checking Accuracy of Water Analyses. *Anal. Chem.* **1949**, *21*, 631. [CrossRef]
35. Legnaioli, S.; Botto, A.; Campanella, B.; Poggialini, F.; Raneri, S.; Palleschi, V. Univariate Linear Methods. In *Chemometrics and Numerical Methods in LIBS*; Palleschi, V., Ed.; Wiley: Hoboken, NJ, USA, 2022; pp. 259–276, ISBN 978-1-119-75961-4.
36. Guezenoc, J.; Gallet-Budynek, A.; Bousquet, B. Critical Review and Advices on Spectral-Based Normalization Methods for LIBS Quantitative Analysis. *Spectrochim. Acta Part B At. Spectrosc.* **2019**, *160*, 105688. [CrossRef]
37. IUPAC. Nomenclature, Symbols, Units and Their Usage in Spectrochemical Analysis—II. Data Interpretation. *Pure Appl. Chem.* **1976**, *45*, 99–103. [CrossRef]
38. Schäffer, R.; Dietz, A. Standardized Schoeller Diagrams—A Matlab Plotting Tool. *ESS Open Arch.* **2022**, 1–17. [CrossRef]
39. Ma, S.; Tang, Y.; Zhang, S.; Ma, Y.; Sheng, Z.; Wang, Z.; Guo, L.; Yao, J.; Lu, Y. Chlorine and Sulfur Determination in Water Using Indirect Laser-Induced Breakdown Spectroscopy. *Talanta* **2020**, *214*, 120849. [CrossRef]
40. Tang, Z.; Hao, Z.; Zhou, R.; Li, Q.; Liu, K.; Zhang, W.; Yan, J.; Wei, K.; Li, X. Sensitive Analysis of Fluorine and Chlorine Elements in Water Solution Using Laser-Induced Breakdown Spectroscopy Assisted with Molecular Synthesis. *Talanta* **2021**, *224*, 121784. [CrossRef] [PubMed]
41. Poggialini, F.; Legnaioli, S.; Campanella, B.; Cocciaro, B.; Lorenzetti, G.; Raneri, S.; Palleschi, V. Calculating the Limits of Detection in Laser-Induced Breakdown Spectroscopy: Not as Easy as It Might Seem. *Appl. Sci.* **2023**, *13*, 3642. [CrossRef]
42. Hark, R.R.; Harmon, R.S. Geochemical Fingerprinting Using LIBS. In *Laser-Induced Breakdown Spectroscopy*; Musazzi, S., Perini, U., Eds.; Springer Series in Optical Sciences; Springer: Berlin/Heidelberg, Germany, 2014; Volume 182, pp. 309–348, ISBN 978-3-642-45084-6.
43. Samek, O.; Beddows, D.C.S.; Kaiser, J.; Kukhlevsky, S.V.; Liska, M.; Telle, H.H.; Whitehouse, A.J. Application of Laser-Induced Breakdown Spectroscopy to In Situ Analysis of Liquid Samples. *Opt. Eng.* **2000**, *39*, 2248. [CrossRef]
44. Palleschi, V. Avoiding Misunderstanding Self-Absorption in Laser-Induced Breakdown Spectroscopy (LIBS) Analysis. *Spectroscopy* **2022**, *37*, 60–62. [CrossRef]

45. Rao, A.P.; Jenkins, P.R.; Auxier, J.D.; Shattan, M.B.; Patnaik, A.K. Analytical Comparisons of Handheld LIBS and XRF Devices for Rapid Quantification of Gallium in a Plutonium Surrogate Matrix. *J. Anal. At. Spectrom.* **2022**, *37*, 1090–1098. [CrossRef]
46. Tang, Y.; Ma, S.; Chu, Y.; Wu, T.; Ma, Y.; Hu, Z.; Guo, L.; Zeng, X.; Duan, J.; Lu, Y. Investigation of the Self-Absorption Effect Using Time-Resolved Laser-Induced Breakdown Spectroscopy. *Opt. Express* **2019**, *27*, 4261. [CrossRef] [PubMed]

Disclaimer/Publisher's Note: The statements, opinions and data contained in all publications are solely those of the individual author(s) and contributor(s) and not of MDPI and/or the editor(s). MDPI and/or the editor(s) disclaim responsibility for any injury to people or property resulting from any ideas, methods, instructions or products referred to in the content.

MDPI AG
Grosspeteranlage 5
4052 Basel
Switzerland
Tel.: +41 61 683 77 34

Chemosensors Editorial Office
E-mail: chemosensors@mdpi.com
www.mdpi.com/journal/chemosensors

Disclaimer/Publisher's Note: The title and front matter of this reprint are at the discretion of the Guest Editors. The publisher is not responsible for their content or any associated concerns. The statements, opinions and data contained in all individual articles are solely those of the individual Editors and contributors and not of MDPI. MDPI disclaims responsibility for any injury to people or property resulting from any ideas, methods, instructions or products referred to in the content.

www.ingramcontent.com/pod-product-compliance
Lightning Source LLC
LaVergne TN
LVHW072251110526
838202LV00106B/2545